Misinformation and Disinformation

Victoria L. Rubin

Misinformation and Disinformation

Detecting Fakes with the Eye and AI

Victoria L. Rubin
Western University
London, ON, Canada

ISBN 978-3-030-95658-5 ISBN 978-3-030-95656-1 (eBook)
https://doi.org/10.1007/978-3-030-95656-1

This Springer imprint is published by the registered company Springer Nature Switzerland AG
The registered company address is: Gewerbestrasse 11, 6330 Cham, Switzerland

Introduction

Abstract How do we detect, deter, and prevent the spread of mis- and disinformation with the human eye and AI? How does theory inform the practice, and how do the evidence-based research and best practices in lie-catching and truth-seeking professions—inform AI? The book looks into well-established human practices such as the routines and processes used in detective work, journalism, and scientific inquiry, and how they contribute toward innovative AI solutions. The book explains the principles, inner workings, and recent evolution of five types of state-of-the-art AI technologies suitable for curtailing the spread of mis- and disinformation: automated deception detectors, clickbait detectors, satirical fake detectors, rumor debunkers, and computational fact-checking tools.

Keywords Misinformation; Disinformation; Fake news; False news; Falsehoods; Infodemic; Satire; Clickbait; Information manipulation; Online deception; Library and Information Science; LIS; Journalism; Psychology; Communication; Social Sciences; Infodemiology; Natural Language Processing; NLP; Artificial Intelligence; AI; Machine Learning; ML; Algorithms; Automated Deception Detection; News verification; Automated fact-checking; Satire detection; Clickbait identification

Have you ever been duped? Do you tend to believe what you see or hear? Do you naturally trust others? If so, like many of us, you may be an easy target for sinister pranks or serious fraud. The world is full of phone scammers, email spammers, and other con-artists who work hard to influence what we think, what choices we make, and what we ultimately do. The classic definition of deception is a message intentionally transmitted to bolster a false impression, idea, or belief in someone else's mind (Buller & Burgoon, 1996). Why do we fall for such messages? Being a mark for an online con-artist is an unenviable position but avoiding it by staying on top of the latest online tricks and false claims is not easy. So, how can we get to the truth and avoid being fooled? What shapes do manipulative influences take? How do we guard ourselves against unwanted deceptive attempts? With so much of our daily lives spent online, we are learning how to be more vigilant, how to fight our unconscious biases, and how to be more critical of online sources. Are there any assistive technologies that we can rely on to alleviate these burdens? Or, is the battle ultimately in our own minds?

This book shares the results of my 15-year quest to understand deceptive behaviors: the shapes and forms of deceptive messages, how they are expressed in language, and how they can be spotted using both the human eye and AI (Artificial Intelligence). I focus on how theory informs practice and how the evidence-based research and best practices from multiple professions (e.g., journalists, detectives, social and computer scientists) can help us to systematically separate lies from truths, mistaken beliefs from facts, and myths from reality. How do we amplify and complement human intelligence with artificial ones? What have the leading research and development (R&D) labs achieved thus far toward the goal of scalable AI-solutions to curtail the spread of mis- and disinformation? How successful have their efforts been? Is AI mimicking the procedures and know how of the experts, or does it require entirely new systematic approaches?

My plunge into researching the capabilities of AI deception detection systems automatically forced me to scrutinize our innately human behaviors. I consider the very essence of human intelligence, by contrasting it with AI, and make observations about our manipulative use of language, as well as human ineptitude in spotting lies. While the human mind is the ultimate built-in detector, it requires preparation, awareness, and practice to achieve better thinking. As Hemingway once said to a reporter from *The Atlantic*, "every man should have a built-in automatic crap detector operating inside him. It also should have a manual drill and a crank handle in case the machine breaks down" (Manning, 1965). I review what may cause such breakdowns: the circumstances that make us susceptible to being fooled and manipulated. I also speculate about how to extricate ourselves from the powers of persuasive propaganda.

Driving Forces Behind This Book

What drove me to write this book is my desire to endorse the value of education and show its concrete application in the fight against the infodemic. Without any distinct political or partisan agenda, I promote rational thinking, well-informed decision-making, and deeper self-awareness. Of the two modes within which we process information, known in social and cognitive psychology as the dual processing model (Chaiken & Ledgerwood, 2007), one mode is to think in a quick, associative, relatively automatic, and superficial way. The other way is an effortful, reflective, systematic mode of thinking. Which mode we are in influences our outcomes such as the conclusions, judgments, and attitudes we form and commit to memory. I warn against the dangers of superficial thinking and emphasize the need to be critical of the information we encounter in digital environments.

My main claim in this book is that in the fight against the infodemic, some technological assistance is inevitable and likely to come from AI-enabled applications. In other words, our human intelligence can be, at least in part, enhanced with an artificial one. We need systematic analyses that can reliably and accurately sift through big data that comes at us in large volumes, with velocity, and in a variety of formats. Since opinions, claims, thoughts, news, requests, promises, and so forth are often expressed in words (or textual format), we are talking in part about linguistic behaviors.

I have always been passionate about languages. I was born, raised, and educated in a Russian-speaking part of Ukraine. My working language for the last 25 years has been English. I have spent over half of my life in North America; progressed from one graduate degree to another in the States. By now, I have been a professor and the director and principal investigator of the Language and Information Technology Research Lab (LIT.RL) at Western University in Canada for about 15 years.

As a multilingual, I converse in up to six different languages including French, Spanish, and Japanese. I also occasionally get by in a few Slavic languages due to their similarities to my native Russian and Ukrainian. Eager to discover new cultures and use my linguistic gifts, I have relocated from country to country, and traveled extensively in Europe, Mexico, and Japan. How we use languages in the context of our everyday social lives has been my perpetual curiosity.

What specifically fascinates me is how we use language under challenging circumstances, especially when what we think and what we say does not exactly match. During my travels, I kept running into differences in attitudes about lying across languages and cultures. For instance, Japanese speakers puzzled me with their apparent inability to refuse face-to-face requests. It would force them at times to invent fake excuses or be vague, instead of simply saying "No" (as described earlier in Rubin (2014).) Spanish speakers in Mexico would rather give me wrong driving directions than admit that they simply did not know the route. Many improvized freely with their knowledge of local autostradas.

Lying and deception may be distinctly cultural, yet universal, in the sense of their relevance to our human condition, and our ability to share ideas through language. Most cultures deeply frown upon serious prevarications but seem to excuse white lies such as lying about surprise birthday parties. Some cultures disagree on whether to disclose the terminal diagnosis to fatally sick patients (see, for example, Blum (2007)). Universally, people lie to avoid hurting their loved ones, to get themselves out of predicaments, and for the sake of self-preservation, image-management, or other personal gains. The more I studied deceptive behaviors, the more I was confronted with variations in deception's forms, formats, motives, and justifications. I looked to put the pieces of the truth-deception puzzle together.

Intended Readers

If you are interested in issues concerning truth and deception, digital fake news, information disorder, mis- or disinformation, and the use of AI to curtail the info-demic, this book is for you. If you have a persistent online presence and generate online content (as a blogger, "citizen journalist," or other website creator), or if you simply consume large volumes of social media daily, you may be looking for this book as a primer on deception research in the humanities and social sciences, with an additional technological digest of "know-hows." You can build on these foundations for educational campaigns, policy making, or for your personal educational, financial, or political advantage.

Anyone that deals with large amounts of information, professionally or otherwise, probably realizes that AI can help solve the problem of the infodemic, but they may lack the mathematical or computer science background needed to understand how this can be done. I am a computational linguist and information scientist by training. Over the past 15 years, I have taught future information professionals (librarians, archivists, record managers, metadata specialists, etc.) in our library and information science (LIS) programs. In my experience, more LIS students are comfortable reading in the humanities and social sciences than in the more technical computational disciplines. Some graduate students are certainly more tech-savvy than others, but most share a curiosity for innovative technological solutions. They often wonder what promise intelligent systems hold in solving the problem of the mis- and disinformation. How can AI be combined with instruction and training that information professionals offer at their workplaces? Even if we gloss over some technicalities, we can still discuss the inner workings of the AI systems. This conversation can reveal what otherwise may seem like magic tricks to those who do not read specialized AI literature. Considering the pros and cons of AI systems can stimulate intelligent conversations about their capabilities, the principles that guide their research and development, and the rationale behind the adoption of these systems.

If you are a programmer or an expert in information retrieval or other areas within computer science broadly, you can still benefit from the overview of the human side of language computing. I explore psychological, philosophical, and communication models that account for human linguistic behaviors: the human factor behind the data-driven world of AI computing. For more technical readers who are well versed in AI and natural language processing (NLP), my overview may feel a bit lightweight, but it will lead you to the primary literature for more technical details.

To bridge the disciplinary boundaries, I combine psychological, philosophical, and linguistic insights into the nature of truth and deception, trust and credibility, cognitive biases, and logical fallacies. I then translate these insights into practical terms by drawing on the professional practices of expert lie detectors and truth seekers. I explain how their expert routines can be automated and augmented by AI methods.

Information professionals and technology users more broadly will find this book useful because it accumulates multiple perspectives on mis- and disinformation, otherwise scattered across diverse professional journals, into one volume. A few books about "fake news" prioritize the political and historical perspective (e.g., Bennett & Livingston, 2020). Others are filled with real-life examples of fraudulent news reports to advise practicing journalists and reporters (e.g., Silverman, 2021). Books from a psychological viewpoint focus on the human susceptibility to misinformation. Media literacy infographics instruct you on how to fight fake news (IFLA, 2021) and are popularly available in thin report brochures (e.g., Cooke, 2018), which were produced in the wake of the 2016 U.S. Presidential Election, and more recently, the COVID-19 pandemic (e.g., Ostman, 2020). An occasional chapter in library and information literature explains AI countermeasures to digital attacks like phishing, spamming, and social bots, but their skepticism about an eventual AI solution is palpable (Dalkir & Katz, 2020). In my research and writing, I am cautiously optimistic about the success of technological solutions, especially when AI is applied in combination with heightened critical thinking. None of the books I have reviewed so far combine aspects of human and artificial intelligence as two necessary parts for the resolution of the infodemic problem as definitively as I have here. That is the main contribution of my book: it explains how these two intelligences fit together.

Book Contributions

This book translates descriptions of the AI detection of mis- and disinformation into digestible portions about the principles, processes, techniques, analyses, and other research considerations. One of the aims is to facilitate the dialog between tech developers and less tech-savvy digital media users. The secrecy of technological "know-hows" often keeps users in the dark, preventing us from understanding how algorithms make choices. There are, however, hundreds of research labs and institutions that publish findings and innovative solutions in scientific journals and conference proceedings. Barriers arise from the complexity of specialized jargon or the background knowledge required to access and understand AI literature (e.g., Shu & Liu, 2019).

This book is an overview and culmination of over 10 years of NLP R&D in my LiT.RL lab[1]—from the earliest features for discerning deception automatically to more successful prototypes of clickbait and satire detectors. My doctoral students and I have published extensively on the topic in specialized journals and conference proceedings, but I felt the need to share it with a wider professional audience across disciplinary boundaries. My doctoral students in the lab are ambitious and capable collaborators with a variety of skills. They assist me in data collection,

[1] See https://victoriarubin.fims.uwo.ca/research/.

management, analysis, programming, and testing. What precipitated this book are years of brain-storming sessions, active discussions in the lab, and our interviews with journalists, satirists, clickbait creators, and usability study participants. We have collected real-world samples of fake news from political campaigns, contrasted fabricated bluff and real news reports, and participated in computer science programming challenges to distinguish the two. It is now time to disseminate what we have learned about the infodemic of mis- and disinformation more widely, both the human factors and the automated ways of detection, since if neither users nor systems are able to filter out rubbish, the accumulation of inaccurate information threatens the very usefulness of information retrieval.

Book Structure at a Glance

Between the introductory and conclusive chapters, the book contains two parts of four chapters each. **Part I** of the book seeks to understand the nature of deception and why we fall for it, how the human mind conceives of the truth and how we distinguish it from deception. This section establishes the theoretical footing for our understanding of how we interact with new information. To do this, I use evidence-based research from interpersonal, social, and cognitive psychology, computer-mediated communication, and insights from library and information science, as well as discussions in philosophy.

Chapter 1 frames the problem of mis- and disinformation as the proliferation of deceptive, inaccurate, and misleading information in digital media and information technologies. Discussing divergent terms and their nuances, the chapter narrows the field down to several types of fakes that can be identified with the naked eye and AI. My infodemiological model, the (Rubin, 2019) Disinformation and Misinformation Triangle, serves as a starting point in which we recognize the three minimal interacting causal factors that fuel the infodemic—susceptible digital media users, virulent fakes, and toxic digital environments. Three corresponding interventions—automation, education, and regulation—are proposed to interrupt the interaction among these factors.

Chapter 2 surveys psychological studies on lying and what distinguishes various kinds of deception from truths. I explore the cognitive biases that predispose us to being manipulated into believing untruths. Chapter 3 taps into library and information science insights into trust and credibility, their perceived components, and their connection to other markers of high-quality information. Chapter 4 establishes that truth may be seen from different philosophical perspectives, and how your view of reality impacts how you establish facts and build up your knowledge.

Each chapter in **Part I** brings your attention to central chapter concepts and their interconnections. I invite you to ponder key questions before you proceed to read the research synthesis that addresses each question. For instance, consider what truth, facts, and reality actually mean to you, and how they relate to each other in your understanding before you read about how philosophical worldviews guide us in establishing truths and presuming reality in Chap. 4.

Part II focuses on how the theoretical and empirical knowledge from **Part I** can be applied in practice. Both human and automated stepwise procedures can be put to good service by truthfully informing the public, using the best knowledge of experts in their fields. Alternatively, there are well-documented techniques for influencing the public mind, for commercial or political gains, regardless of what is known by the experts.

Chapter 5 discusses law enforcement, scientific inquiry, and investigative reporting as examples of well-established traditions for truth-seeking. Each applied field has its own systematic ways of collecting strong, supportive evidence and conducting thorough inquiries to reach valid conclusions. Ideally, the best practices of such inquiries lead to establishing facts and reliable knowledge. When experts are not well-trained, diligent, or honest, mistakes and missteps may happen despite these established systems, as exemplified by cases of wrongful convictions, scientific dishonesty, and journalistic fraud.

By contrast, many practices in advertising, public relations, and marketing, canvassed in Chap. 6, have the intent to persuade and manipulate the public opinion from the outset. Advertising techniques used in marketing campaigns and political propaganda frequently engage in truth-bending. Their persuasion mastery often exploits human biases and logical fallacies. Manipulative techniques need to be recognized first before we can resist their powers. I describe activities to identify biases, logical fallacies, manipulative advertising techniques, and other propaganda tricks.

Chapter 7 culminates this book with a thorough review of the AI systems that can help our human eyes identify and call out fakes of various kinds for the benefit of the public good. I explain, in plain language, the principles behind automated deception detectors, rumor debunkers, satire and clickbait detectors, and automated fact-checkers, both how they work and where they tend to fail.

In the **Conclusion**, I offer suggestions on how to incorporate the lessons from each chapter of the book into media and information literacy education to help curtail the infodemic. Chapter 8 aggregated the key arguments and claims about the use of automated ways of detecting various online fakes and puts forward ten recommendations for educational, AI-based, and regulatory interventions, which are articulated summatively, as a package of countermeasures to control the infodemic. Reverting to rigorous systematic thinking is the way to inoculate the public against the mindless and unquestioning consumption of manipulated content.

Some of the AI solutions described in this book have user interfaces and can be installed on personal computing devices. I encourage proactive citizens, librarians, and information professionals to download and experiment with such sample systems. Collaborations across disciplines are now needed to assess how effective AI solutions are when compared to traditional instruction in classrooms, in libraries, or at home. Now that AI can help to identify and label fakes, the public should be more widely informed of the pros and cons when considering its adoption. Further research should evaluate how to best present AI predictions to digital media users to build confidence in their use. Decisions about which AIs to adopt and how to interact with them have societal level impacts, so these choices should not be left solely to giant tech companies. This is especially true when deciding how to incorporate automation with another society-wide measure—the legislative regulation of digital environments to disrupt mis- or disinformation cycles. Our society should then be able to make better-informed decisions about important matters such as our health care, laws, public policies, finances, and voting preferences. While we debate and decide on how to best regulate toxic social media platforms, information professionals and educators can direct their attention to cultivating more discerning minds and encouraging more tech-savvy digital media use. The book can help you make sense of what AI does and how, possibly, spur conversations on various flavors of fakes and how to identify them with an eye and AI.

References

Bennett, W. L., & Livingston, S. (2020). *Disinformation age: Politics, technology, and disruptive communication in the United States*. Cambridge University Press.

Blum, S. D. (2007). *Lies that bind: Chinese truth, other truths*. Rowman & Littlefield Publishers. Retrieved from http://ebookcentral.proquest.com/lib/west/detail.action?docID=1351099

Buller, D. B., & Burgoon, J. K. (1996). Interpersonal deception theory. *Communication Theory, 6*(3), 203–242.

Chaiken, S., & Ledgerwood, C. (2007). Dual process theories. In *Encyclopedia of social psychology*. SAGE.

Cooke, N. A. (2018). *Fake news and alternative facts: Information literacy in a post-truth era*. ALA Editions.

Dalkir, K., & Katz, R. (Eds.). (2020). *Navigating fake news, alternative facts, and misinformation in a post-truth World*. IGI Global. https://doi.org/10.4018/978-1-7998-2543-2

IFLA. (2021, February 18). How to spot fake news. *The International Federation of Library Associations and Institutions*. Retrieved from https://www.ifla.org/publications/node/11174

Manning, R. (1965, August). Hemingway in Cuba. *The Atlantic*. Retrieved from http://www.theatlantic.com/magazine/archive/1965/08/hemingway-in-cuba/399059/

Ostman, S. (2020, March 20). *Fighting fake news in the pandemic: Using your library's digital reach to thwart misinformation*. American Libraries Magazine. Retrieved from https://americanlibrariesmagazine.org/blogs/the-scoop/covid-19-fighting-fake-news-pandemic/

Rubin, V. L. (2014). Pragmatic and cultural considerations for deception detection in Asian languages. Guest editorial commentary. *TALIP Perspectives in the Journal of the ACM Transactions on Asian Language Information Processing (TALIP), 13*(2), 1–8. https://doi.org/10.1145/2605292

Rubin, V. L. (2019). Disinformation and misinformation triangle: A conceptual model for "fake news" epidemic, causal factors and interventions. *Journal of Documentation, ahead-of-print*. https://doi.org/10.1108/JD-12-2018-0209

Shu, K., & Liu, H. (2019). Detecting fake news on social media. *Synthesis Lectures on Data Mining and Knowledge Discovery, 11*(3), 1–129. https://doi.org/10.2200/S00926ED1V01Y201906DMK018

Silverman, C. (Ed.) (2021). *Verification handbook for disinformation and media manipulation (Online)*. DigitalJournalism.com. Retrieved from https://datajournalism.com/read/handbook/verification-3

Contents

List of Abbreviations

ACL	Association for Computational Linguistics
AFP	Agence France-Presse
AFT	(Wikipedia's) Article Feedback Tool
AI	Artificial Intelligence
AIDA	Awareness-Interest-Desire-Action (in advertising)
AOI	Area of Interest
API	American Press Institute
BBB	Better Business Bureau
BILD	Building Industry and Land Development Association (Toronto, Canada)
3 Cs	Times of "Conflict, Crisis, and Catastrophe" (in research about rumors)
CBC	Canadian Broadcasting Corporation (Canada)
CBCA	Content-Based Criteria Analysis
CIHR	Canadian Institutes of Health Research
CMC	Computer-Mediated Communication
CNE	Cámara Nacional Electoral (National Court on Elections, Argentina)
CTR	Click-through rate
DAGMAR	Defining Advertising Goals for Measured Advertising Results
DRIP	Differentiate, Reinforce, Inform, and Persuade (a marketing model)
EMA	European Medicines Agency (for the scientific evaluation, supervision, and safety monitoring of medicines in the EU)
ERIC	Education Resources Information Center (U.S.)
EU	European Union
FTC	Federal Trade Commission (U.S.)
GIGO	Garbage In, Garbage Out
HCI	Human–Computer Interaction
HTML	HyperText Markup Language
IAB	Interactive Advertising Bureau (Europe, U.S.)
IAMAI	Internet and Mobile Association of India
ICT	Information and Communications Technology
IFCN	Poynter's International Fact-Checking Network

IFLA	International Federation of Library Associations and Institutions
IMC	Integrated Marketing Communications
IPA	Institute for Propaganda Analysis (U.S.)
IR	Information Retrieval
JSTOR	Journal Storage (a database for Internet access of scholarly journals)
LIS	Library and Information Science
LiT.RL	Language and Information Technology Research Lab (Western University, Canada)
LOP	Location of Presence
MAIN	Modality, Agency, Interactivity, and Navigability (a model)
MCQ	Memory Characteristic Questionnaire
MIS	Management Information Systems
ML	Machine Learning
MMS	Multimedia Message Service
mRNA	Messenger Ribonucleic Acid (vaccine technology)
NACOLE	National Association for Civilian Oversight of Law Enforcement (U.S.)
NLP	Natural Language Processing
NSERC	Natural Sciences and Engineering Research Council of Canada
NSF	National Science Foundation (U.S.)
OBA	Online Behavioral Advertising
PKM	Persuasion Knowledge Model
PPC	Pay-Per-Click (paid search marketing)
PR	Public Relations
R&D	Research and Development
RM	Reality Monitoring
RIRO	Rubbish In, Rubbish Out
SEM	Search Engine Marketing
SEO	Search Engine Optimization
SERPs	Search Engine Results Pages
SMS	Short Message Service
SSHRC	Social Sciences and Humanities Research Council of Canada
SVA	Statement Validity Assessment
UI	User Interface
UNESCO	United Nations Educational, Scientific and Cultural Organization
UNICEF	United Nations International Children's Emergency Fund
URAC	Utilization Review Accreditation Commission
URL	Uniform Resource Locator
VDPV	Vaccine-Derived Poliovirus
WHO	World Health Organization
WOM	Word-of-Mouth
WOT	Web of Trust (an online reputation and Internet safety service)

Part I
Human Nature of Deception and Perceptions of Truth

Chapter 1
The Problem of Misinformation and Disinformation Online

There has never been, nor will there ever be, a technological innovation that moves us away from the essential problems of human nature.

(Broussard, 2019, p. 8)

Abstract **Chapter 1** frames the problem of deceptive, inaccurate, and misleading information in the digital media content and information technologies as an infodemic. Mis- and disinformation proliferate online, yet the solution remains elusive and many of us run the risk of being woefully misinformed in many aspects of our lives including health, finances, and politics. **Chapter 1** untangles key research concepts—*infodemic, mis-* and *disinformation, deception, "fake news," false news,* and various types of *digital "fakes."* A conceptual infodemiological framework, the Rubin (2019) Misinformation and Disinformation Triangle, posits three minimal interacting factors that cause the problem—susceptible hosts, virulent pathogens, and conducive environments. Disrupting interactions of these factors requires greater efforts in educating susceptible minds, detecting virulent fakes, and regulating toxic environments. Given the scale of the problem, technological assistance as inevitable. Human intelligence can and should be, at least in part, enhanced with an artificial one. We require systematic analyses that can reliably and accurately sift through large volumes of data. Such assistance comes from artificial intelligence (AI) applications that use natural language processing (NLP) and machine learning (ML). These fields are briefly introduced and AI-enabled tasks for detecting various "fakes" are laid out. While AI can assist us, the ultimate decisions are obviously in our own minds. An immediate starting point is to verify suspicious information with simple digital literacy steps as exemplified here. Societal interventions and countermeasures that help curtail the spread of mis- and disinformation online are discussed throughout this book.

Keywords Infodemic · Misinformation · Disinformation · Fake News · False News · Falsehoods · Satire · Clickbait · Information manipulation · Online deception · Library and Information Science · LIS · Journalism · Psychology · Communication · Social Sciences · Infodemiology · Natural Language Processing · NLP · Artificial Intelligence · AI · Machine Learning · ML · Algorithms · Automated Deception Detection · News Verification · Automated fact-checking · Satire Detection · Clickbait Identification

V. L. Rubin, *Misinformation and Disinformation*,
https://doi.org/10.1007/978-3-030-95656-1_1

1.1 Introduction

1.1.1 Infodemic

Inaccurate, deceptive, and other misleading information is accumulating in large amounts online; it is a Google search away and it easily seeps into our social media newsfeeds. In September 2020, the World Health Organization (WHO), jointly with fellow United Nations organizations like UNICEF and UNESCO, put out a public statement warning the public of a growing global *infodemic*. The infodemic—an overabundance of online and offline information, and deliberate attempts to disseminate wrong information—is undermining governmental responses worldwide and jeopardizing measures to control the COVID-19 pandemic (World Health Organization et al., 2020). The problem of the infodemic is inescapable, atop the minds of many digital media users since gaining the attention of the North American media in relation to the 2020 US Presidential Elections. A vast international research agenda was subsequently published, showing collaboration across 20 professional disciplines in over 35 countries, declaring the need for cross-disciplinary standards, and in essence announcing the birth of *infodemiology*, an emerging transdisciplinary scientific field of managing the infodemic (World Health Organization, 2021a, p. X).

A large-scale systematic analysis of over 126,000 Twitter stories tweeted by about 3 million people more than 4.5 million times (Vosoughi et al., 2018) found that falsehoods diffused considerably faster and more broadly than the truth, with more pronounced effects in false political news than in false news about less-partisan topics such as terrorism, natural disasters, science, urban legends, or financial information. Other empirical social science researchers, monitoring social media trends in the US and worldwide, report that mis- and disinformation online is a common concern among digital media users. A 2019 survey of 6127 US adults by the Pew Research Center, a US-based nonpartisan think tank, found that the creation and spread of made-up news and information was viewed by the majority of the survey participants as causing considerable harm. Namely, "more Americans view made-up news as a very big problem for the country than identify terrorism, illegal immigration, racism, and sexism that way. Additionally, nearly seven-in-ten U.S. adults (68%) say made-up news and information greatly impacts Americans' confidence in government institutions, and roughly half (54%) say it is having a major impact on our confidence in each other" (Mitchell et al., 2019) (Fig. 1.1).

These polling data were echoed by a December 2020 NPR/Ipsos poll, which found that more than 8 in 10 (83%) of their participants report they are concerned (of which 54% are very concerned) about the spread of false information, and specifically, 69% say that the information they receive on social media is not accurate (Newall, 2020, p. 3). The poll also showed that recent misinformation has been establishing itself among some Americans (see Fig. 1.2). Out of 10 knowledge test questions about historic events, most Americans correctly answered questions about past events famed for misinformation (e.g., the moon landing, Barack Obama's

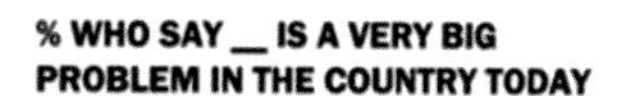
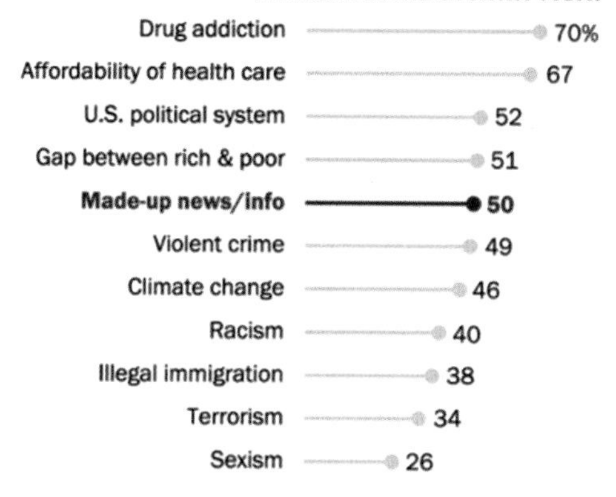

Fig. 1.1 Results of a Pew Research Center 2019 survey of US adults showing US adults reports of concerns about made-up news and its detriment to Americans' confidence in government, each other, and political leaders' ability to get work done (Mitchell et al., 2019)

birthplace, and 9/11), yet showed more ambiguity on recent events (e.g., around origins of COVID, QAnon tenets on politics and media control, and the nature of the 2020 Black Lives Matter protests) (Newall, 2020, p. 3). Such surveys imply that the general public concern about the spread of mis- and disinformation is valid and urgent solutions are needed as the problem is already taking a toll on society at large.

Deceptive or misleading information threatens to interfere with many aspects of our everyday lives: Both the news we get and the information we seek and then use to make choices about health, finances, and politics. Internalizing and acting on

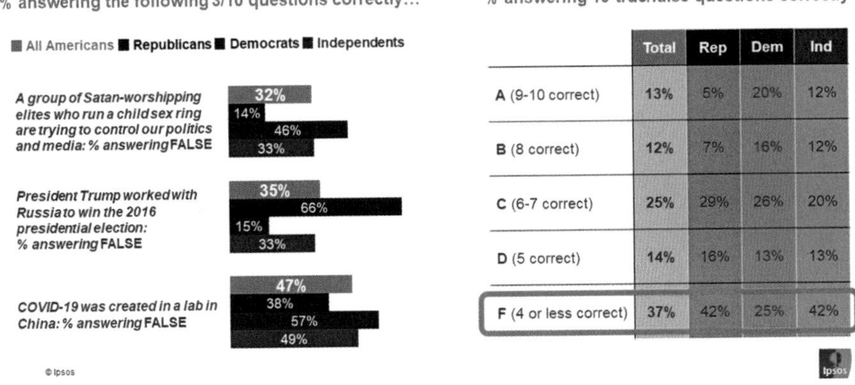

Fig. 1.2 NPR/Ipsos 2020 knowledge test answers show mis–/disinformation is taking roots among Americans, broken down by political demographic (Newall, 2020)

incorrect information often leads to undesirable consequences. For example, misrepresenting scientific findings such as the alleged harms of vaccines may increase vaccine hesitancy in the population, as we have seen in 2021 during the second year of the COVID-19 pandemic. While Moderna's mRNA COVID-19 vaccine was being administered to millions of people around the world, "the DC Dirty Laundry" website, known for promoting conspiracies and exhibiting extreme biases, made false claims about the vaccine's harms. One post (Fig. 1.3), shared thousands of times over social media, misrepresented a TED Talk by Tal Zaks, the chief medical officer of the US pharma firm Moderna, and claimed that he allegedly confirmed that messenger RNA vaccines can alter human DNA. The Agence France-Presse (AFP) Fact Check confirms that Zaks did not make the purported comments and embeds the TED Talk video as proof (AFP Espagne, AFP Hong Kong, 2021).

The debunking article reiterates that scientists had previously rejected similar false claims by explaining the mechanism of the mRNA vaccines which trigger an immune response to repel the active virus with a snippet of genetic code (AFP Espagne, AFP Hong Kong, 2021).

For instance, Reuters had debunked these claims multiple times (since Reuters Staff, 2020), citing experts from the European Medicines Agency (EMA) recommending the authorization of the mRNA-based Moderna vaccine (Reuters Staff, 2021). In January 2021, EMA assured European citizens that Moderna meets European Union (EU) standards for EU-wide vaccination campaigns, categorically stating: "The mRNA from the vaccine does not stay in the body but is broken down shortly after vaccination" (Glanville, 2021). Recent studies of the impact of mis- and disinformation on vaccine hesitancy found that "relative to factual information, recent misinformation induced a decline in intent of 6.2 percentage points (95th percentile interval 3.9–8.5) in the UK and 6.4 percentage points (95th percentile

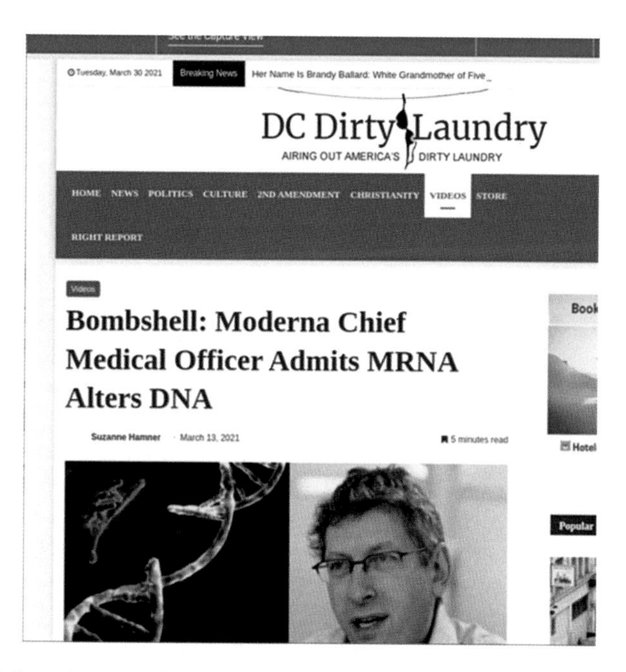

Fig. 1.3 Disinformation example. A March 30, 2021, article by "the DC Dirty Laundry" falsely claims that Tal Zaks confirmed in his TED Talk that "mRNA injection for COVID-19 can change your genetic code or DNA" (https://perma.cc/QRW4-EK4J?type=image)

interval 4.0–8.8) in the USA among those who stated that they would definitely accept a vaccine" (Loomba et al., 2021). The decrease in public willingness to get vaccinated demonstrates the destabilizing effect that false harm claims have on public health matters at the societal scale.

False news can also drive the misallocation of resources during terrorism attacks and natural disasters, or otherwise interfere with the democratic processes of a society. After the 2020 US Presidential election season, 6-in-10 Americans polled about viral online misinformation said it had a big impact on the presidential election, about on par with the portion who say the same of news media coverage (Mitchell et al., 2020).

1.1.2 Mis- and Disinformation

Disinformation is defined by the Library of Congress law specialists and analysts as "false information deliberately and often covertly spread [...] in order to influence public opinion or obscure the truth" in their 2019 Report for US Congress (Global Legal Research Directorate Staff, 2019). Interpersonal psychology and communication studies distinguish disinformation from misinformation by *the intent to deceive*. Inaccurate or misleading information can be spread unintentionally in an act of *misinformation*.

Both terms have persisted into the early 2020s and scholars in this area agree that they are near equivalents, distinguished from each other by the communicator's intent to deceive. One commonly cited explanation goes as follows: "*Disinformation* is content that is intentionally false and designed to cause harm. It is motivated by three distinct factors: to make money; to have political influence, either foreign or domestic; or to cause trouble for the sake of it. When disinformation is shared it often turns into misinformation. *Misinformation* also describes false content but the person sharing doesn't realise that it is false or misleading. Often a piece of disinformation is picked up by someone who doesn't realise it's false, and [who] shares it with their networks, believing that they are helping. The sharing of misinformation is driven by socio-psychological factors. Online, people perform their identities. They want to feel connected to their 'tribe', whether that means members of the same political party, parents that don't vaccinate their children, activists who are concerned about climate change, or those who belong to a certain religion, race or ethnic group" (Wardle, 2019, p. 8).

The art of *deception* is at the core of intentional disinformation. While *digital deception* is a relatively new online phenomenon (Hancock, 2012), various "offline" deceptive strategies have been around since the dawn of time. These strategies have shape-shifted into a variety of online fakes such as outright falsifications in the news, misleading satire, and clickbait. Some of those fakes have taken on viral qualities, such as their ability to replicate using the machinery of social media. Each variety of fakes deserves a careful examination of its characteristics and potential weaknesses as a deceptive strategy (see Chap. 2).

1.2 Framework for Causes and Interventions

In terms of the potential causes of the infodemic, there are at least three necessary contributing factors that are interacting and enabling the spread of mis- and disinformation. I borrow the classic disease triangle model from epidemiology (Scholthof, 2007) and translate it into the context of digital communication. Just as in the epidemiological model, the three factors at the apices of the Mis- and Disinformation Triangle (Rubin, 2019) are compromised hosts, virulent pathogens, and conducive environments (see Fig. 1.4).

First, who are the proverbial hosts of the infodemic? It appears some digital media users are more susceptible to mis- or disinformation than others. We are more susceptible to misunderstandings, misinterpretations, or hasty judgments when we are overloaded with information, pressed for time, unaware of our own biases and cognitive limitations, or if we simply lack media literacy skills. Behavioral sciences research explains what makes us easy marks for con artistry and how the empirical evidence can advise us on how to avoid being fooled (see Chap. 2, Sect. 2.4 "Why Mis-/Disinformation is Effective").

Second, various fakes are the pathogens of the infodemic. They are either newly generated or regurgitated messages that convey deceptive, inaccurate, or misleading

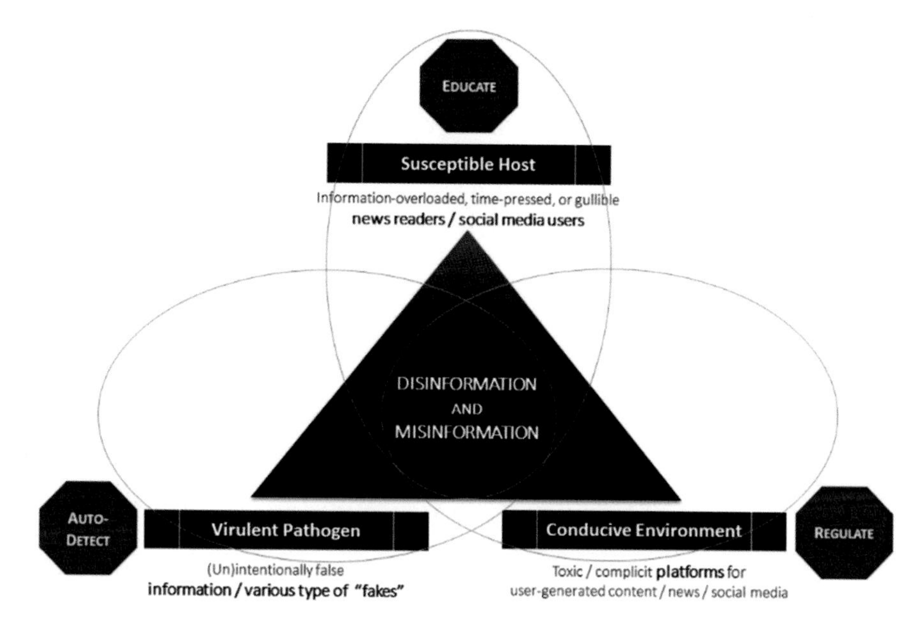

Fig. 1.4 The disinformation and misinformation triangle. This infodemiological model identifies three interacting causal factors responsible for the spread of the mis- and disinformation. Three corresponding interventions are proposed to interrupt the interaction among these factors: automation, education, and regulation (in red "stop sign"). (Redrawn from Rubin, 2019)

information. They persist since their original creators, the agents of disinformation, are highly motivated to continue spawning a variety of fakes for their potential financial or political gain. Online fakes may appear unique to their digital environments but many of them have roots, as you will see, in their time-tested offline equivalents (see Chap. 2, Sect. 2.3 "Many Faces of Deception").

Third, social media platforms are poorly regulated, which makes them conducive to further propagation of the infodemic. They are complacent in allowing fakes to reach their susceptible users. Promotional content, at the heart of the ad revenue model, is meagerly vetted for veracity (see Chap. 6 for the ad revenue model and advertising paradigms in marketing communication). Ads may carry mis- and disinformation without any repercussions to their creators or to the social media platforms that host the ads and benefit from the ad revenues. Digital advertising revenue across all digital entities (beyond just news) continues to grow, with technology companies playing a large role in the flow of both news and revenue (Pew Research Center Report, 2019). Digital content moderation and sporadic labeling have thus far been a weak barrier to the penetration of mis- and disinformation, as the ad revenues of social media platforms take precedence over their moral obligations to uphold truth for the public good. Thus, more stringent legal measures are necessary.

From the epidemiological perspective, interactions between the fakes as virulent pathogens, users as susceptible hosts, and digital platforms as conducive environments is what creates the infodemic. The metaphor of digital deception as a disease fits within the classic epidemiological explanation for the spread of infections. Having a conceptual model like this is useful because we can borrow the traditional epidemiological control measures and apply them in this new context. To be effective, these countermeasures must interrupt the interactions of the three minimal causal factors, and should be applied simultaneously, consistently, and relentlessly. I advocate for three concrete countermeasures.

First and foremost, as a society, we need to have the public will and means to better educate susceptible digital media users. This reduces the susceptibility of hosts. Recent efforts have shown success in neutralizing the influence of misinformation through inoculating messages as a form of misconception-based education. These "mis-information vaccines" explain the flawed argumentation used by disinformers and highlight the scientific consensus on topics like climate change (Cook et al., 2017). Social engineering tactics call for improving the presentation, confidence, and integration of information, and overcoming social forces, for instance, by reducing group-think by using messages framed with group values in mind, which increases acceptance on even controversial issues (Endsley, 2018). One recent study's preliminary results show that older adults (a demographic group especially susceptible to disinformation online) are able to judge the veracity of news headlines more accurately after participating in a 2-month-long digital media literacy training program (Dyakon, 2020). (See also Chap. 8, Sect. 8.2. "Concrete Recommendations for Educational Interventions"). Cultivating awareness and super-vigilance and training the eye to distinguish deceptive attempts turn out to be effective and important methods to pursue, especially within vulnerable demographics.

Second, we need to develop accurate automatic mis- and disinformation identification systems and promote their adoption as assistive tools. Much of this book's content ties back to current advances in AI. The future of content moderation and verification at the necessary scale will rely on successful technological implementations (see also Chap. 8, Sect. 8.3. "Automated Interventions" and Chap. 7 for a detailed review of AI-based systems for combatting mis- and disinformation). The challenges of this automation work also make us consider the challenges of governance, accountability, censorship, and the right to free speech.

The third important intervention is the establishment of more stringent regulation of toxic media platforms. Many people are now realizing this need is urgent. Some countries' legislatures have acted to break up the giant tech monopolies who have been historically reluctant to assume any responsibility for the content they host, but who are now pressured to moderate it. Platforms make attempts at labelling content (Rosen, 2020) but should not be the sole arbitrators of truth. Their efforts have been criticized for policing public opinions too much and yet not enough. (See Chap. 8, Sect. 8.4. "Regulatory and Legislative Interventions").

In short, these three types of interventions—education of susceptible hosts, automation of deceptive content recognition, and environmental regulation—are the most promising tools we currently have to control the spread of mis- and disinformation. The content of this book is a synthesis of multidisciplinary research and technological advances that can help us curtail the infodemic.

1.3 Immediate Digital Literacy Steps

Most of us prefer making informed decisions—rather than misinformed ones—be it about health, investments, environment, or politics. There are immediate steps you can take now, if you are suspicious about any information you receive. You can verify the source, date, and author by yourself, or consult experts or reliable fact-checking websites to make sure it is not a joke or a rumor. This advice is given consistently across reputable organizations in many fields such as health (by the World Health Organization, see Fig. 1.5), librarianship (by the International Federation of Library Associations and Institutions, see Fig. 1.6), and journalism (by the Canada's National Observer, see their website with five-step instructions.[1]).

These guidelines may be intuitive, amounting to what is known as the "stop-and-think" tactics or "think-before-you-click" approach to slowing down misinformative social media resharing. Yet they may be tricky to remember, so such iconographics are good to have in front of your eyes as reminders.

Aside from these simple steps, you can also deepen your awareness of the stumbling blocks, persistent arguments, and major findings in deception research. Beyond the impact on you as an individual, the societal objective is to prevent mass disinformation of the population, which requires robust measures in information technologies and greater involvement by interdisciplinary communities of scientists, researchers, educators, and other enthusiasts.

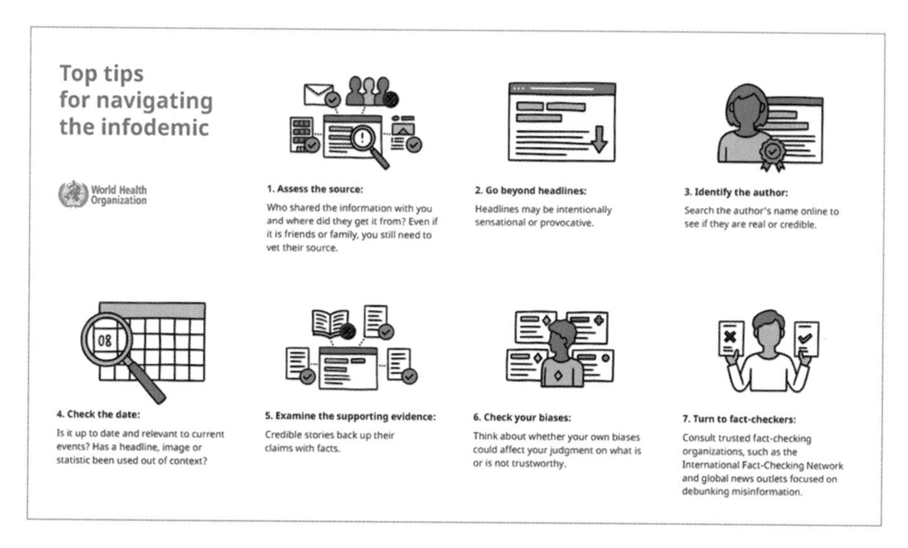

Fig. 1.5 The World Health Organization top tips infographic to identify mis- or disinformation in navigating the infodemic (World Health Organization, 2021b)

[1] See Canada's National Observer (2019) "Five Step Guide: How to Spot Fake News" via https://www.nationalobserver.com/spot-fake-news (accessed on March 16, 2021).

Fig. 1.6 The International Federation of Library Associations and Institutions infographic on how to spot fake news (IFLA, 2021)

The problem of pervasive deception is felt acutely in 2022, but as you will see, it dates back to the times of ancient philosophers (see Chap. 4). The present has added the computing part, of course (see Chap. 7). So, this book seeks answers to how our cumulative knowledge from the past and the newest technological advancements can be combined into modern interventions to bring the current online infodemic under control.

1.4 Terminology in Mis- and Disinformation Research

Intensive experimental deception detection research in psychology goes back to 1950–1970, and most studies during that period were about people communicating face-to-face in pre-digital settings. For instance, in addition to being studied in everyday communication, personal, and workplace relations, deception was also explored in forensic and criminal settings (Granhag et al., 2004; Vrij, 2000) as well as in power and politics (Galasinski, 2000; Hancock, 2012). Chapter 2 overviews fundamental psychological questions such as what motivates people to deceive (Sect. 2.2), what people deceive about, and what forms deception takes (Sect. 2.3).

The late 1990s gave rise to technologically enabled deception in new online contexts. Digital media—computer-mediated communication in blogs, e-mails, wikis, tweets, and all sorts of posts in social media—are "easy venues for deception ranging from harmless white-lies and evasions to blatant misrepresentations" (Rubin, 2010, p. 2). Many varieties of deception strategies are accomplished via technological means. Some types—like omission or equivocation—simply appear relatively unchanged in their new digital guises, while others—like doctored images or videos, *deepfakes*—are born digitally and owe their manipulative powers to online affordances.[2]

1.4.1 Digital Deception

With much of our professional and leisure lives conducted via our devices (phones, tablets, laptops, etc.), it should come as no surprise that the control and manipulation of information is enacted through technologically mediated deceptive messages. Jeff Hancock, the Stanford Social Media Lab director and one of the current leading researchers in the field, extended Buller and Burgoon's (1996) earlier definition of face-to-face deception to *digital deception*, marking "the intentional control of information in a technologically mediated message to create a false belief in the receiver of the message" (Hancock, 2012, p. 2).

As of 2012, two potentially overlapping digital deception types were prevalent and primarily discussed in the literature: *identity-based digital deception* is the false manipulation or display of a person or organization's identity (e.g., an e-mail designed to look like it originated from a Nigerian prince in need of distributing vast sums of money); and *message-based digital deception* is a message exchanged between interlocutors that is manipulated or controlled to be deceptive (e.g., a mobile phone call to excuse lateness with a fictional traffic jam) (Hancock, 2012).

1.4.2 "Fake News"

Neither variety of digital deception captured the general public's imagination as much as the term "fake news" did around the 2016 US Presidential Elections. A handful of academic papers prior to that time used the term to roughly mean digital deception or false reporting, in some cases referring to satirical fakes, like those

[2]What I mean by *affordances* here is actions that are possible within a given environment, referring to what users can do within the parameters of a particular technology, like what the features of a cellphone afford you to do.

published by satirical newspapers *The Onion* and *The Harvard Lampoon*. Of the 34 academic articles (between 2003 and 2017) that were found to use the term "fake news" in their meta-analysis study, Molina and colleagues from the Penn State University (2021) were able to disambiguate the nature of seven different types of online content related to the label of "fake news"—false news, polarized content, satire, misreporting, commentary, persuasive information, and citizen journalism—to be contrasted with the "real news." Real or truthful reporting in digital journalism is a norm in well-established print or radio "legacy journalism" outlets,[3] and reputable news sources can be considered genuine unless proven otherwise, retracted, or corrected (see also Chap. 5, Sect. 5.3.1 "Journalistic Goals and Aspirations in Pursuit of Facts").

Since around 2016, the term "fake news" has been "irredeemably polarized" and co-opted by politicians to refer to any information put out by sources that do not support their partisan positions (Vosoughi et al., 2018, p. 1146). Right-wing US politicians and commentators used it to castigate left-wing reporting and critical news organizations (Molina et al., 2021). After such political weaponization of the term, many scholars swiftly distanced themselves from its use and adopted alternatives such as mis- and disinformation, or more broadly, problematic information and information disorder.

1.4.3 Information Disorder and Other Problematic Content

Information disorder is an overarching concept, originally introduced and popularized by Claire Wardle and Houssein Derakhshan in their (Wardle & Derakhshan, 2017) Report for the Council of Europe, that encompasses both mis- and disinformation, as well as a third category of "malinformation." *Malinformation* is explained as "genuine information that is shared with an intent to cause harm. An example of this is when Russian agents hacked into emails from the Democratic National Committee and the Hillary Clinton campaign and leaked certain details to the public to damage reputations" (Wardle, 2019, p. 8). An influential article in *Science Magazine* (Lazer et al., 2018) frames mis- and disinformation as multiple varieties of the disorder. At times, *dis*-information is drawn as the middle of the two overlapping circles, like a Venn diagram (see Fig. 1.7).

The Data & Society Research Institute out of New York City collects all of the above information disorders under *problematic information* online, pointing out that the boundaries between these slippery terms are blurry. They add a variety of systematic persuasion campaigns into the mix: "advertising (companies trying to persuade the public to buy goods and services), public relations (companies, non-profit organizations, or other non-governmental groups trying to persuade the public

[3] See, for example, www.nytimes.com in the US, www.bbc.co.uk in the UK, or www.cbc.ca in Canada.

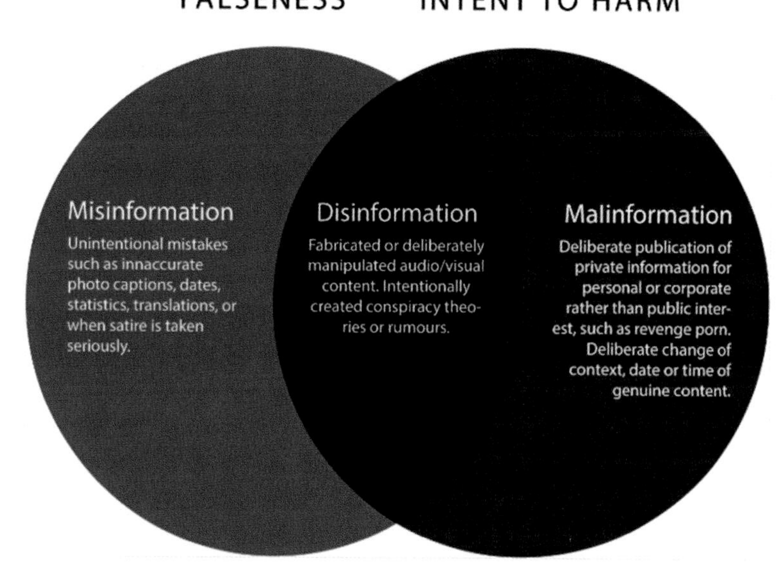

Fig. 1.7 Three types of information disorder (Wardle & Derakhshan, 2017)

to view them more positively), and public diplomacy/public affairs (countries trying to improve their public reputations in other nations)" (Jack, 2017, p. 15). These campaigns, depending on the perspective, can "rightly be referred to as *propaganda*—systematic information campaigns that are deliberately manipulative or deceptive" (Jack, 2017, p. 15) (see also Chap. 6). Other categories—*hoax, satire, parody,* and *culture jamming*—are treated separately due to their use of critique or cultural commentary (Jack, 2017). The spectrum of harm allows one to arrange various types from least to most harmful: "clickbait content, misleading content, genuine content reframed with a false context, imposter content when an organisation's logo or influential name is linked to false information, to manipulated and finally fabricated content" (Wardle, 2019, p. 12) (see Fig. 1.8).

An alternative multipart typology was proposed in *Digital Journalism* by Tandoc et al. (2018), based on levels of deception, as orthogonal dimensions to facticity (i.e., the degree of reliance on facts): news satire, news parody, fabrication, manipulation, advertising, and propaganda.

Many definitions and classifications are "a useful starting point in enhancing our understanding of the phenomenon" but are not easy to implement in practice for AI: "we need more such distinguishing characteristics and dimensions, especially those that can be usefully incorporated in automated detection algorithms" (Molina et al., 2021, p. 181). As people say, sometimes less is more, and in the case of AI R&D having fewer but clearly defined tangible categories is beneficial.

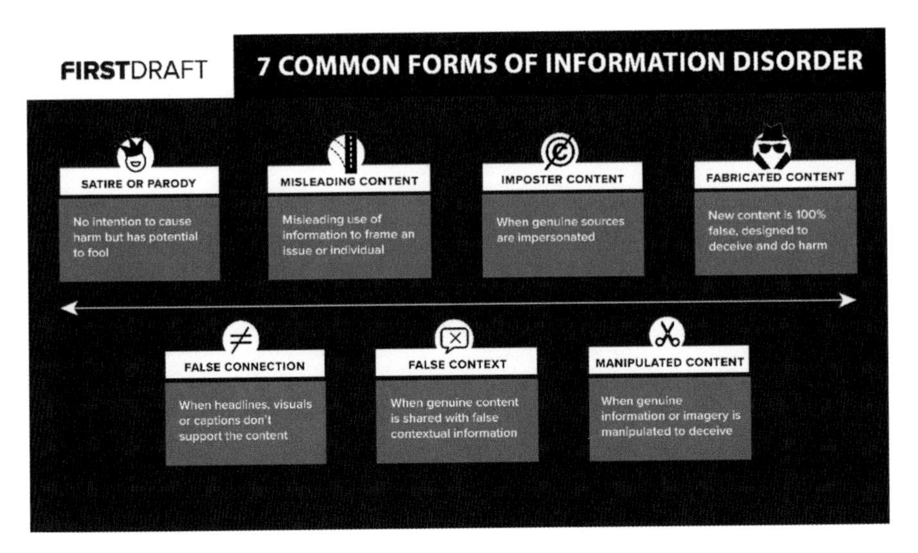

Fig. 1.8 Seven common forms of information disorder (Wardle, 2019)

1.5 Computational Fields

In the early 2000s, online news, social media, and dedicated "fake news" websites started to display a range of deceptive strategies—from equivocation to exaggeration, from misattribution to outright falsification. The more serious and prevalent deceptive attempts became, the more urgent is the need to figure out a mechanism for spotting various kinds of lies online. How do deceivers express outright falsifications? What do the words of an online deception attempt look like? Are there patterns in the similarities and differences between truths and lies? How apparent are they to everyone?

Both sides of deceptive communication needed to be studied—the deceiver and the deceived, the con and the mark. Deceptive intentions are rarely openly stated, they could be rather covert and so subtle that they are barely perceivable. In the legal system, for instance, figuring out intentions is one of the biggest challenges. But are there linguistic cues to deceptiveness? Perhaps there are linguistic "footprints" left by deceivers in what they state and how. Studying practices in lie detection (see Chap. 5, Sect. 5.2) makes us aware of common deceptive strategies and some palpable language cues such as evidence of preplanned or spontaneous execution. Are there differences in their language expression along any of those lines? There was an acute need to find methods to identify a "fake" by analyzing language online, with the hope of implementing it in a computerized identification system. Computational linguistics is best suited for such pursuits in language-related technological innovation.

1.5.1 *AI, Computational Linguistics, and Natural Language Processing*

The field in which the phenomena of digital deception, mis-, and disinformation is studied is referred to by two names—computational linguistics (CL) and natural language processing (NLP)—often used interchangeably. The field is indeed at the intersection of human and artificial intelligence, one informing and enhancing the other. Undoubtedly, other research areas study the phenomena from various angles but what singles out NLP in AI is the promise of computational language technology to assist and enhance human capabilities. At the intersection of computer sciences and linguistics, the research on automated detection of mis- and disinformation requires both a computational and a linguistic background. Additional knowledge of human nature and behavior also typically comes from empirical research in social and interpersonal psychology, studies of computer-mediated communication, human–computer interaction, and information sciences.

Computational linguists find ways to decipher language and draw insights from language patterns and regularities. We create artificially intelligent computer programs, or algorithms, that can "understand" and use language like humans do (Liddy et al., 1999). NLP is a subset of AI that enables computers to "understand," interpret, translate, and otherwise manipulate human language. We can say that AI (more specifically, NLP and ML) can be used to curb mis- and disinformation, to an extent. What this means is that we can create stepwise instructions (or algorithms) or predictive language models that analyze textual data as artifacts of human behavior, and then render automated decisions to the system's users (e.g., rumor or not).

NLP algorithms can now do many kinds of operations with languages: They mimic our human ability to read, write, speak, question and answer, and translate. Machine algorithms can "read in" textual data as input, first character by character, then recognizing strings of characters as words, and sequences of words as sentences. They store everything that was learned from texts and then analyze it, looking for what is already known or predefined from other resources, connecting new information to previously established meanings. Word senses are inferred from the context, or looked up and matched to preexisting lists, libraries, and dictionaries, or are gleaned from other sample texts. NLP algorithms reinterpret and reassign meaning, building from smaller language units to more abstract terms to achieve human-like understanding.

You are likely already using some rudimentary NLP applications such as spell-checkers and grammar checkers. Your search keywords may be autocompleted based on your prior searches or patterns in other people's searches. Statistics often tell us the chances of one solution being more likely than another, say "Donald Trump" may be more commonly used than "Donald Duck" in combination with "press conference." Predictions of more likely word sequences are enabled by preliminary analyses of patterns in volumes of search queries. When NLP predictions coincide with what you had in mind, it may indeed feel like magic: "How does Google know?!"

When AI capabilities serve you right, they are often taken for granted. For instance, we may turn to machine translation applications—Google Translate, Microsoft Translator, or IBM Watson Language Translator—when we need to figure out an unknown sequence of words. These NLP tools may recognize the language and give you an idea of what it roughly means in your native tongue. We are quick to notice the flaws when the results of NLP and AI predictions are incorrect or feel unnatural. Yet, I still marvel at AI when my teen shows off his casual morning conversation with Siri on his phone, literally asking it about the weather for the day. Such speech-to-text and text-to-speech NLP technologies, as Siri or Alexa, are so seamlessly integrated with natural language search engines that even a toddler can interact with them.

Such examples demonstrate just a bit of what AI-driven NLP technologies can do. It should come as no surprise that AI algorithms can be harnessed for more sophisticated analyses as well. Combining the power of observation from large amounts of data with clever feature engineering helps machines learn more about human language and its uses. Machine-learned predictions can be made about the presence of such complex human phenomena as subjectivity, sarcasm, and emotional charge. Thus, at the core of the AI that "understands" human language and makes intelligent predictions are NLP algorithms that analyze, label, and restructure textual data by finding, matching, extracting, and augmenting meaningful pieces of information as specified by their developers.

1.5.2 Shift from Factuality to Subjectivity Analysis

How did we get from a basic NLP analyses like spellchecking to the automated prediction of deceptive news and fact-checking? In broad strokes, it took about 70 years or so of intense R&D efforts. The first 50 years of NLP were focused on obtaining objective information from texts, so that we can retrieve, analyze, sort, and use it. The early field was concerned with finding out what was objectively and factually said: Who did what, when, and where, or what is stated about all of that. The main driving force behind the science of information searching (Salton & McGill, 1983) from around the mid-1950s till the late 1990s was the retrieval of objectively relevant documents as a whole to reduce information overload.

Since the early 2000s, with the growth of the Internet and social communication platforms, NLP researchers became interested in how subjectivity is expressed in human languages. It was a stark contrast to the earlier presumption of the objectivity of information as used in the more classic information retrieval task. The original intention of putting documents on the Internet was, above all, to inform others. Then NLP shifted to identifying opinions, emotions, attitudes, sarcasm, and other subtle properties of language (Wiebe et al., 2005). Consequently, these various forms of automated sentiment analysis flourished (Liu, 2012; Pang & Lee, 2008). The rise of social networking technologies allowed researchers and developers to access a greater number of varieties of language.

Since the dawn of "weblogging" in the early 2000s, people started keeping personal diaries online and sharing their stories publicly. Early bloggers, the precursors of social media users, were publicly reasoning and commenting on events in their lives and the lives of others. Their thoughts and opinions, openly stated intentions, laments, and confessions disclosed their subjective realities. Early blogs, just like any other user-generated content, offered linguistic data with insights into human behaviors and attitudes and quickly became an important source of research data.

As a graduate student at Syracuse in the early 2000s, I studied the properties of blogs for NLP applications. I was specifically tasked with finding posts that showed emotion. I found emotionally loaded words could reveal our inner states ("happy," "sad," "terrified") and thus set the tone of an overall positive or negative attitude toward a person, object, or event (Rubin et al., 2004). In my early blog-harvesting journeys, I also noticed assertive words that signaled confidence ("probably," "must be," "definitely") while hedges showed doubt and hesitation in how people expressed ideas ("possibly," "might not happen," "I doubt that it could") (Rubin et al., 2005). I observed and documented significant differences in the prevalence of such subjective markers in news reports and editorial pieces (Rubin, 2006). My dissertation documented the early markers of (un)certainty, giving rise to the novel NLP task of *certainty identification* and consequently the related task of *factuality (or facticity) identification* in which events that "were judged as not having happened, or as being only possible, are different from those derived from events evaluated as factual" (Saurí & Pustejovsky, 2012, p. 261).

Following my interest in deception in everyday life, I started wondering whether blogs could show subtle language cues that may reveal the author's deceptive intentions. When we tell a story from our subjective points of view, do we simply state what we know to be true, or do we also happen to reveal our hidden intentions? Can people, for instance, unintentionally leak their intent to deceive? The pervasiveness of deception and its damaging powers are obvious to most of us, but it is equally puzzling why it is so hard to figure out when someone is lying.

In spite of the many advantages of publicly available research data (ethical data use issues aside), they lack important controls. To identify the true objective cues of deception, we needed to compare samples of lies and truths. In subsequent studies, study participants were asked to provide subjective narratives: Some intentional lies, others were truths. The analysis of differences across the truthful and deceptive samples revealed curtain tendencies, for instance, in the use of pronouns (they, them, his, my, etc.) (Rubin & Conroy, 2012a). Consistent with prior research, deceptive language indeed shows its own unique properties that contrast with the language of truth-tellers (Mihalcea & Strapparava, 2009; Zhou & Zhang, 2008). Such fundamental insights in these early work results, though not uncontroversial, spurred the development of AI systems that can automatically seek out and identify various "fakes" in textual data.

1.6 Varieties of Fakes Identifiable with AI

My doctoral students and I have consistently drawn attention to deceptive patterns in online language since 2010 (Rubin, 2010; Rubin & Chen, 2012; Rubin & Conroy, 2012a, 2012b; Rubin & Lukoianova, 2014). Other early works in the fields of AI, NLP, and ML were in search of distinguishable features of tangible varieties of digital fakes to develop systems that could somehow capture such fakes automatically (Feng et al., 2012; Zhang et al., 2012; e.g., Zhou et al., 2004). In 2015, my lab conducted a survey of the existing content-based and network-based automated methods at the time (Conroy et al., 2015), and defined the newly formulated task of *fake news detection* as the categorization of news along a continuum of veracity from legitimate to fraudulent news. By 2015, we argued that digital deception needed close examination and put forward a typology of "fake news," suggesting methods for recognizing them computationally using NLP and other AI techniques (Chen et al., 2015; Conroy et al., 2015; Rubin et al., 2015).

Needless to say, since the 2016 US Presidential Election and its popularization of the term "fake news," the problem has been on the radar of many tech companies, social media users, and computer scientists in academe. As a result, the field of automated deception detection and fact-checking has flourished in the past 5 years or so. By 2019, my lab developed a proof-of-concept News Verification Browser that uses NLP and ML algorithms to find misleading and deceptive news—outright falsehoods, satire and clickbait (Rubin et al., 2019).[4]

At the same 2015 conference, we conceived of three tangible varieties of digital fakes that were feasible to detect with AI: serious journalistic fabrications, large-scale multiplatform hoaxes, and humorous fakes (Rubin et al., 2015), each discussed next in turn.

1.6.1 Serious Journalistic Fraud, or Falsehoods

At the time when early automated deception detection efforts were made, data were scarce since instances of *serious journalistic fraud* rarely became public knowledge. Up until the early 2010s, traditional news media were generally quick to retract fraudulent reporting emanating from their newsrooms, if and when they were uncovered. Cases of outright fabrications by celebrity hosts or journalists caused scandals. When journalistic fraudsters were exposed, they faced harsh consequences for their dishonest reporting (Compton & Benedetti, 2015; Shingler, 2015). For instance, a high-profile French-Canadian foreign correspondent, Francois Bugingo, was investigated by the Professional Federation of Quebec Journalists on a report of fabricated information (Montreal CTV News, 2015a), and he admitted to "errors in

[4] This suite of working proof-of-concept applications is freely accessible on GitHub for anyone to download and experiment with.

Fig. 1.9 Three types of fake news form three subtasks in fake news detection (Rubin et al., 2015): (**a**) Bugingo's fabrications exposed (Shingler, 2015); (**b**) large-scale hoaxes (Matt, 2015); (**c**) news satire (The Onion, 2015)

judgement" in his plagiarized reporting a week later (Montreal CTV News, 2015b) (see Fig. 1.9a, cited in Rubin et al. (2015)). News fabrications have provided rich data to computational linguists for automated detection of mis- and disinformation in digital settings since the early 2010s, much like false testimonies in the justice system have interested psychologists since the 1970s. (For contrast, see Chap. 5, Sect. 5.3 on standards and norms in high-quality investigative reporting.)

Pre-2016 fabrications were still hard to collect, verify, and aggregate into databases that could be used as comparative sample data for AI R&D. But to the AI developers who set out to detect lies, such narratives have become valuable since they are likely to exhibit cues of deception akin to "verbal leakages" (see Chap. 2, Sect. 2.5). Databases of sample fake news in politics, entertainment, business, and other domains are now available, curated, and distributed for use in R&D (e.g., see Dogo et al.' (2020), who use several of them comparatively to study the coherence of fake news).

In a later R&D effort (Rubin et al., 2019), we refined the original (Rubin et al., 2015) typology and implemented a proof-of-concept suite of technologies, the News Verification (NV) Browser, with one of the three detection technologies identifying outright falsehoods. Conceptually, we equated falsehoods to serious journalistic fraud, digital deception, as well as disinformation, since all three concepts reflect the deceptive intent of the message creator. (See also Chap. 7, Sect. 7.2 "Automated Deception Detectors").

1.6.2 Rumors or Hoaxes

The second tangible variety of fakes is related to *rumors or hoaxes*, either specific to one platform (such as Twitter), or appearing across social media platforms. Multiplatform attacks can reinforce rumors because people tend to go to alternate sources if they hear a rumor. Well-coordinated and well-timed rumors can mobilize populations to act in the interest of the "rumor engineers" or generally cause some confusion in the public. Infamously, the multiplatform attack #ColumbianChemical plant hoax used an identifiable communication pattern of text messages to the residents' neighboring the plant in Centerville, Louisiana, and social media venues (see Fig. 1.9b, cited in Rubin et al. (2015)). The rumor of the alleged explosion and toxic gas release at the plant convinced the surrounding citizens to evacuate quickly, caused chaos and traffic jams on the highways when panicked citizens tried to leave the area, and created public outrage. Such highly coordinated attacks are sometimes referred to nowadays as "social media hacks" or instances of disinformation campaigns and they require time and effort to orchestrate. In simple terms, they are viral rumors. In the AI community, *rumor debunking* is a specific task for identifying (or busting) such rumors, with methodologies usually tailored to the specifics of social media platform formats and their networking affordances. (For an overview of automated rumor debunkers, see Chap. 7, Sect 7.5; Rubin (2017), with an example for Twitter; Liu et al. (2015) and for Sina Weibo Wu et al. (2015)).

1.6.3 Satirical Fakes

Another tangible category is *satirical fakes* (or humorous fakes, news satire, parody, satirical game shows), which are reports of plainly obvious nonexistent, surreal, alarming events which can be taken seriously by mistake. For instance, Jack Warner, the former FIFA vice president, "has apparently been taken in by a spoof article from the satirical website *The Onion*" (Topping, 2015) after *The Onion* had suggested that the FIFA corruption scandal would result in the 2015 Summer Cup moving to the US (see Fig. 1.9c, cited in Rubin et al. (2015). Satire seems to be experiencing a renewed level of popularity in the 2000s. Some notable examples in the English- and French-speaking worlds include *Saturday Night Live*, *The Daily Show*, and *The Colbert Report* in the US; *Have I Got News for You* and *Mock the Week* on the BBC in the UK, *CBC's This is That* and *The Rick Mercer Report* in Canada; and *Rendez-vous avec Kevin Razy* on Canal+ in France. Satirical fakes rely heavily on irony and deadpan humor to emulate a genuine news source, mimicking credible news sources and stories. Mock newscast segments provide a steady stream of data to be analyzed, but their writers' intentions to entertain, mock, and be absurd may cause algorithms to pick up cues of unbelievability, sensationalism, or humor instead of cues for deception.

Since satires often achieve wide distribution outside of the context provided by their home site, they should be explicitly distinguished from legitimate news, as well as from serious fabricated news and malicious hoaxes. If readers are aware of the humorous intent, they may no longer be predisposed to take the information at face value. Social media technologies should identify humor and prominently display the originating sources to alert users, especially on decontextualized news aggregators or social media platforms. AI researchers have also devised ways to identify the presence of absurdity and humor in order to pinpoint satire (see Chap. 7, Sect. 7.4; Burfoot & Baldwin, 2009; Rubin et al., 2016).

1.6.4 *Deepfakes*

Deepfakes have been found to be persuasively disinformative, but "no more so than equivalent misinformation conveyed through existing news formats like textual headlines or audio recordings" (Barari et al., 2021, p. 1). Note that this book explicitly ignores image and video fakes formats due to their features being distinctly different from text- and speech-based cues for automated deception detection with NLP.[5]

1.6.5 *Resulting AI Technologies*

As dark creative minds continue to invent new types and formats of digital deception, conceptual work on distinguishing those types has to keep up and precede any R&D efforts in AI. What is clear is that simply referring to *fake news detection* as the task in AI is insufficient and vague. What does a resulting system, or any mobile *fake news app*, claim to identify precisely? Many subvarieties of fakes require different methods of detection.

There are at least five types of NLP and ML technologies in AI, with multiple implementation methodologies and subtasks in each because each of the five targets a corresponding variety of fakes (see Chap. 7 for the discussion of algorithmic details). *Automated deception detectors* (see Chap. 7, Sect. 7.2) were developed based on years of deception detection research in experimental psychology (see Chap. 2). *Clickbait detectors*, *satirical fake detectors*, and *rumor debunkers* (see Chap. 7, Sect. 7.3, 7.4, and 7.5) rely more heavily on the more recent explosion of problematic content in false and misleading news, and each warrants its own ways of identification. Computational fact-checking tools (see Chap. 7, Sect. 7.6) attempt to recombine several verification subtasks and bring them into a single user interface, mimicking investigative journalism practices and manual fact-checking (see Chap. 5, Sect. 5.3).

1.7 Conclusions

How do we detect, deter, and prevent the spread of various kinds of digital fakes manually (using the critical human eye) and automatically (using NLP- and ML-enabled AI technologies)? To date, there is no complete and effective solution to halt mis- or disinformation from propagating, leaving many of us wondering what can be done to change the situation. While researchers, scientists, developers, educators,

[5] Current trends and advances in (non-text based) creation and detection of *deepfakes* (Mirsky & Lee, 2021) and their surrounding controversy (Barari et al., 2021) may still be of interest to some readers, but are outside of my book's scope.

and legislators are looking for elusive solutions, many of us remain at risk of being woefully misinformed. Your personal risks can be decreased by exercising awareness of the situation, causal factors, and potential solutions to the online infodemic.

The central problem is human interaction with information: How we go about handling and comprehending the information we encounter. We need to start with the premise that the problem at hand is not purely technological. AI cannot possibly be the exclusive tool to debunk rumors or to convince anyone to be critical of any news. AI assists people by leading them on the verification path: It can alert us to the possibility of foul play and suggest that we be skeptical of it. The ultimate decisions are made by each individual—to believe or not, and to act on that belief or not. Solutions to the mis- and disinformation problem must incorporate our behaviors, cognition, inherent beliefs, and acquired attitudes in our interactions with novel information. We have to be aware of the possible kinds of deception that we may encounter, in particular, unverified information, rumors, propaganda, and other potential lies. The ultimate countermeasures and defense systems have to be erected within our critical minds. The mis- and disinformation problem is framed as a socio-technological one, and any solutions to this problem have to start from a deeper understanding of the human mind within the context of information exchanges and the deceptive intentions of individuals in society. As the epigraph of this chapter suggests in the words of software developer and data journalist, Meredith Broussard (2019), we should not mistakenly believe that technology is a solution to every problem.

References

AFP Espagne, AFP Hong Kong. (2021, April 12). Moderna boss did not say "vaccines change your DNA." *AFP Fact Check*. Retrieved from https://factcheck.afp.com/moderna-boss-did-not-say-vaccines-change-your-dna

Barari, S., Lucas, C., & Munger, K. (2021, January 13). Political deepfakes are as credible as other fake media and (sometimes) real media. OSF Preprints. doi: https://doi.org/10.31219/osf.io/cdfh3

Broussard, M. (2019). *Artificial unintelligence: how computers misunderstand the world*. MIT Press. Retrieved from https://mitpress.mit.edu/books/artificial-unintelligence

Buller, D. B., & Burgoon, J. K. (1996). Interpersonal deception theory. *Commun Theory, 6*(3), 203–242.

Burfoot, C., & Baldwin, T. (2009). Automatic satire detection: are you having a laugh?, *Proceedings of the ACL-IJCNLP 2009 Conference Short Papers*, pages (pp. 161–164). Association for Computational Linguistics.

Canada's National Observer. (2019, July 19). *How to spot fake news*. National Observer. Retrieved from https://www.nationalobserver.com/spot-fake-news

Chen, Y., Conroy, N. J., & Rubin, V. L. (2015). News in an online world: the need for an automatic crap detector, *Proceedings of the Association for Information Science and Technology, 52*(1), 1–4. https://doi.org/10.1002/pra2.2015.145052010081

Compton, J. R., & Benedetti, P. (2015). *News, lies and videotape: the legitimation crisis in journalism*. Rabble.ca. Retrieved from http://rabble.ca/news/2015/03/news-lies-and-videotape-legitimation-crisis-journalism

Conroy, N. J., Rubin, V. L., & Chen, Y. (2015). Automatic deception detection: methods for finding fake news. *Proceedings of the Association for Information Science and Technology, 52*(1), 1–4. https://doi.org/10.1002/pra2.2015.145052010082

Cook, J., Lewandowsky, S., & Ecker, U. K. H. (2017). Neutralizing misinformation through inoculation: exposing misleading argumentation techniques reduces their influence. *PLoS One, 12*(5), e0175799. https://doi.org/10.1371/journal.pone.0175799

Dogo, M. S., Deepak, P., & Jurek-Loughrey, A. (2020). Exploring thematic coherence in fake news. In I. Koprinska, M. Kamp, A. Appice, C. Loglisci, L. Antonie, A. Zimmermann, R. Guidotti, Ö. Özgöbek, R. P. Ribeiro, R. Gavaldà, J. Gama, L. Adilova, Y. Krishnamurthy, P. M. Ferreira, D. Malerba, I. Medeiros, M. Ceci, G. Manco, E. Masciari, et al. (Eds.), *ECML PKDD 2020 workshops* (pp. 571–580). Springer International Publishing. https://doi.org/10.1007/978-3-030-65965-3_40

Dyakon, T. (2020, December 14). *Poynter's MediaWise training significantly increases people's ability to detect disinformation, new Stanford study finds*. Poynter. Retrieved from https://www.poynter.org/news-release/2020/poynters-mediawise-training-significantly-increases-peoples-ability-to-detect-disinformation-new-stanford-study-finds/

Endsley, M. R. (2018). Combating information attacks in the age of the internet: new challenges for cognitive engineering. *Human Factors: The Journal of the Human Factors and Ergonomics Society, 60*(8), 1081–1094. https://doi.org/10.1177/0018720818807357

Feng, S., Banerjee, R., & Choi, Y. (2012). Syntactic stylometry for deception detection. In *50th Annual meeting of the association for computational linguistics, ACL 2012—Proceedings of the conference* (pp. 171–175).

Galasinski, D. (2000). *The language of deception: a discourse analytical study*. Sage Publications.

Glanville, D. (2021, January 6). *EMA recommends COVID-19 vaccine Moderna for authorisation in the EU* [text]. European Medicines Agency. Retrieved from https://www.ema.europa.eu/en/news/ema-recommends-covid-19-vaccine-moderna-authorisation-eu

Global Legal Research Directorate Staff, L. L. of C. (2019). *Government Responses to Disinformation on Social Media Platforms* [Web page]. Retrieved from https://www.loc.gov/law/help/social-media-disinformation/index.php

Granhag, P. A., Andersson, L. O., Strömwall, L. A., & Hartwig, M. (2004). Imprisoned knowledge: criminals beliefs about deception. *Leg Criminol Psychol, 9*(1), 103.

Hancock, J. T. (2012). Digital deception: when, where and how people lie online. In A. N. Joinson, K. McKenna, T. Postmes, & U.-D. Reips (Eds.), *The Oxford handbook of internet psychology* (p. 20). Oxford University Press. Retrieved from http://www.oxfordhandbooks.com/10.1093/oxfordhb/9780199561803.001.0001/oxfordhb-9780199561803-e-019

IFLA. (2021, February 18). *How to spot fake news*. The International Federation of Library Associations and Institutions. Retrieved from https://www.ifla.org/publications/node/11174

Jack, C. (2017). *Lexicon of lies: Terms for problematic information*. Data & Society Research Institute. Retrieved from https://datasociety.net/pubs/oh/DataAndSociety_LexiconofLies.pdf

Lazer, D. M. J., Baum, M. A., Benkler, Y., Berinsky, A. J., Greenhill, K. M., Menczer, F., Metzger, M. J., Nyhan, B., Pennycook, G., Rothschild, D., Schudson, M., Sloman, S. A., Sunstein, C. R., Thorson, E. A., Watts, D. J., & Zittrain, J. L. (2018). The science of fake news. *Science, 359*(6380), 1094–1096. https://doi.org/10.1126/science.aao2998

Liddy, E. D., Paik, W., McKenna, M. E., & Li, M. (1999). *Natural language information retrieval system and method* (United States Patent No. US5963940A). Retrieved from https://patents.google.com/patent/US5963940A/en.

Liu, B. (2012). *Sentiment analysis and opinion mining*. Morgan & Claypool Publishers. Retrieved from http://www.cs.uic.edu/~liub/FBS/SentimentAnalysis-and-OpinionMining.html

Liu, X., Nourbakhsh, A., Li, Q., Fang, R., & Shah, S. (2015). *Real-time rumor debunking on twitter*. 1867–1870. doi: https://doi.org/10.1145/2806416.2806651.

Loomba, S., de Figueiredo, A., Piatek, S. J., de Graaf, K., & Larson, H. J. (2021). Measuring the impact of COVID-19 vaccine misinformation on vaccination intent in the UK and USA. *Nature Human Behaviour, 5*(3), 337–348. https://doi.org/10.1038/s41562-021-01056-1

Matt (2015). *#ColumbianChemicals hoax: Trolling the Gulf coast for deceptive patterns*. Retrieved from https://www.recordedfuture.com/columbianchemicals-hoax-analysis/
Mihalcea, R., & Strapparava, C. (2009). The lie detector: explorations in the automatic recognition of deceptive language. *Proceedings of the ACL-IJCNLP 2009 Conference Short Papers* (pp. 309–312). Association for Computational Linguistics.
Mirsky, Y., & Lee, W. (2021). The creation and detection of Deepfakes: a survey. *ACM Computing Surveys, 54*(1)., 7:1–7:41. https://doi.org/10.1145/3425780
Mitchell, A., Gottfried, J., Stocking, G., Walker, M., & Fedeli, S. (2019, June 5). *Many Americans say made-up news is a critical problem that needs to be fixed*. Pew Research Center's Journalism Project. Retrieved from https://www.journalism.org/2019/06/05/many-americans-say-made-up-news-is-a-critical-problem-that-needs-to-be-fixed/
Mitchell, A., Jurkowitz, M., Oliphant, J. B., & Shearer, E. (2020, December 15). *Concerns about made-up election news are high, and both parties think it is mostly intended to hurt their side*. Pew Research Center's Journalism Project. Retrieved from https://www.journalism.org/2020/12/15/concerns-about-made-up-election-news-are-high-and-both-parties-think-it-is-mostly-intended-to-hurt-their-side/
Molina, M. D., Sundar, S. S., Le, T., & Lee, D. (2021). "Fake news" is not simply false information: a concept explication and taxonomy of online content. *The American Behavioral Scientist (Beverly Hills), 65*(2), 180–212. https://doi.org/10.1177/0002764219878224
Montreal CTV News. (2015a, May 23). *Quebec foreign correspondent suspended over allegations of false information*. Montreal. Retrieved from https://montreal.ctvnews.ca/quebec-foreign-correspondent-suspended-over-allegations-of-false-information-1.2387746
Montreal CTV News. (2015b, May 29). *Montreal journalist Francois Bugingo admits to falsifying stories*. Montreal. Retrieved from https://montreal.ctvnews.ca/montreal-journalist-francois-bugingo-admits-to-falsifying-stories-1.2398611
Newall, M. (2020, December 30). *More than 1 in 3 Americans believe a 'deep state' is working to undermine trump*. Ipsos. Retrieved from https://www.ipsos.com/en-us/news-polls/npr-misinformation-123020
Pang, B., & Lee, L. (2008). Opinion mining and sentiment analysis. *Foundations and Trends in Information Retrieval, 2*(1–2), 1–135. https://doi.org/10.1561/1500000001
Pew Research Center Report. (2019). *Trends and facts on online news: state of the news media*. Pew Research Center's Journalism Project. Retrieved from https://www.journalism.org/fact-sheet/digital-news/
Reuters Staff. (2020, May 19). *False claim: a COVID-19 vaccine will genetically modify humans*. Reuters. Retrieved from https://www.reuters.com/article/uk-factcheck-covid-19-vaccine-modify-idUSKBN22U2BZ
Reuters Staff. (2021, January 13). *Fact check: genetic materials from mRNA vaccines do not multiply in your body forever*. Reuters. Retrieved from https://www.reuters.com/article/uk-factcheck-genetic-idUSKBN29I30V
Rosen, G. (2020, August 18). *Community standards enforcement report*. About Facebook. Retrieved from https://about.fb.com/news/2020/08/community-standards-enforcement-report-aug-2020/
Rubin, V. L. (2006). Identifying certainty in texts. Thesis. In *School of information studies*. Syracuse University
Rubin, V. L. (2010). On deception and deception detection: Content analysis of computer-mediated stated beliefs, *Proceedings of the Association for Information Science and Technology, 32*: 1–10. Retrieved from http://dl.acm.org/citation.cfm?id=1920377
Rubin, V. L. (2017). Deception detection and rumor debunking for social media. InSloan, L. & Quan-Haase, A. (Eds.) *The SAGE Handbook of Social Media ResearchMethods*, London: SAGE: (pp. 342–364). https://uk.sagepub.com/en-gb/eur/the-sage-handbook-of-social-media-research-methods/book245370
Rubin, V. L. (2019). Disinformation and misinformation triangle: a conceptual model for "fake news" epidemic, causal factors and interventions. *Journal of Documentation*, 75 (5), 1013-1034. https://doi.org/10.1108/JD-12-2018-0209

Rubin, V. L., & Chen, Y. (2012). Information manipulation classification theory for LIS and NLP. *Proceedings of the American Society for Information Science and Technology, 49*(1), 1–5. https://doi.org/10.1002/meet.14504901353

Rubin, V. L., & Conroy, N. (2012a). Discerning truth from deception: human judgments and automation efforts. *First Monday, 17*(3) Retrieved from http://firstmonday.org/ojs/index.php/fm/article/view/3933/3170

Rubin, V. L., & Conroy, N. (2012b). The art of creating an informative data collection for automated deception detection: A corpus of truths and lies. *Proceedings of the American Society for Information Science and Technology, 49*, 1–11.

Rubin, V. L., & Lukoianova, T. (2014). Truth and deception at the rhetorical structure level. *Journal of the Association for Information Science and Technology, 66*(5), 12. https://doi.org/10.1002/asi.23216

Rubin, V. L., Stanton, J. M., & Liddy, E. D. (2004). *Discerning emotions in texts*. AAAI Symposium on Exploring Attitude and Affect in Text, Stanford, CA. https://www.aaai.org/Papers/Symposia/Spring/2004/SS-04-07/SS04-07-023.pdf

Rubin, V. L., Liddy, E. D., & Kando, N. (2005). Certainty identification in texts: categorization model and manual tagging results. In J. G. Shanahan, Y. Qu, & J. Wiebe (Eds.), *Computing attitude and affect in text: theory and applications* (pp. 61–76). Springer-Verlag.

Rubin, V. L., Chen, Y., & Conroy, N. J. (2015). *Deception detection for news: three types of fakes. Proceedings of the Association for Information Science and Technology, 52*(1), 1–4. https://doi.org/10.1002/pra2.2015.145052010083

Rubin, V. L., Conroy, N. J., Chen, Y., & Cornwell, S. (2016). Fake news or truth? Using Satirical Cues to Detect Potentially Misleading News. *Proceedings of the Second Workshop on Computational Approaches to Deception Detection*, 7–17, San Diego, California. Association for Computational Linguistics. http://aclweb.org/anthology/W/W16/W16-0800.pdf

Rubin, V. L., Brogly, C., Conroy, N., Chen, Y., Cornwell, S. E., & Asubiaro, T. V. (2019). A news verification browser for the detection of clickbait, satire, and falsified news. *Journal of Open Source Software, 4*(35), 1208. https://doi.org/10.21105/joss.01208

Salton, G., & McGill, M. J. (1983). *Introduction to modern information retrieval*. McGraw-Hill.

Saurí, R., & Pustejovsky, J. (2012). Are you sure that this happened? Assessing the factuality degree of events in text. *Computational Linguistics, 38*(2), 261–299. https://doi.org/10.1162/COLI_a_00096

Scholthof, K.-B. G. (2007). The disease triangle: pathogens, the environment and society. *Nature Reviews Microbiology, 5*(2), 152–156. https://doi.org/10.1038/nrmicro1596

Shingler, B. (2015, May 23). *Foreign correspondent suspended by media outlets after report he fabricated stories*. CBC News. http://www.cbc.ca/news/canada/montreal/françois-bugingo-foreign-correspondent-suspended-by-media-outlets-1.3085118

Tandoc, E. C., Lim, Z. W., & Ling, R. (2018). Defining "Fake News". *Digit Journal, 6*(2), 137–153. https://doi.org/10.1080/21670811.2017.1360143

TheOnion. (2015). *FIFAfranticallyannounces2015summerworldcupinUnitedStates*. Retrievedfrom http://www.theonion.com/article/fifa-frantically-announces-2015-summer-world-cup-u-50525

Topping, A. (2015, May 31). *Ex-FIFA vice president Jack Warner swallows Onion spoof*. The Guardian. Retrieved from http://www.theguardian.com/football/2015/may/31/ex-fifa-vice-president-jack-warner-swallows-onion-spoof

Vosoughi, S., Roy, D., & Aral, S. (2018). The spread of true and false news online. *Science, 359*(6380), 1146–1151. https://doi.org/10.1126/science.aap9559

Vrij, A. (2000). *Detecting lies and deceit*. Wiley.

Wardle, C. (2019). *Understanding information disorder*. Retrieved from https://firstdraftnews.org/wp-content/uploads/2019/10/Information_Disorder_Digital_AW.pdf?x76701

Wardle, C., & Derakhshan, H. (2017). *Information disorder: toward an interdisciplinary framework for research and policy making, Council of Europe DGI*. EU DisinfoLab. Retrieved from https://www.disinfo.eu/academic-source/wardle-and-herakhshan-2017/

Wiebe, J., Wilson, T., & Cardie, C. (2005). Annotating expressions of opinions and emotions in language. *Lang Resour Eval, 39*(2), 165–210. https://doi.org/10.1007/s10579-005-7880-9

World Health Organization. (2021a). *WHO public health research agenda for managing infodemics*. World Health Organization. Retrieved from https://apps.who.int/iris/handle/10665/339192

World Health Organization. (2021b, March 16). *Let's flatten the infodemic curve*. WHO Newsroom Spotlight. Retrieved from https://www.who.int/news-room/spotlight/let-s-flatten-the-infodemic-curve

World Health Organization, UN, UNICEF, UNDP, UNESCO, UNAIDS, ITU, UN Global Pulse, & IFRC. (2020, September 23). *Managing the COVID-19 infodemic: promoting healthy behaviours and mitigating the harm from misinformation and disinformation.* Joint statement. Retrieved from https://www.who.int/news/item/23-09-2020-managing-the-covid-19-infodemic-promoting-healthy-behaviours-and-mitigating-the-harm-from-misinformation-and-disinformation

Wu, K., Yang, S., & Zhu, K. Q. (2015). *False rumors detection on Sina Weibo by propagation structures*. IEEE International Conference on Data Engineering, ICDE, 651-662.

Zhang, H., Fan, Z., Zheng, J., & Liu, Q. (2012). An improving deception detection method in computer-mediated communication. *Journal of. Networks, 7*(11). https://doi.org/10.4304/jnw.7.11.1811-1816

Zhou, L., & Zhang, D. (2008). Following linguistic footprints: automatic deception detection in online communication. *Communications of the ACM, 51*(9), 119–122. https://doi.org/10.1145/1378727.1389972

Zhou, L., Burgoon, J. K., Nunamaker, J. F., & Twitchell, D. (2004). Automating linguistics-based cues for detecting deception in text-based asynchronous computer-mediated communications. *Group Decision and Negotiation, 13*(1), 81–106.

Chapter 2
Psychology of Misinformation and Language of Deceit

If falsehood, like truth, had only one face, we would be in better shape.
For we would take as certain the opposite of what the liar said.
But the reverse of truth has a hundred thousand shapes and a limitless field.

(Michel de Montaigne, 1572–1580/1957)

Abstract **Chapter 2** focuses on deception as a communicative behavior and establishes, in broad strokes, what deceptive strategies can be used by deceivers and mass disinformers, and what motivates deceptive communication. We consider definitions of deception and typological distinctions, assuming some deceptive strategies have found their way into the digital realm. What do people deceive about? Why can digital media users be susceptible to mis- and disinformation? Interpersonal social psychology and communication research on deception informs us about characteristics, content, and language of deceit that can be identified with analytical thinking or by automated means. Fact-checking and content moderation are typical responses to mis- and disinformation. Other coordinated countermeasures include preemptive inoculation, warning labeling, and literacy campaigns.

Keywords Computer-mediated communication · Interpersonal psychology · Communicative act · Uncooperative communication · Deceptive messages · Misinformation · Disinformation · Digital Deception · Digital fakes · Deceptive behavior · Deceiver · Deceived · Susceptible digital media users · Digital media users bias · Psychological propensities · Truth bias · Confirmation bias · Lazy thinking · General gullibility · Pseudo-profound bullshit · Politically biased thinking · Identity-protective thinking · Dual processing model · Intuitive thinking · Effortful analytical thinking · Cognitive miserliness · Reasoning abilities · Inference · Information behavior · Reading tendencies · In-depth reading · Information snacking · Selective reading · Countermeasures · Coordinated public interventions · Preventative measures · Fact-checking · Content Moderation · Labeling · Inoculation · Targeted messaging campaigns · Raising public awareness · How-to-spot-fake-news tips · Media literacy · News literacy · Digital literacy · Information literacy

V. L. Rubin, *Misinformation and Disinformation*,
https://doi.org/10.1007/978-3-030-95656-1_2

2.1 Introduction: Deception Is Uncooperative Communication

Mis- and disinformation are versions of unintentional (or intentional) deceptive communicative acts in which a false, inaccurate, or misleading message is mass-delivered over digital channels for the purpose of creating a false impression in someone's mind. This view emphasizes the tripart nature of mis- or disinformation: It is procedurally a communicative act, which transmits a particular message (with its content and form of expression), and there are certain behaviors that both the sender (i.e., the deceiver) and the receiver of the message (i.e., the deceived) engage in. Studies in computer-mediated communication and interpersonal psychology reveal some patterns in deceptive behaviors and provide some explanations of their nature.

The act of deceptive communication is considered uncooperative (McCornack, 1992) in the sense that it covertly or blatantly violates the principles that govern conversational exchanges. Most of us enter in communication with each other expecting to cooperate and to be successful at our communicative goals. We typically proceed with an assumption of truthfulness, honesty, and trust as essential to both communicating parties. Linguists working in the field of pragmatics observe that we adhere to one or more conventions when we interact with each other, though we are often unaware of them until they are broken. Paul Grice, for instance, articulated our mutual expectations in his Cooperative Principle of Communication (Grice, 1975), postulating his famous Four Maxims: Say what you believe to be true (Maxim of Quality), do not say more than needed (Maxim of Quantity), stay on the topic (Maxim of Relevance), and do not be vague (Maxim of Manner). Deception breaks these norms: One or more of the Grice's Maxims are violated and cooperation is undermined, precisely because the deceiver's communicative goals and intentions have little to do with cooperation. The deceiver is guided by ulterior motives and uses deceptive communication to achieve them.

2.2 Motivations to Deceive and Disinform

Those who deceive a friend, coworker, relative, or stranger in small "person-to-person" circumstances and those who disinform many people on a large scale share communicative goals. In both contexts—that of deceivers and disinformers—these agents have deliberate intentions to instill falsehoods in someone else's mind. Their motivations, if not identical, bear a resemblance to one another. Experimental behavioral research has shown that people prefer to achieve their goals truthfully whenever possible and only resort to lying if they cannot achieve their goals through honest means (Levine et al., 2010). These findings are consistent with the *veracity principle*, put forward by Sissela Bok, a philosopher and ethicist who in the late 1970s noted a moral asymmetry between honesty (which requires no justification) and deceit (which does) (Bok, 1999). (See also Chap. 4 on philosophy of truth and deceit.) So, why do people deceive in the first place? What drives the deceivers? And, how, if at all, do these reasons to deceive translate to problematic content online?

Assuming that people deceive for the same basic reasons as they disinform, we will look at the body of deception research at the person-to-person magnitude and see how it can shed light on the scaled-up version of person-to-group magnitude. These insights are important in the context of finding automated AI solutions to the problem of mis- and disinformation because they provide detection cues—such as the pragmatic contexts of language use—that can be taken into account by algorithms. While AI attempts to solve the problem by identifying and flagging falsehoods and misleading information, it must be clearly acknowledged that much work has to happen "upstream of the problem." This means understanding what necessitates people to deceive and disinform on a large scale. This context is instrumental for providing prevention measures and regulating environments by eliminating incentives and thus constricting the spawning of falsehoods in the first place.

In interpersonal communication and psychology research, deception (and lies, i.e., prevarication, specifically) is stereotypically seen as selfish, disruptive, and hurtful to the deceived, but there are additional nuances. Aldert Vrij, social psychologist and internationally recognized expert in deception, says that in addition to the commonly thought of motives (to either gain material advantage or avoid materialistic loss or punishment), people also lie for psychological reasons such as to avoid embarrassment or to present a better "edited self" (Vrij, 2008). Psychologists affirm that people deceive to achieve their social interaction goals, much like what linguists in pragmatics call communicative goals. In everyday interactions, people lie to influence others, make a good impression, to reassure and support others (DePaulo et al., 1996, 2003), or to accomplish their social interaction goals in romantic relationships (Ennis et al., 2008). People deceive when their communication goals are not attainable through honest means (Levine et al., 2010).

In the 1990s, experimental researchers divided the reasons to deceive into "three categories: "Self-centered" lies are told to protect the self (e.g., "I did not do it"). "Other-oriented" lies are told to protect another (e.g., "your hair looks nice today"). Finally, "altruistic" lies are told to protect a third party (e.g., "Julie could not have done that as she was at home with me at 9 pm") (DePaulo et al., 1996; Vrij, 2000, p. 200). Although, other-oriented and altruistic lies may also protect one's own well-being (e.g., to help maintain a relationship or protection from hassles and arguments stemming from the truth) (Ennis et al., 2008 p. 106).

By 2008, Vrij reviewed his distinctions between the motivations to deceive in the face of mounting experimental evidence and study participants' self-reports. He explains in his well-cited book that some combinations of the following three motivational dimensions are possible: (1) self- or other-orientated benefits, (2) materialistic advantage gain or cost avoidance, and (3) psychological reasons (Vrij, 2008). An example of a self-oriented material advantage includes glossing over the shortcomings of one's own house when presenting it to a potential buyer. Self-oriented psychological advantage gain may motivate exaggerating one's success on exams to build positive impressions in others. Denying guilt, or a child's reluctance to admit misdeeds, can be interpreted as self-oriented avoidance of psychological costs or punishments. Making a flattering comment about a friend's attractiveness can boost her self-confidence. And an innocent father owning up to his son's crime is motivated by the father's desire to avoid his son's punishment. People may also lie socially to please each other at work or in romantic relationships, to maintain a better mutual relationship, or for psychological reasons like to save face or avoid damaging one's self-esteem (Vrij, 2008).

2.2.1 Deceptive Content

What do people generally lie about? You can imagine a lie about almost anything, but research has been done that could help us gain a bird's eye view of the typical content of lies. Studies in information sciences have poured over user-generated content online to scrutinize stories that describe incidents of deception, the surrounding circumstances, and the ways in which lies were discovered. A content analysis of 324 blog posts containing descriptions of deceptive incidents identified 11 self-reported deception varieties, listed here from most common to the least: *mis/dis-informing, scheming, lying (deliberately prevaricating), misleading, misrepresenting, cheating, (mostly) showing frustration about deception, white-lying, name-calling, plagiarizing,* and "*bs*"-*ing* (Rubin, 2010). These blog posts covered nine broad domains in life: environment, finances and insurance, law enforcement, media and entertainment, medicine and health, personal relations, politics, religion, and science. This wide-ranging span of domains confirms that there is practically no life sphere that has not been touched by someone's deceptive intent.

2.2.2 Serious and Everyday Lies

As an alternative to distinguishing deception by its type, context, or domain, we can also consider the consequences of deceptive acts when answering the question of when and how people lie. Some lies seem to be more easily dismissed than others.

Researchers distinguish between serious lies and everyday lies. Psychologists directed study participants to report what they lie about, and later systematized their responses. In their study of 77 students and 70 community members, DePaulo et al. (1996) identified five groups of *everyday lies:* about *feelings and opinions* (e.g., exaggerating about being hurt by a comment); about *actions, plans, and whereabouts* (e.g., lying that a check was sent); about *knowledge, achievements, and failings* (e.g., lying about poor performance on a test while acing it); about *explanations for behaviors* (e.g., being late for work because of car troubles); and about *facts and personal possessions* (e.g., lying about father being an ambassador).

The authors' follow-up study of *serious lies* (DePaulo et al., 2004) collected autobiographical narratives about lies from 128 students and 107 community participants. These more-substantial falsehoods were categorized into eight groups indicative of most significant life tasks in the study participants' lives. The eight content-themed groups included lies about *affairs or other romantic cheatings*, about *misdeeds* (e.g., lying on the college application about having been suspended for giving out drugs on a high school field trip); about *personal facts or feelings* (e.g., lying about a miscarriage); about *forbidden socializing* (e.g., describing plans to spend the night babysitting instead of attending the forbidden school dance); about *money or jobs* (e.g., investing money in the stock market after promising to save it for a down payment on a home); about *death, illness, or injury* (e.g., a parent or grandparent's serious illness is misrepresented to a teenager); about *identity* (e.g., lies about frequenting a bar for gay men); and about *violence or danger* (e.g., a lie told by a commanding officer who claimed that there were no enemies in a village he knew to be heavily defended) (DePaulo et al., 2004). Interestingly, most serious lies were told by or to the participants' closest relationship partners—not strangers—in order to get what they wanted, or to do what they felt they were entitled to do. Other motives were to avoid punishment, to protect themselves from confrontation, to appear to be the type of person they wished they were, to protect others, and to hurt others (DePaulo et al., 2004). It would be reasonable to assume that most of these motivations still hold for putting out intentionally deceptive content online.

2.2.3 Incentives to Disinform

How directly applicable is this interpersonal deception research to the motivations underpinning the proliferation of mis- and disinformation online? Do the reasons to deceive online fit within these typologies? The ad revenue model of social platforms has been implicated as the main motivator for generating falsehoods. Facebook, Twitter, and Google's targeted advertising business models, their lack of transparency, and the lack of accountability in their opaque algorithmic systems have been called "the root cause of their failure to staunch the flow of misinformation" and "despite these companies' commitment to take unprecedented steps to control the problem, they are failing" (Maréchal et al., 2020, p. 5). Two types of algorithms have been used to propagate or to prohibit various forms of online discourse: "(1) content-shaping algorithms that determine what individual users see when they use a company's online services (including those that target ads) and (2) content moderation algorithms that help human reviewers identify (and sometimes remove) content that violates the company's rules" (cited in Maréchal et al., 2020, p. 11; Maréchal & Biddle, 2020). Borrowing a metaphor from the oil industry, this report contends that it is impossible to "clean up downstream pollutants like misinformation or dangerous speech without tackling the upstream processes—targeted advertising and algorithmic systems—that make this speech so damaging to our information environment in the first place" (Maréchal et al., 2020, p. 10).

The structural incentives of the "targeted ads" business model and their content propagation mechanisms are responsible for the spread and abundance of mis- and disinformation online, but the motivations of the agents of disinformation—the originators of falsehoods—are still simplistically understood. Not unlike the early 1990s research in deception, we readily acknowledge two selfish veins: material (money) or political gains (influence and popularity). Disinformers may also promote their own positive "self-edited" image or an improved image of the organization they have vested interests in (aligning with the concept of other-oriented gains). More research is needed to see if the other person-to-person pedestrian motivations also make people ambitious enough to disinform at scale. It is reasonable to assume so, but to the best of my knowledge such assumptions have rarely been questioned in research, perhaps because the monetary and political gains are so self-evident.

Occasional media interviews with those who make their living producing "fake news" confirm the assumption of strong financial incentives, at least anecdotally. In the wake of the spike of fake news associated with the 2016 US Presidential Election, BuzzFeed News was the first to bring to light a story about a cluster of faux sites and viral made-up stories which all originated from the Macedonian town of Veles and which were all written by youths. The reporters' source, a university student in Veles who agreed to speak anonymously, described his US politics site: "Yes, the info in the blogs is bad, false, and misleading but the rationale is that if it gets the people to click on it and engage, then use it." Interviewers Craig Silverman and Lawrence Alexander expanded: "These sites open a window into the economic incentives behind producing misinformation specifically for the wealthiest advertising markets and specifically for Facebook, the world's largest social network, as well as within online advertising networks such as Google AdSense" (Silverman & Alexander, 2016).

Later in 2016, NPR correspondent Laura Sydell, tracked down the renter of a single Amazon server space that put up sites like NationalReport.net, USAToday.com.co, and WashingtonPost.com.co: Jestin Coler, a Los Angeles resident and former magazine subscriptions salesman, database administrator, and freelance writer for magazines. Listed online as the founder and CEO of Disinfomedia and christened by NPR "a sort of godfather of the industry," Coler built a fake news "empire" by employing about 20–25 writers (Sydell, 2016). In November 2016, he aggregated over half a million Facebook reshares on the fabricated story "FBI Agent Suspected in Hillary Email Leaks Found Dead In Apparent Murder-Suicide" which was published by a fictitious newspaper. According to Snopes.com, "there was no truth to this story. The Denver Guardian is simply a fake news website masquerading as the online arm of a (non-existent) big city newspaper" (Mikkelson, 2016). During the NPR interview, Coler bragged about how easily people believe fake news and said that if he left his business, "dozens, maybe hundreds of entrepreneurs will be ready to take his place" because "they know now that fake news sells and they will only be in it for the money" (Sydell, 2016). Coler reported making "good money" from ads on fake news websites and he fit into a pattern of fake news writers who especially targeted Trump supporters: "He wouldn't give exact figures, but he says stories about other fake-news proprietors making between $10,000 and $30,000 a month apply to him." Coler insisted that his behavior was not about money but rather "about showing how easily fake news spreads" but then admitted that "the money gave him a lot of incentive to keep doing it regardless of the impact" (Sydell, 2016).

A 2018 BBC exposé profiled another prolific fake news writer, this time a resident of the U.S. East Coast around Portland, Maine: Christopher Blair, formerly a construction worker and liberal political blogger. In the interview with Anisa Subedar, he exhibits an almost sardonic arrogance and enjoyment in describing which buttons he would push on any given day in his false news, describing gun control, police brutality, and feminist policies as one set of options or making up "a controversial incident, a crime, a new law or a constitutional amendment" as an alternative. The stories he fabricated included "rumours about Hillary Clinton's health problems and illegal dealings, stories about luminaries like Pope Francis lining up behind the Republican candidate, and other false news sure to either please or rile Trump's supporters" (Subedar, 2018). His tactic was to write sensational and sometimes offensive, bigoted, or racist stories that would provoke an emotional response and get clicks and reshares, so that "he was able to use Google's advertising platform to convert page views into money." His original motivation was a search for a more lucrative and easier lifestyle than construction work. He claims that he places enough disclaimers to demonstrate that his writing is all a satire because his raison d'être is to "trick conservative Americans into sharing false news, in the hope of showing what he calls their 'stupidity'" and then exposing the most extreme and racist people that take his bait (Subedar, 2018). In terms of psychological motivations to deceive, there are clearly stated ideological underpinnings, with aspirations to politically motivated "prankster-ism" as a defense.

When political aspirations take on a more organized effort and turn into concerted political disinformation campaigns, the links to propaganda become apparent. Techniques used in mass persuasion campaigns may aim to create chaos, confusion, and information overflow, or to undermine trust in previously well-established institutions, norms, and traditions. Motivations may be further removed from that of pure financial gain but the psychological cravings for power, influence, and the desire to present "an edited self" may still be at play. The communicative goal of reaching and influencing large groups of people can be better achieved with well-known methods of propaganda, persuasive communication, and advertising. Carefully orchestrated and coordinated campaigns can be reinforced with automated technologies to deliver messages in multiple formats such as visual and audio memes. "The purveyors of disinformation prey on the vulnerability or partisan potential of recipients whom they hope to enlist as amplifiers and multipliers. In this way, they seek to animate us into becoming conduits of their messages by exploiting our propensities to share information for a variety of reasons" (Ireton & Posetti, 2018, p. 8). Moreover, what may be read as a disinformation attack with the intention to tarnish someone's reputation may have the financial backing of commercial competitors. Further connections to propaganda, mass persuasion, marketing, and advertising campaigns are discussed in Chap. 6.

In short, while deceivers in interpersonal everyday communication are motivated along three dimensions—self-oriented or altruistic benefits, materialistic gain or cost avoidance, and psychological reasons—the independent disinformers who fabricate stories are primarily motivated by the financial and political gains found by exploiting the current structures of online information environments. Meanwhile, other powerful psychological motives such as boosting self-confidence or avoiding damage to self-esteem are likely still at play and deserve further probing and elaboration.

2.3 Many Faces of Deception

We have thus far considered why people lie in everyday communications and compared those motivations to incentives to disinform at larger scales. The next section starts with definitions of deception and its characteristics, and then gets into typological distinctions. I assume that some types of deceptive strategies have found their way into the digital realm.

2.3.1 Definitions of Deception

Deception is both a communicative act and a message itself. In the context of interpersonal psychology and communication, deception specifically refers to a message that is knowingly and intentionally transmitted by a *sender* to foster such a false belief or conclusion by the *perceiver* (Buller & Burgoon, 1996; cited in Rubin, 2010). This definition has been widely adopted in the deception research community. Paul Ekman (2001) prefers to define deception as "a deliberate choice to mislead a target without giving any notification of the intent to do so," which is still consistent with Bella DePaulo et al.' (2003) wording: "a deliberate attempt to mislead others." Notice that both definitions exclude self-deceptions, eliminate honest mistakes (which are naturally unintentional), emphasize the deliberate nature of the act, as well as acknowledge a minimum of two communicators. Hancock's definition adds that the deceiver's goal is to "convince someone else to believe something that the deceiver believes to be false" (Hancock, 2012, p. 2). Deception does not necessarily have to be malevolent or criminal in nature. Even harmless dishonest praise is still deceptive. At the same time, even unintentionally misleading or erroneous messages (typically not considered deceptive) can harmfully misinform. Thus, humor, jokes, irony, parody, sarcasm, or satire can be controversially considered misinformation in the sense that they are easily misinterpreted, even when the sender does not intend to deceive. Both types of unintentionally misleading messages—errors disseminated unknowingly and misconstrued humor—contribute to the spread of problematic information and need to be counteracted to curb the problem.

2.3.2 *Deception Is Pervasive*

Deception as a phenomenon is widespread in society. Research has long established a simple truth about lying: Most of us lie and do so frequently. In fact, deception often serves as a social lubricant that facilitates our communication. "People likely have lied to one another for roughly as long as verbal communication has existed. Deceiving others can offer an apparent opportunity to gain strategic advantage, to motivate others to action, or even to protect interpersonal bonds" (Southwell et al., 2017, p. 1). Lying is thus said to be a frequent and fundamental part of everyday life (DePaulo et al., 1996; Ennis et al., 2008). Trivial and harmless white lies account for the high frequency of lying, yet this high frequency remains notable considering that deception is generally frowned upon in most societies. In everyday communication, people deceive more frequently than they care to admit. An opinion poll conducted by Miller and Stiff (1993) suggested that lying is more pervasive in everyday life than had been previously assumed: Only 12% of respondents claimed they never told a lie, and 24% said that they lied at least once during the previous day.

I use *lying* and *deceiving* throughout the book almost interchangeably, but I need to note that deception in its inventory of strategies is much broader than narrowly construed "lying" (i.e., prevarication). Deceivers manipulate the minds of those who pay attention to them in many ways: they falsify, redirect attention, equivocate, skirt issues, or plainly distort the reality to suit the deceiver's ends to a means. When we exchange information, in whatever format, we do not simply inform but often use persuasion: we forewarn, we predict, we promise and so forth. We communicate for pragmatic reasons—to get attention, gain fame, raise sales, sell ideology, sway voters, preach a way of life, or incite civil unrests. If the message that we communicate does not match our own beliefs or knowledge, it is then intentionally disinformative, or following the definitions above, it is deceptive. In other words, at the core of intentional disinformation is the art of deception, with its various forms and strategies. What are these forms and strategies? Now we can consider the various mechanisms of deception in greater detail, with an eye on how some varieties may be easier or more difficult to spot.

2.3.3 *Deception Varieties and Typologies*

How many varieties of deceptive strategies are there? Typologies describing deception varieties are comprised of as few as 2 categories and as many as 46. The distinctions between them are often based on comparative and contrasting features. For the purpose of developing AI detection applications, the understanding of these phenomena should be very precise so that we can discriminate types by their subtleties. Instances of each variety have to be documented, analyzed, and profiled in detail so that distinctions can be made automatically based on the unique features of each type.

The most minimal typology splits deception into two fundamental categories by how passive or active the deceiver is: *deception by commission* (purposeful and conscious erroneous communication) and *deception by omission* (consciously allowing a person to believe something untrue) (Chisholm & Feehan, 1977).

Alternatively, David Buller and Judee Burgoon's influential Interpersonal Deception Theory distinguishes three deception varieties based on seven differentiating features: amount of information, sufficiency of information, degree of truthfulness, clarity, relevance, ownership, and intent. The three varieties are *falsification* (lying or describing "preferred reality"), *concealment* (omitting material facts), and *equivocation* (dodging, skirting issues by changing the subject, or offering indirect responses) (Buller et al., 1996; Buller & Burgoon, 1996). Falsification was found to be most deceptive and least readily detected, since it is most prevalent and most practiced. The absence of information (either through concealment or intentional omission) is also hard to identify, either by the eye or AI. Equivocation offers the least amount of clarity, completeness, directness, and often induces suspicion and so it is most readily detected by humans (Buller et al., 1996), and most likely, by automated means as well (Rubin, 2010).

Studying deception in the context of close personal relationships, Sandra Metts (1989) diverges from Buller and Burgoon's threefold typology primarily in terms and scope. While Metts' *falsification* refers to asserting information contradictory to the true information or explicitly denying the validity of the true information, her *omission* is about withholding all references to the relevant information, not unlike Burgoon and Buller's concealment. Her third category, *distortion* refers to manipulation of the true information through exaggeration, minimization, and equivocation, such that a listener would not know all relevant aspects of the truth or would logically misinterpret the information provided. The latter category is often referred to as misleading and can be readily found in satirical news (Rubin et al., 2016), where it has the potential to be taken out of context and be misinterpreted. Interestingly, the type of communicator relationship helped predict the frequencies of different types of deceptive strategies used by the participants in Metts' study. Married study participants reported proportionately more instances of omission and fewer instances of explicit falsification relative to other relationship types; they also reported more psychological motivations which focused on avoiding threats to their partner's self-esteem (Metts, 1989). Respondents who were dating reported proportionately more motivations focused on protecting their resources and avoiding stress or abuse from their partner, with more reasons focused on avoiding relational trauma or termination (Metts, 1989). In other words, information about the context of deceptive communications is important when assessing a communicator's choice of strategies.

Other typologies of deception isolate one or another distinguishing attribute, and sometimes regroup narrower subtypes under broader ones, while overall prioritizing the significance of the main distinctions described above (falsification, omission, equivocation). For instance, Holly Payne (2008) analyzed workplace relationships through student employees' logs of a two-work-shift period and found that multiple forms of deception were inevitable. She used O'Hair and

Cody's (1994) fivefold typology of deceptive acts: *lies* (direct acts of fabrication), *evasion* (redirecting communication away from sensitive topics), *concealment* (hiding or masking true feelings or emotions), *overstatement* (exaggerating or magnifying facts), and *collusion* (where the deceiver and the target cooperate in allowing deception to take place). The findings indicated that employees overwhelmingly concealed information and lied primarily to supervisors and customers in order to protect emotions, evade work, cover their mistakes, emotions, or policy violations, and to mislead customers in order to increase sales, commissions, or gratuities (Payne, 2008). In organizational contexts, employees do not just lie for self-benefit, there are also competitive pressures (to meet objectives) and social pressures such as role conflicts (Payne, 2008). Payne's recommendations to workplace managers—to set up workplace procedures, policies, and reward structures to minimize deception and to investigate policy violations—are directly applicable to creating less toxic digital environments. Table 2.1 provides an approximate alignment of deception categories from extensively used taxonomies in interpersonal psychology research listed from that with the fewest categories to the most numerous.

Society clearly disapproves of deception; we teach our children not to lie from the time they are able to articulate their thoughts. Some forms of deception are, however, more socially acceptable than others. White lies, for instance, lack malevolent intentions and may be more readily recognized and excused. The pro-social goals of benevolent deception may include keeping others from harm, preventing hurt feelings, or genuinely meaning to do good for others such as by concealing plans for a surprise party (Walczyk et al., 2008). It has been empirically established since the 1970s and 1980s that people use white lies a lot (to save face, avoid

Table 2.1 Alignment of deception varieties across taxonomies in interpersonal psychology and communication research. (Adapted from Rubin & Chen, 2012)

Chisholm and Feehan (1977)	Buller and Burgoon (1996)	Metts (1989)	O'Hair and Cody (1994) Payne (2008)	Hopper and Bell (1984)
Commission	Falsification		Fabrication	Lies
	Equivocation	Distortion		
			Over-statement	
			Collusion	
			Evasion	
Omission	Concealment	Omission	Concealment	Masks
				Fictions
				Playings (e.g., jokes, bluff, hoaxes)
				Crimes (e.g., forgeries, con, conspiracies)
				unlies (e.g., distortion, misrepresentation)

tensions or conflicts, guide social interaction, affect interpersonal relationships, or achieve personal power), and they justify their white lies as the means to an end with much ease (Camden et al., 1984).

The most extensive typology of deception that I am aware of was conducted by Hopper and Bell (1984), who collected 46 deception-related terms and asked their study participants to judge each on whether the term was considered: harmful, socially unacceptable, immoral, premediated, prolonged in enactment, or hard to detect. Their study did not restrict its inventory of terms to narrowly construed malevolent concepts such as criminal disguise and forgery but also acknowledged pretense and broader playful terms (such as hoaxes, teasing, jokes, and even theater in which the deceiver and the deceived agree to cooperate). By clustering the features of these varieties by likeness, the researchers arranged them into six families of deception: *fictions, playings, lies, crimes, masks,* and *unlies* (Rubin & Chen, 2012).

Settling on just one typology of deceptive types for identification with the eye or AI may be difficult because each offers its own nuances. Unearthing such typologies is important since they demonstrate variation in degrees of "untruth," a multitude of ways in which a false impression can be created in one's mind, and underscore the importance of the situational contexts, motivations, and incentives to deceive.

If the ultimate goal is to be immune to such tricks, we need to be able to recognize them in their various shapes by their unique features. For AI-enabled identification of deception, it is also important to spell out the features that constitute a particular variety. Some of the features may be hard or impossible to grasp digitally. For instance, in case of omissions, there is literally no digital footprint to uncover in terms of the language in use at any particular time. However, AI may be able to scrutinize prior events, statements, or posts to provide situational context and lead the system on a trail of discovery. In the case of equivocation tactics, there are tangible expressions such as hedging, restatements, or exaggerations which can be searched and analyzed—and if found in abundancy—the system can make conclusions about the presence of this type of deception. (See also, varieties of digital fakes in Chap. 1, Sect. 1.4–1.6 and how to detect them with AI in Chap. 7.)

2.4 Why Mis-/Disinformation Is Effective

Now that we have established what people typically do to deceive, and what deceptive strategies they use, we will turn to the minds of the deceived ones. Why do we fall for what is just not true? Why do we believe mis- or disinformation? As previously established, deception is a communicative act involving at least two actors: the sender and the receiver. In other words, to be deceived online you have to receive a deceptive message in some way. This could be very minimal; perhaps reading a misleading headline, glancing at a doctored picture, noticing a snippet of online rumor, or opening a spam e-mail. The next section considers the psychological underpinnings of being mis- or disinformed. What do interpersonal, social, and cognitive psychologies tell us about the ease of being fooled? What blinds the deceived? What expectations do people typically hold when encountering new information? Let us examine what makes us susceptible to mis- or disinformation.

2.4.1 External Circumstances and Human Propensities

Often, explanations of being fooled boil down to the availability, convenience, and ease of access to free but poor-quality information. Journalists identify "people who cannot afford to pay for quality journalism, or who lack access to independent public service news media" as populations who are especially vulnerable to mis- and disinformation (Ireton & Posetti, 2018). In other words, if some baloney comes up as the first result for a quick internet search and it is free, it may be the only immediate option for people to view. Authoritative sources and some legacy news outlets have greater expertise, credibility, and some fact-checking procedures in place to eliminate fraudulent information, but they may at times be pay-walled in. A fee of $1 (Canadian) a week for *the New York Times* or *the Washington Post* creates an extra accessibility hurdle and a financial barrier, especially for marginalized communities or for those who may be reluctant or unwilling to provide credit card information online. However, the explanation that "high-quality reporting is costly" seems to be an oversimplification. Even when the financial and accessibility hurdles are removed on the path to getting truthful reporting and genuine high-quality information, other factors play a role in how people process the new information they encounter.

Some external circumstances can predispose us to believing untruths or hinder us from evaluating information more critically. In addition, internal convictions and psychological propensities can bias our judgments. Those external circumstances include some pressures that we experience during information seeking, whether we are pointedly searching, distractedly surfing, or scrolling a news feed. The sheer volume and speed with which information is pushed at us and the mountain of options which face us limit our cognitive capacity to process, select, and retain important pieces of information. The problem of *information overload*, also called *information overflow*, has been recognized in multiple scientific communities including business studies, social sciences, computing, and information sciences. The abundance of information has paradoxically resulted in difficulties obtaining high-quality information when it is needed (Edmunds & Morris, 2000). Social media users can rarely spare the time and energy to fact-check every single piece of news that they come across in their newsfeed (Chen et al., 2015; Rubin, 2019). People cope by cutting corners: We tend to select the first suggested search result, we skim the news headlines, we look for images in posts and tweets to get a quicker idea about their content, and we develop other strategies to further trim the load. If we simply pay less attention to what we are reading or if we are reluctant to spend time analyzing what we read, we can come to rushed, incomplete, erroneous, or altogether incorrect conclusions.

A recent study (by Pennycook & Rand, 2019b) suggests that susceptibility to blatantly inaccurate political headlines is driven by "lazy thinking" (meaning that people simply fail to think), and that it is not driven by the alternative "politically biased" or "identity-protective thinking." In other words, those susceptible to mis- or disinformation are in the habit of not examining the information they encounter in much depth.

The literature on reading behaviors has recognized new norms when reading on digital screens as a way of handling information overflow. A survey of 113 participants (by Liu, 2005) reported more time was spent on browsing and scanning, spotting keywords, one-time reading, nonlinear reading, and reading selectively, while less time was spent on in-depth reading and concentrated reading with an associated decrease in sustained attention. "We don't have time to grasp complexity, to understand another's feelings, to perceive beauty, and to create thoughts of the reader's own" (Wolf, 2018). Such essential slower, time-demanding "deep reading" processes like inference from internalized knowledge, analogical reasoning, perspective-taking and empathy, critical analysis, and the generation of insight, "all of which are indispensable to learning at any age," may be under threat in many parts of the world, caution experts in education via *the Guardian* (Wolf, 2018).

We often do not read much beyond headlines. A 2016 French study (by Gabielkov et al., 2016) found that 59% of all links shared on social networks were not clicked on at all, implying the majority of article shares are not based on reading the actual content behind the link. In an apparent response to such findings, in summer 2020 Twitter introduced a new feature to prompt users to first open the link if they were about to reshare it without reading its contents. This prompt was meant to nudge users to slow down (Hatmaker, 2020). Researchers in philosophy of computing refer to "information snacking" on the web as a type of information behavior with decreased demands on cognitive effort (Taraborelli, 2008). People interact with information in short frequent bursts throughout the day, often reading carelessly or impatiently. All of the information-related behaviors discussed above contribute to the problem of being misinformed.

2.4.2 *Cognitive Biases and Individual Traits*

Psychologists have also identified several human biases and individual traits which, in part, explain our predispositions to be fooled and misinformed. Studies of interpersonal communication and psychology tell us that humans are not good lie detectors: We are fairly ineffective at recognizing deception (DePaulo et al., 1997; Frank & Feeley, 2003; Rubin, 2019, p. 201; Rubin & Conroy, 2012; Vrij, 2000) and this is for two potential reasons. First, most people tend to assume that the information they receive is true and reliable. We are inherently *truth biased*, meaning that we believe that the majority of people are honest most of the time (Buller & Burgoon, 1996). Cooperation in communication pays off and we have been socialized to value honesty in society.

The medical research community has also uncovered evidence that our truth bias is an important adaptive behavior with biological underpinnings: Having a truth bias requires less cognitive effort and is associated with fewer long-term health problems (Reardon et al., 2019). In addition, people also exhibit *general gullibility:* They are receptive to "pseudo-profound bullshit" or ideas that they do not fully understand (Pennycook et al., 2015). In other words, we tend to believe each other even when we might not fully understand what we are being asked to believe.

Confirmation bias, another common pitfall, causes people to simply see only what they want to see or to arrive at the conclusions that they initially favored. Our beliefs and expectations influence the selection, retention, and evaluation of evidence. This can be broken down by four types of cognition patterns: hypothesis-determined information seeking and interpretation, failure to pursue a falsificationist strategy in contexts of conditional reasoning, a resistance to change a belief or opinion once formed, and overconfidence (illusion of validity) in one's own view (Peters, 2020). In terms of news literacy, confirmation bias manifests itself in peculiar ways. For example, conservative viewers of the news satire program *the Colbert Report* tended to believe that the comedian's statements were sincere, while liberal viewers tended to recognize the satirical elements (LaMarre et al., 2009; Rubin, 2019). In addition to these cognitive biases, news readers may lack the information literacy skills required to interpret news critically (Hango, 2014). University educators report a substantial overlap of the requirements for information literacy and the discerning disposition needed to identify "fake news" (Delellis & Rubin, 2020). Being information literate requires not just the ability to find information, but also the ability to critically evaluate it through language skills and the use of diverse perspectives. Detecting mis- or disinformation is just an extension of these skills (Delellis & Rubin, 2020). Overall, people just do not think critically enough about the information they encounter (Pennycook & Rand, 2019a). Plus, some may also lack the historical or cultural context needed to interpret the message correctly, for instance, for understanding satire in a second language or within a foreign culture (Rubin, 2019).

The illusory truth effect is a well-recognized phenomenon whereby the familiarity with a concept, idea, or an object makes it easier to think about it. *Processing fluency* is one of the cognitive mechanisms that refers to the subjective ease experienced while mentally processing any statement. In other words, fluency is about how easily and quickly our minds process information once it is picked up by our senses, and how easy it is to retrieve it from our memory. Fluency seems to speed up inferences, for instance, fluent words seem more familiar, fluent names seem more famous, and fluent paintings are more appreciated (Wang et al., 2016, p. 739). Now, neuropsychological behavioral evidence also suggests that fluency underlies the illusion of truth, with "repetition of a statement increases its likelihood of being judged true" (Wang et al., 2016, p. 739). Hearing something 20 times does not mean that we have 20 new pieces of information, but the more we hear it, the more likely we are to believe it as true. Repetitions makes statements more fluent in our minds. That is how we come to believe oft-repeated information, especially if they are falsehoods that sound plausible but go unquestioned (Giles, 2010). In addition, complicated abstract language is harder to process and remember than simpler more concrete wording, notoriously exploited in commercial and political ads (see Chap. 6 for marketing and advertising practices). The more we hear simply worded ads, the more familiar they seem, and the more powerful and true they appear. Such repetition impacts our decisions in our daily lives.

According to the dual processing model in psychology, humans have two distinct ways of thinking—analytical and intuitive, and while there is no consensus in the psychological community on what triggers the switch from one mode to another, the role of detecting conflict has been hypothesized (Pennycook, 2017). Our natural tendency is to use as little cognitive effort as necessary to solve our problems, an evolutionary psychological trait of being *cognitive misers* (Shane, 2020).

Pennycook and Rand (2019a, 2019b) are emphatic in suggesting that cultivating or promoting our *reasoning abilities* should be part of the solution for the mis- and disinformation problem. They found that people who engaged in more reflective reasoning did better at telling true from false, regardless of whether the headlines aligned with their political views, and regardless of their demographics such as levels of education and political leaning. "People who think more analytically (those who are more likely to exercise their analytic skills and not just trust their 'gut' response) are less superstitious, less likely to believe in conspiracy theories, and less receptive to seemingly profound but actually empty assertions (like 'Wholeness quiets infinite phenomena')" (Pennycook & Rand, 2019a).

Effortful analytical thinking to overcome this cognitive laziness or miserliness is required to arrive at accurate beliefs and conclusions and to avoid being fooled. Instead of relying on pure intuition, we (should) use effortful credibility judgments and assessments of information quality (Chap. 3) when dealing with online information that informs our important decisions or actions. Our beliefs and worldviews (Chap. 4) also have formative influences on our decisions and cognitive heuristics.

2.5 Language of Deceit and Cues to Its Detection

When facing so many difficulties—information overload, time pressures, and our general tendency to cut corners mentally when encountering new information—we should not be surprised that some deception can go undetected. So, when we are given the time and opportunity to think and assess the veracity of a statement or a story, how good are we, humans, on average, at spotting a lie?

2.5.1 Human Inability to Detect Deception

"A large body of research indicates that most individuals are unable reliably to detect when others are lying" (Miller & Stiff, 1993; cited in Wiseman, 1995). This tendency has been researched in social psychology and communications studies where, typically, participants are presented with a lie and a truth and are asked to distinguish the two in what is called a lie–truth discrimination task (Wiseman, 1995). When findings from multiple studies of this type were analyzed comparatively, researchers concurred that the human ability to detect deception in a lie–truth discrimination task is only slightly better than chance. Typical lie detection accuracy rates are in the 55–58% range (Frank & Feeley, 2003), with the mean detection accuracy only reaching 54% across over 100 experiments with over 1000 participants in total (DePaulo et al., 1997). Basically, lay people, if not alerted to the presence of deception, are poor detectors (Twitchell et al., 2004). With training, professional lie-catchers such as law enforcement personnel or witness credibility assessment experts, typically show only a slight improvement over the detection success rate of untrained people (Frank & Feeley, 2003). For a discussion of the time-tested professional practices used for establishing truth by the justice system, as well as in journalism and the sciences, see Chap. 5.

2.5.2 *Cues to Lie Detection by Mode of Perception*

Let us now consider what could alert us to the presence of deception in a message. Are there any cues to watch out for, and if so, how reliable are they? Thus far, we have looked at deceptive communication actors on both sides—the sender who encodes the message, and the receiver who decodes the message in the information exchange. It is now time to turn to the code itself: How the information is packaged and presented to the receiver as the actual content of the communicated message. Each one of us, in a way, is a lie detector. To detect lies, we need to recognize valid cues that may indicate deceit.

Empirical research has never uncovered any "Pinocchio response," in the sense that there is no one single deceptive cue that is indicative of deception, but scientists continue to examine the combinations of cues that can be associated with deception (Carr et al., 2019). These cues to deception can be distinguished by their mode of perception. Visual cues are observable sights during the communication, for example, eye contact, body movements, and facial expression, while auditory or vocal cues include the way in which words are said, for example, voice pitch, pauses, and hesitation. Verbal cues are part of the language used in the message and include simple measures such as the number of words written or the length of sentences, as well as more complex linguistic cues identified by morphological, lexical, syntactic, semantic, discourse, and pragmatic analyses. People often mistakenly believe that having more information available to us (audiovisual cues with images, audio, video formats) makes us more likely to spot a lie by noticing that something is off. "In experimental studies, people are willing to infer deception from odd behaviors such as arm raising, head tilting, and staring that violate normative expectations but that never accompany real-world lies (Bond et al., 1992)" (cited in Bond et al. (2013), p. 219). Contrary to these popular misconceptions, a meta-analysis demonstrated that no nonverbal or speech variable is strongly and consistently related to lying (DePaulo et al., 2003). The 1995 Megalab Truth Test recruited over 41,471 UK participants (a huge response by 1995 standards) to test which mode of perception showed higher lie detection rates: radio listeners detected the lies over 73% of the time (i.e., using audio cues), newspaper readers over 64% (i.e., verbal cues), and television viewers about 52% (i.e., all three categories of cues) (Wiseman, 1995). This test determined that logically assessed verbal cues are likely among the strongest and that the presence of visual cues in video may be more distracting than helpful.

It is intriguing that "cues to truthfulness are not particularly appealing; it is deception cues that captivate the human imagination" (Bond et al., 2013, p. 218). However remarkably attractive this idea is to some of us, there is simply no simplistic solution in existence. Paul Ekman's (2001) subtle and fleeting facial micro-expressions, for instance, captured the imagination of viewers of *Lie To Me*, an American criminal investigation television series based around this method of lie detection. Deception researchers also took a lot of interest in micro-expressions and other behavioral leakages, but these measures either turned out not to be strongly indicative of deceit (Bond et al., 2013; DePaulo et al., 2003) or difficult to apply in practice.

2.5.3 Verbal Cues

Verbal cues have been found to be more valid deception indicators than nonverbal behavior (Reinhard, 2010), which is a good reason to use a combination of such cues in AI applications. Verbal cues tend to be more reliable (Wiseman, 1995), possibly since richer visual and auditory media distract perceivers' attention from the verbal essence (Vrij, 2004).

So, what are these verbal cues and patterns that objectively differentiate deceptive messages from truthful ones? Compared to truth-tellers, deceivers, for instance, tend to make more negative statements, give more indirect and less detailed answers (Granhag et al., 2004), use fewer self-references but more negative emotions (Zhou et al., 2008), and use more sense-based words (e.g., seeing, touching) (Hancock et al., 2004). Such patterns are too complex for people to identify without specialized analytical tools. These automated deception detection tools using NLP and ML are discussed in detail in Chap. 7.

The type of deception defines its content and form, and thus affects the recognizable features which can be detected. For instance, the language used in exaggeration is expressed differently than in equivocation. When looking for the omission of facts, we need to primarily consider the surrounding context, while in falsifications, how and what is said is as important as the context. Misleading information like clickbait is marked by prototypical phrasing, for example, "13+ Trending Pics That Give Us A Unique View Of The World" (Zimmer & Diply.com, 2021) or "12+ Wedding Cake Designs That We Would Definitely Question Eating" (written for Diply.com by Mikolajczak, 2021).

For automated identification, the use of linguistic features (like the use of numerals at the start of headlines in clickbait) can be easy patterned and applied in detection of previously unseen instances. For example, this numeral cue, when combined with others, resulted in a detection accuracy rate as high as 94% in Brogly and Rubin's automated clickbait detection study (Brogly & Rubin, 2019). Some types of disinformation have more readily recognizable characteristic features than others, no matter which typology you adhere to. See also Chap. 7, Sect. 7.1.3. "Chapter Roadmap: AI Detection by Type of Fake." Outright fabrication—the most prolific form of disinforming online—is a more challenging type, since it is less formulaic and has no predictable forms of expression. As is widely acknowledged, humans have deceived each other for thousands of years, yet, our understanding of deception still remains nebulous—deception detection is a difficult task (Carr et al., 2019).

2.5.4 Incentives to Deceive

Research has also shown that knowing contextual cues—such as the circumstances and incentives of deceivers—can help human lie detectors identify deception. Charles Bond and colleagues found near perfect accuracy in uncovering deceit (97% in their lie–truth discrimination tasks) when people were given information about incentives that the liar faced (Bond et al., 2013). This revelation points both humans and AI into a slightly different direction than taken thus far—away from looking for cues solely within the content of potentially deceptive messages toward looking at the cues indicating the circumstances which incentivize deceit. We apparently tend to underestimate the power of circumstances, which are forces strong enough to make most individuals lie regardless of their internal traits or scruples. The presence of such circumstances might be an almost-perfect deception cue to enter into the equation for an incentive-based lie detection algorithm which could hypothetically be capable of fairly accurate a priori prediction (Levine et al., 2010).

2.6 Putting Psychological Research to Practice: Interventions

Considering our propensity for being fooled, and in the absence of digital savviness and media literacy, there is a growing need to step up measures to counter the influence of mis- and disinformation and provide some assistance to digital media users of social and information technologies.

2.6.1 Fact-Checking and Content Moderation

When it comes to potentially fraudulent news or false claims in public discourse, fact-checking immediately comes to mind as a quality-control strategy. Fact-checking, as an activity outsourced from independent newsrooms, started in 2003 with the appearance of Factcheck.com based out of the United States. In 2009, the

Pulitzer Prize was awarded to PolitiFact to recognize fact-checking as a highly valued form of journalism (Mantzarlis, 2016), which in turn raised the visibility of similar organizations. According to the database of fact-checking organizations maintained by the Reporters' Lab at Duke University, there were 304 initiatives in 84 countries counted in October 2020 (Stencel & Luther, 2020). Why an increase in these organizations? The boundaries between news production, content generation, and information sharing have been gradually blurred in online news and social media environments (Chen et al., 2015). Digital media users nowadays carry the burden of discerning truth from deception and verifying information for its quality and correctness, reliability of sources, and contextual relevance. The journalism community has long been alarmed by this shift in responsibility, in the rise of "a citizen media culture" in which "the loudest or most agreeable voice wins and where truth is the first casualty" (Kovach & Rosenstiel, 2010, p. 7). According to the 2019 Pew Research Center State of the Media Report, in the U.S., a vast majority (93%) of adults get at least some news online (either via a mobile device or desktop/laptop), and online spaces have begun to host the digital homes of both legacy news (print, radio, and broadcast) outlets and new, "born on the web" news outlets. Digital media users are thus increasingly getting their information from online sources and may simply require more assistance with making sense of it.

Social media platforms, as publishers of information, have reluctantly taken on the task of content verification by hiring armies of human content moderators to filter out undesirable speech, and additionally by applying algorithms to help. The enormous amount of content challenges "show[s] the futility of moderating content across networks with more than 3 billion [Facebook] users" (Thurm, 2021). In addition to deception detection, challenges identified by the Facebook Oversight Board include the complexity of moderation decisions in ethical and political contexts; algorithmic shortcomings; and a lack of consistency in standards, policies, or appeal procedures. A comprehensive solution remains elusive.

Meanwhile, setting AI solutions aside (more in Chap. 7), three countermeasures to mis- and disinformation have been developed: "inoculating," labeling, and educating about media literacy more broadly. Inoculation is used in specific domains with targeted messaging campaigns. For example, promoting facts about vaccination to new parents in order to decrease rates of vaccination hesitancy. Labelling misinformation is popular with social media platforms and professional fact-checkers: Various labels are used to reveal the result of an independent assessment of information quality, whether this content verification was done automatically or by human content moderators. And last, but not least, is some form of direct education about mis- and disinformation. These educational campaigns instruct people on how to discern multiple varieties of fakes and verify information. Although there have been few large-scale trials, several initiatives include short-term campaigns and a more permanent embedding of various literacies (media, news, information, and digital literacy) in the educational curricula, from early grade schools through to graduate studies.

2.6.2 Inoculating

One strategy—sometimes conceptualized as "thought inoculation," or "sociopsychological attitudinal inoculation"—has been proven effective in the context of specific domains such as health messaging, political campaigns, and polarized issues. Following the epidemiological analogy that I used in this book's Chap. 1, mis- or disinformation spreads in the population like an infectious disease due to the interaction of three necessary factors—susceptible users, conducive environments, and persistent viral attacks (Rubin, 2019). The infection in this analogy is conceptualized as "thought contagions" (Lynch, 1998), "memes," or ideas, behaviors, and styles that spread from person to person in a culture and which have a pattern of cultural information transmission. The goal is to slow the cultural transmission of such an infection by exposing the population to a weakened version of the virus, following the epidemiological practice of vaccination. This counter-messaging has been referred to in many ways: inoculation, vaccination, refutational preemption, pre-bunking, or conferring resistance to persuasive influence. Such measures are typically short term, domain specific, and targeted to a specific misconception (such as the efficacy of fad diets or symptoms of COVID). In essence, inoculations are a set of measures that forewarn people and help them to develop attitudinal resistance. They do this by activating people's awareness of currently circulating false claims, stimulating their willingness to think more critically about these claims, and by refuting potential counterarguments. Essentially, they equip people with accurate information as a relevant counterbalance to misinformation.

Van der Linden et al.' (2017) study found that "communicating the scientific consensus on human-caused climate change significantly increases public perception of the expert consensus" (p. 6). In the study, participants were preemptively forewarned that politically or economically motivated actors may seek to undermine the findings of climate science and had the nature of such disinformation campaigns explained to them. This study's findings emphasized that "communicating a social fact, such as the high level of agreement among experts about the reality of human-caused climate change, can be an effective and depolarizing public engagement strategy" (van der Linden et al., 2017 p. 6), provided that inoculators have knowledge about both how people attend to, process, and organize new information, and about the structure of the information environment in which people are forming their judgments and opinions.

The experiences of health promotion practitioners who aim to create positive attitudes toward desirable health behaviors (such as physical activity, dietary patterns, safer sex, avoidance of harmful substances) can be adapted to other themes. The practice of health communication is to craft messages that communicate positive health attitudes and which protect these positive attitudes against potential challenges. "A conventional inoculation message begins with a forewarning of impending challenges to a held position, then raises and refutes some possible challenges that might be raised by opponents. For example, an inoculation message designed to discourage teen cigarette smoking (e.g., Pfau et al., 1992) might begin with a warning that peer pressure will strongly challenge their negative attitudes toward smoking, then follow this forewarning with a handful of potential counterarguments they might face from their peers (e.g., 'Smoking isn't really bad for you') and the refutations to these counterarguments (e.g., 'Actually, smoking is harmful in a number of ways …')" (Compton et al., 2016, p. 2). This type of message design can weaken the effects of counterarguments that may be employed by disinformation agents.

An innovative approach to inoculating their citizens was demonstrated by the Taiwanese Government in 2020–2021. Taiwan, with a population of over 23 million, of whom about 1.2 million work or reside in China, was expected to have the second highest number of cases of coronavirus disease (COVID-19) due to the proximity to and the number of flights with China (Wang et al., 2020, p. 1). Instead, according to the John Hopkins University tracking system, Taiwan has only had 10 COVID-related deaths with 990 COVID cases in total by mid-March 2021 (Johns Hopkins Coronavirus Resource Center, 2020). By any comparison worldwide, it is an exemplary record of a minimal number of COVID-19 infections. Taiwan has clearly managed the crisis effectively, with one creative strategy that is of particular interest to the discussion of countermeasures to mis- and disinformation here.

The Taiwanese Digital Ministry implemented their 2–2–2 "humour over rumour" strategy to quash the "infodemic" associated with COVID. To decode the triple two digits, the Ministry provides a response to mis-/disinformation within 20 min, in 200 words or fewer, alongside 2 fun images. The strategy capitalizes on the ability of satire to point out and cure societal follies. *The Guardian* reported on how Taiwan handled a rumor that toilet paper was being used in manufacture of face masks in order to stop panic-buying during toilet paper shortages. "The Taiwanese Premier, Su Tseng-chang, released a cartoon of himself wiggling his bum, with a caption saying: "We only have one pair of buttocks." It sounds silly, but it went viral. "Humour can be far more effective than serious fact-checking," comments *the Guardian* reporter (Mahdawi, 2021) (Fig. 2.1). Other digital campaigns that used viral memes and animal mascots have been similarly effective.

2.6.3 Labeling

Another, admittedly simple, way to signal mis- or disinformation is to label any online content with a system of warning labels. Human fact-checkers use various gradations of a truth-to-deception continuum. For instance, PolitiFact, a project run out of the nonprofit Poynter Institute For Media Studies, provides one-of-six ratings of "the Truth-O-Meter," which reflect the relative accuracy of a statement based on a reporter's fact-checking research and a combined vote by three editors.

The Truth-O-Meter decreases in the level of truthfulness from true to pants-on-fire: TRUE (the statement is accurate and there's nothing significant missing); MOSTLY TRUE (the statement is accurate but needs clarification or additional

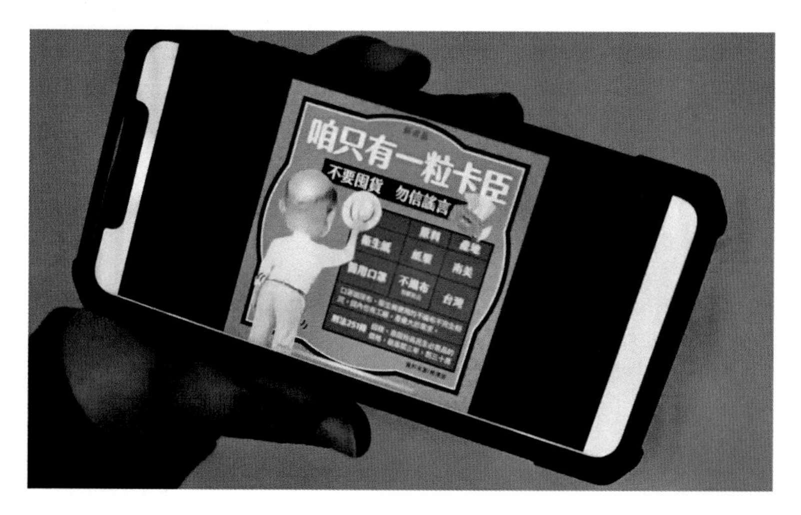

Fig. 2.1 "Butt of the joke … Taiwan's premier, Su Tseng-chang, fighting misinformation with his backside." Sam Yeh/AFP/Getty Images (Mahdawi, 2021)

information); HALF TRUE (the statement is partially accurate but leaves out important details or takes things out of context); MOSTLY FALSE (the statement contains an element of truth but ignores critical facts that would give a different impression); FALSE (the statement is not accurate); PANTS ON FIRE (the statement is not accurate and makes a ridiculous claim)" (Holan, 2020). A ratings is typically applied to the fact-checked content and shown to the readers, followed by explanations. For an example, see how a false claim was debunked (Kim, 2021a) with a confirmed reason (Kim, 2021b) for the isolation of the Texas power grid in the February 2021 winter storm, as fact-checked by the PolitiFact.

The PolitiFact Truth-O-Meter labeling system, as one of the more established ones, has been used extensively in experimental datasets that are collected for the purposes of automating and evaluating the accuracies of the task (see Chap. 7).

Since May 2020, Twitter has been using its own system of three broader categories to typify misinformation by its propensity to harm. The three Twitter categories—misleading information, disputed claims, and unverified claim—were color-coded by their propensity of harm (moderate or severe) and subsequent actions (from no action to warning and removal) (Roth & Pickles, 2020).

These labels may also lead to external additional information sources. According to a blog statement by Twitter's Head of Site Integrity and Director of Global Public Policy Strategy and Development (Roth & Pickles, 2020), these categories were clarified as follows:

- "Misleading information—statements or assertions that have been confirmed to be false or misleading by subject-matter experts, such as public health authorities.

- Disputed claims—statements or assertions in which the accuracy, truthfulness, or credibility of the claim is contested or unknown.
- Unverified claims—information (which could be true or false) that is unconfirmed at the time it is shared."

Roth and Pickles (2020) provided sample screenshots of Twitter warnings applied to real tweets in practice, although this threefold classification itself is far from being crystal clear.

Other giant tech companies such as YouTube and Facebook also place warning labels on their content, but have been more reluctant to do so. "Among the tech companies, Twitter seems to be the most willing to at least experiment with product changes to slow the spread of information, even if it cuts our engagement with its service" (Fowler, 2020). The fear is that the desired effect of such warnings may be reversed by attracting additional attention and thus reshares of undesirable content, therefore increasing the polarization and hate speech around posts labeled as "false." Another fear is that having the majority of content unlabeled may erroneously imply that their content is verified and truthful, or alternatively that false headlines which fail to be tagged are considered validated and thus are seen as more accurate, in a fallacy known as the "implied truth effect" (Pennycook et al., 2020). Little information has been released by the big tech companies on the efficacy of their warning labels and much controversy still exists around them. In part, it is due to big tech companies' secrecy around their data, likely due to fears that greater transparency may hurt their vested interests in user engagement by decreasing sensationalism and the polarization of opinions. This situation is concerning since the results of AI interventions are likely to be communicated to digital media users through such simple warning labels or through automatically crafted messages, inoculation-style, but the efficacy of these labels is largely untested. As a result, this issue remains unresolved. Research is needed on how to most effectively counter false claims and communicate the results of either automated or human content verification in these contexts without distorting belief in true information.

Preliminary evidence shows that "false headlines are perceived as less accurate when people receive a general warning about misleading information on social media or when specific headlines are accompanied by a 'Disputed' or 'Rated false' tag" with relatively modest magnitudes of these effects. This study also found that using a "Rated false" tag "lowers its perceived accuracy more than adding a 'Disputed' tag (Facebook's original approach) relative to a control condition" (Clayton et al., 2020). The presence of these warnings was found to cause untagged headlines to be seen as more accurate, and the researchers suggested "to attach verifications to some true headlines—which removes the ambiguity about whether untagged headlines have not been checked or have been verified—[this] eliminates, and in fact slightly reverses, the implied truth effect" (Pennycook et al., 2020, p. 4944).

2.6.4 Literacy Education

The third strategic countermeasure to mis- or disinformation is to step up education in various kinds of literacies related to the use of new information online. Susceptible groups of the population can be identified, and short-term direct training can be deployed to hone their skills and provide the knowledge needed to identify disinformation online. Various initiatives are underway for such coaching, especially through online resources and self-directed study. For example, the Poynter Institute has partnered with the Stanford Social Media Lab to teach "Americans of all ages how to sort fact from fiction online" (Dyakon, 2020). In collaboration with MediaWise, a nonprofit journalism organization, the group ran a digital media literacy program for older adults. This age group was chosen by the researchers because prior research found that they were two to seven times more likely to reshare fake news. This was perhaps due to a lack in their digital literacy skills. Preliminary study results showed a significant increase in the ability to accurately judge the veracity of news headlines among the MediaWise enrollee group, when compared to the control group (Dyakon, 2020; Moore & Hancock, 2021). The success of the program was attributed to their "secret sauce" by the program managers—"teaching people how to spot mis- and disinformation using real-life examples of inaccurate information that has gone viral on social media platforms—is central to the effectiveness of our training and courses" (Dyakon, 2020).

Other initiatives that emphasize the various forms of literacy have recently sprung up. For example, the News Literacy Project, an American educational nonprofit organization, focuses on working with educators to teach another highly susceptible group of media users—school-age students—to be critical consumers of media (About Our Organization Page, 2021). The Project provides programs and resources including an e-learning platform, an app, a podcast, shareable tips and tools, and news literacy events. They distinguish between several fields that they draw on: "***Media literacy*** generally refers to a broad discipline that seeks to teach students how to access, analyze, evaluate, create, and take action using all forms of communication (including entertainment media). ***News literacy*** is focused on helping students understand the role that credible information and a free press play in their lives and in a robust democracy and seeks to help them determine the credibility of news and other information. ***Information literacy*** is aligned with library sciences and seeks to help students find, evaluate, and use information effectively. ***Digital literacy*** aims to teach students how to use information and communications technologies in effective, responsible and ethical ways" (About Our Organization Page, 2021).

Similarly, in Canada, the CIVIX Digital Information Literacy Initiative is "dedicated to building the skills and habits of active and informed citizenship among young Canadians" and "aims to support teachers in empowering students with the knowledge and skills required to locate the information they can trust" (About Us Page, 2021).

When talking about the full array of news content, it may not always be possible to inoculate the population with preemptive refutation or warning labels on every single message. Instead, the focus should shift to the cultivation of the mind. A large-scale study of digital media literacy intervention in the U.S. and India used "tips" on how to spot false news (similar to those by WHO, IFLA, and Canada's National Observer; see Chap. 1, Sect. 1.3). "The intervention improved discernment between mainstream and false news headlines among both a nationally representative sample in the United States (by 26.5%) and a highly educated online sample in India (by 17.5%)" (Guess et al., 2020, p. 15,536). The authors underscored that their intervention was brief and inexpensively disseminated at scale, but it helped users gauge the credibility of the news content that they encounter on a variety of different topics or issues. Even the simplest approach of cultivated self-awareness can slow down the spread of misinformation. The simplest literacy approach is consistent with Twitter's latest initiative to remind users to read the articles that they intend to reshare: "stop-and-think" to resist our natural urge to "click-and-reshare." This basic approach is by no means novel but, if consistently practiced, it could serve as an effective "mental hygiene."

2.7 Summary

Chapter 2 frames deception as a communicative behavior and discusses its underlying mechanisms, strategies, forms, content, and expression, as well as digital varieties of deception online. Given below is a summary of the key assertions and a reiteration of how this knowledge can be operationalized.

The main premise is that mis- and disinformation should be viewed not only as messages but also as information exchanges with at least two actors—the sender and the receiver of the information. This perspective takes into consideration the circumstances around information encounters and potentially deceptive communicative acts. We can start by analyzing the influences: the motivations and incentives of the deceiver as well as the shortcomings and vulnerabilities of the receiver in context. The circumstantial context of a communication act can be particularly useful for any human assessing the potential veracity of statements. For AI approaches to identifying potential fakes, it is crucial to have clear definitions of each specific phenomenon that the algorithms may be seeking to identify.

Chapter 2 lays out what types of digital fakes there are, what characterizes each type, and provides an estimate of how accessible for processing and analysis the features may be. The overarching claim here is that understanding the psychological underpinnings of deceptive communicative acts sheds light of the variety of digital fakes we encounter online, and the associated circumstances on both sides of the information exchange. Such rich knowledge facilitates the identification of particular kinds of mis- and disinformation with both a naked eye and AI. Here, suffice it to say, the way forward in the identification process is in combining two approaches: "divide and conquer" and "know thy enemy." Namely, R&D in AI has been proceeding with a careful detailing of individual phenomena to demystify vague terms like "fake news." Having a unified typology of digital fakes is useful because it provides an overview of the many subtasks involved in specific AI solutions and acknowledges that each type requires its own methods of identification. The work of detailing, implementing, and testing characteristic indicators has to progress at the level of individual subtasks, for instance, AI programs that are detecting satire, clickbait, or outright falsehoods.

The near future goal of AI veracity assessment is to mimic the reasoning, inference, and rational decision-making that is performed by expert fact-checkers, lie-spotters, or informed analysts, and to provide explanations for those assessments. For a layperson, the goal is to be information literate and aware of mis- and disinformation in general terms and with some nuances, as was described in this chapter.

Chapter 2 also discusses fact-checking and content moderation, preemptive messaging to inoculate the public, the placement of warning labels on fact-checked content, and broader literacy efforts as measures to counter mis- and disinformation. Reiterating that our brain is the ultimate tool for detecting fakes, I suggest that to fight the infodemic at today's scale, we still need assistance of AI-enabled detectors to complement our built-in human ones.

References

About Our Organization Page. (2021, February 18). News Literacy Project. *News Literacy Project*. Retrieved from https://newslit.org/about/

About Us Page. (2021, February 16). *CIVIX News Literacy*. Retrieved from https://newsliteracy.ca/about-us/

Bok, S. (1999). *Lying: Moral choice in public and private life* (2nd Vintage Books ed.). Vintage Books.

Bond, C. F., Omar, A., Pitre, U., Lashley, B. R., Skaggs, L. M., & Kirk, C. T. (1992). Fishy-looking liars: Deception judgment from expectancy violation. *Journal of Personality and Social Psychology, 63*(6), 969–977. https://doi.org/10.1037/0022-3514.63.6.969.

Bond, C., Howard, A., Hutchison, J., & Masip, J. (2013). Overlooking the obvious: Incentives to lie. *Basic and Applied Social Psychology, 35*, 212–221. https://doi.org/10.1080/01973533.2013.764302

Brogly, C., & Rubin, V. L. (2019). Detecting clickbait: Here's how to do it/Comment détecter les pièges à clic. *Canadian Journal of Information and Library Science, 42*, 154–175.

Buller, D. B., & Burgoon, J. K. (1996). Interpersonal deception theory. *Communication Theory, 6*(3), 203–242.

Buller, D. B., Burgoon, J. K., Buslig, A., & Roiger, J. (1996). Testing interpersonal deception theory: The language of interpersonal deception. *Communication Theory, 6*(3), 268–288.

Camden, C., Motley, M. T., & Wilson, A. (1984). White lies in interpersonal communication: A taxonomy and preliminary investigation of social motivations. *Western Journal of Speech Communication, 48*(4), 309–325. https://doi.org/10.1080/10570318409374167

Carr, Z. M., Solbu, A., & Frank, M. G. (2019). Why methods matter: Approaches to the study of deception and considerations for the future. In T. Docan-Morgan (Ed.), *The Palgrave handbook of deceptive communication*. Palgrave Macmillan.

Chen, Y., Conroy, N. J., & Rubin, V. L. (2015). News in an online world: the need for an automatic crap detector, *Proceedings of the Association for Information Science and Technology, 52*(1), 1–4. https://doi.org/10.1002/pra2.2015.145052010081

Chisholm, R. M., & Feehan, T. D. (1977). The intent to deceive. *Journal of Philosophy, 74*(3), 143–159.

Clayton, K., Blair, S., Busam, J. A., Forstner, S., Glance, J., Green, G., Kawata, A., Kovvuri, A., Martin, J., Morgan, E., Sandhu, M., Sang, R., Scholz-Bright, R., Welch, A. T., Wolff, A. G., Zhou, A., & Nyhan, B. (2020). Real solutions for fake news? Measuring the effectiveness of general warnings and fact-check tags in reducing belief in false stories on social media. *Political Behavior, 42*(4), 1073–1095. https://doi.org/10.1007/s11109-019-09533-0

Compton, J., Jackson, B., & Dimmock, J. A. (2016). Persuading others to avoid persuasion: Inoculation theory and resistant health attitudes. *Frontiers in Psychology, 7*, 1–9. https://doi.org/10.3389/fpsyg.2016.00122

Delellis, N. S., & Rubin, V. L. (2020). 'Fake news' in the context of information literacy: A Canadian case study [chapter]. In *Navigating fake news, alternative facts, and misinformation in a post-truth world.* https://doi.org/10.4018/978-1-7998-2543-2.ch004

DePaulo, B. M., Ansfield, M. E., Kirkendol, S. E., & Boden, J. M. (2004). Serious lies. *Basic and Applied Social Psychology, 26*(2–3), 147–167. https://doi.org/10.1080/01973533.2004.9646402

DePaulo, B. M., Charlton, K., Cooper, H., Lindsay, J. J., & Muhlenbruck, L. (1997). The accuracy-confidence correlation in the detection of deception. *Personality and Social Psychology Review, 1*(4), 346–357.

DePaulo, B. M., Kashy, D., Kirkendol, S., & Wyer, M. (1996). Lying in everyday life. *Journal of Personality and Social Psychology, 70*, 979–995.

DePaulo, B. M., Lindsay, J. J., Malone, B. E., Muhlenbruck, L., Charlton, K., & Cooper, H. (2003). Cues to deception. *Psychological Bulletin, 129*(1), 74–118.

Dyakon, T. (2020, December 14). Poynter's MediaWise training significantly increases people's ability to detect disinformation, new Stanford study finds. *Poynter*. Retrieved from https://www.poynter.org/news-release/2020/poynters-mediawise-training-significantly-increases-peoples-ability-to-detect-disinformation-new-stanford-study-finds/

Edmunds, A., & Morris, A. (2000). The problem of information overload in business organisations: A review of the literature. *International Journal of Information Management, 20*(1), 17–28. https://doi.org/10.1016/S0268-4012(99)00051-1

Ekman, P. (2001). *Telling lies: Clues to deceit in the marketplace, politics, and marriage* (3rd ed.). Norton.

Ennis, E., Vrij, A., & Chance, C. (2008). Individual differences and lying in everyday life. *Journal of Social and Personal Relationships, 25*(1), 105–118. https://doi.org/10.1177/0265407507086808

Fowler, G. A. (2020, November 9). *Twitter and Facebook warning labels aren't enough to save democracy*. Washington Post. Retrieved from https://www.washingtonpost.com/technology/2020/11/09/facebook-twitter-election-misinformation-labels/

Frank, M. G., & Feeley, T. H. (2003). To catch a liar: Challenges for research in lie detection training. *Journal of Applied Communication Research, 31*(1), 58–75.

Gabielkov, M., Ramachandran, A., Chaintreau, A., & Legout, A. (2016). Social clicks: What and who gets read on twitter? *Proceedings of the 2016 ACM SIGMETRICS international conference on measurement and modeling of computer science* (pp. 179–192). doi: https://doi.org/10.1145/2896377.2901462.

Giles, J. (2010). Giving life to a lie. *New Scientist, 206*(2760), 42–43. https://doi.org/10.1016/S0262-4079(10)61213-4

Granhag, P. A., Andersson, L. O., Strömwall, L. A., & Hartwig, M. (2004). Imprisoned knowledge: Criminals beliefs about deception. *Legal and Criminological Psychology, 9*(1), 103.

Grice, H. P. (1975). Logic and conversation. In P. Cole & J. Morgan (Eds.), *Syntax and semantics 3: Speech acts* (pp. 41–58). Academic Press.

Guess, A. M., Lerner, M., Lyons, B., Montgomery, J. M., Nyhan, B., Reifler, J., & Sircar, N. (2020). A digital media literacy intervention increases discernment between mainstream and false news in the United States and India. *Proceedings of the National Academy of Sciences, 117*(27), 15,536–15,545. https://doi.org/10.1073/pnas.1920498117

Hancock, J. T. (2012). Digital deception: When, where and how people lie online. In A. N. Joinson, K. McKenna, T. Postmes, & U.-D. Reips (Eds.), *The Oxford handbook of internet psychology* (p. 20). Oxford University Press. Retrieved from http://www.oxfordhandbooks.com/10.1093/oxfordhb/9780199561803.001.0001/oxfordhb-9780199561803-e-019

Hancock, J. T., Curry, L. E., Goorha, S., & Woodworth, M. T. (2004). Lies in conversation: An examination of deception using automated linguistic analysis. *Proceedings of the 26th Annual Conference of the Cognitive Science Society.* (K. Forbus, D. Gentner, & T. Regier, Eds.) Chicago, Illinois

Hango, D. (2014). *University graduates with lower levels of literacy and numeracy skills* (75–006-X). Statistics Canada. Retrieved from http://www.statcan.gc.ca/pub/75-006-x/2014001/article/14094-eng.htm

Hatmaker, T. (2020). Twitter tests a feature that calls you out for RTing without reading the article. *TechCrunch.* Retrieved from https://social.techcrunch.com/2020/06/10/twitter-retweet-prompt-android/

Holan, A. D. (2020, October 27). PolitiFact - The Principles of the Truth-O-Meter: PolitiFact's methodology for independent fact-checking. *@politifact.* Retrieved from https://www.politifact.com/article/2018/feb/12/principles-truth-o-meter-politifacts-methodology-i/

Hopper, R., & Bell, R. A. (1984). Broadening the deception construct. *Quarterly Journal of Speech, 70*(3), 288–302.

Ireton, C., & Posetti, J. (2018). *Journalism, fake news & disinformation: Handbook for journalism education and training—UNESCO Digital Library.* Retrieved from https://unesdoc.unesco.org/ark:/48223/pf0000265552

Johns Hopkins Coronavirus Resource Center. (2020, March 18). Taiwan—COVID-19 overview—Johns Hopkins. *Johns Hopkins Coronavirus Resource Center.* Retrieved from https://coronavirus.jhu.edu/region/taiwan

Kim, N. Y. (2021a, February 18). PolitiFact—Tucker Carlson falsely blames Green new Deal, wind energy for Texas power outage. *@politifact.* Retrieved from https://www.politifact.com/factchecks/2021/feb/17/tucker-carlson/tucker-carlson-falsely-blames-green-new-deal-wind–/.

Kim, N. Y. (2021b, February 18). PolitiFact—Yes, aversion to federal regulation drove Texas to isolate its power grid. *@politifact.* Retrieved from https://www.politifact.com/factchecks/2021/feb/18/facebook-posts/yes-aversion-federal-regulation-drove-texas-isolat/

Kovach, B., & Rosenstiel, T. (2010). *Blur: How to know what's true in the age of information overload: Vol. 1st U.S.* Bloomsbury.

LaMarre, H. L., Landreville, K. D., & Beam, M. A. (2009). The irony of satire: Political ideology and the motivation to see what you want to see in the Colbert report. *The International Journal of Press/Politics, 14*(2), 212–231. https://doi.org/10.1177/1940161208330904

Levine, T. R., Kim, R. K., & Hamel, L. M. (2010). People lie for a reason: Three experiments documenting the principle of veracity. *Communication Research Reports, 27*(4), 271–285. https://doi.org/10.1080/08824096.2010.496334

Liu, Z. (2005). Reading behavior in the digital environment: Changes in Reading behavior over the past ten years. *Journal of Documentation, 61*, 700–712. https://doi.org/10.1108/00220410510632040

Lynch, A. (1998). *Thought contagion: How belief spreads through society: The new science of memes.* Basic Books.

Mahdawi, A. (2021, February 17). Humour over rumour? The world can learn a lot from Taiwan's approach to fake news. *The Guardian.* Retrieved from http://www.theguardian.com/commentisfree/2021/feb/17/humour-over-rumour-taiwan-fake-news

Mantzarlis, A. (2016, June 7). There's been an explosion of international fact-checkers, but they face big challenges. *Poynter.* Retrieved from https://www.poynter.org/fact-checking/2016/theres-been-an-explosion-of-international-fact-checkers-but-they-face-big-challenges/

Maréchal, N., & Biddle, E. R. (2020, March 17). It's not just the content, it's the business model: Democracy's online speech challenge. *New America.* Retrieved from http://newamerica.org/oti/reports/its-not-just-content-its-business-model/

Maréchal, N., MacKinnon, R., & Dheere, J. (2020, May 27). Getting to the source of Infodemics: It's the business model. *A Report from Ranking Digital Rights*. New America. Retrieved from http://newamerica.org/oti/reports/getting-to-the-source-of-infodemics-its-the-business-model/

McCornack, S. A. (1992). Information manipulation theory. *Communication Monographs, 59*, 1–16.

Metts, S. (1989). An exploratory investigation of deception in close relationships. *Journal of Social and Personal Relationships, 6*(2), 159–179. https://doi.org/10.1177/0265407589006002002

Mikkelson, D. (2016, November 5). FALSE FBI agent suspected in Hillary email leaks found dead in apparent murder-suicide. *Snopes.Com*. Retrieved from https://www.snopes.com/fact-check/fbi-agent-murder-suicide/

Mikolajczak, K. (2021, February 10). 12+ wedding cake designs that we would definitely question eating. *Diply.Com*. Retrieved from https://diply.com/c/32969/wedding-cake-designs-would-definitely-question-eating

Miller, G. R., & Stiff, J. B. (1993). *Deceptive communication*. Sage.

Montaigne, M. (1572). *"Of liars" in the complete essays of Montaigne, translated by Donald M. Frame*. Stanford University Press. Retrieved from https://cep.calvinseminary.edu/reading-for-preaching/the-complete-essays-of-montaigne-trans-donald-frame/

Moore, R. C., & Hancock, J. T. (2021). *White paper: The effects of online disinformation detection training for older adults*. Retrieved from https://sml.stanford.edu/ml/2020/12/mediawise-white-paper.pdf

O'Hair, H. D., & Cody, M. J. (1994). Deception. In W. R. Cupach & B. H. Spitzberg (Eds.), *The dark side of interpersonal communication* (pp. 181–213). Erlbaum.

Payne, H. J. (2008). Targets, strategies, and topics of deception among part-time workers. *Employee Relations, 30*(3), 251–263. https://doi.org/10.1108/01425450810866523

Pennycook, G. (2018). A Perspective on the Theoretical Foundation of Dual Process Models [Chapter]. (p. 34). In *Dual Process Theory 2.0*. Routledge. London. https://doi.org/10.4324/9781315204550

Pennycook, G., Bear, A., Collins, E. T., & Rand, D. G. (2020). The implied truth effect: Attaching warnings to a subset of fake news headlines increases perceived accuracy of headlines without warnings. *Management Science, 66*(11), 4944–4957. https://doi.org/10.1287/mnsc.2019.3478

Pennycook, G., Cheyne, J. A., Barr, N., Koehler, D. J., & Fugelsang, J. A. (2015). On the reception and detection of pseudo-profound bullshit. *Judgment and Decision making, 10*(6), 549–563.

Pennycook, G., & Rand, D. (2019a, January 19). Why do people fall for fake news? *The New York Times*. Retrieved from https://www.nytimes.com/2019/01/19/opinion/sunday/fake-news.html

Pennycook, G., & Rand, D. G. (2019b). Lazy, not biased: Susceptibility to partisan fake news is better explained by lack of reasoning than by motivated reasoning. *Cognition, 188*, 39–50. https://doi.org/10.1016/j.cognition.2018.06.011

Peters, U. (2020). What is the function of confirmation bias? *Erkenntnis*. https://doi.org/10.1007/s10670-020-00252-1

Pfau, M., Bockern, S. V., & Kang, J. G. (1992). Use of inoculation to promote resistance to smoking initiation among adolescents. *Communication Monographs, 59*(3), 213–230. https://doi.org/10.1080/03637759209376266

Reardon, R., Folwell, A. L., Keehr, J., & Kauer, T. (2019). Effects of deception on the deceiver: An interdisciplinary view. In T. Docan-Morgan (Ed.), *The Palgrave handbook of deceptive communication*. Palgrave Macmillan.

Reinhard, M.-A. (2010). Need for cognition and the process of lie detection. *Journal of Experimental Social Psychology, 46*(6), 961–971. https://doi.org/10.1016/j.jesp.2010.06.002

Roth, Y., & Pickles, N. (2020, May 11). *Updating our approach to misleading information*. Retrieved from https://blog.twitter.com/en_us/topics/product/2020/updating-our-approach-to-misleading-information.html

Rubin, V. L. (2010). On deception and deception detection: Content analysis of computer-mediated stated beliefs, *Proceedings of the Association for Information Science and Technology, 32*: 1–10. Retrieved from http://dl.acm.org/citation.cfm?id=1920377

Rubin, V. L. (2019). Disinformation and misinformation triangle: a conceptual model for "fake news" epidemic, causal factors and interventions. *Journal of Documentation*, 75 (5), 1013–1034. https://doi.org/10.1108/JD-12-2018-0209

Rubin, V. L., & Chen, Y. (2012). Information manipulation classification theory for LIS and NLP. *Proceedings of the American Society for Information Science and Technology, 49*(1), 1–5. https://doi.org/10.1002/meet.14504901353

Rubin, V. L., & Conroy, N. (2012). Discerning truth from deception: Human judgments and automation efforts. *First Monday, 17*(3). Retrieved from http://firstmonday.org/ojs/index.php/fm/article/view/3933/3170

Rubin, V. L., Conroy, N. J., Chen, Y., & Cornwell, S. (2016). Fake news or truth? Using Satirical Cues to Detect Potentially Misleading News. *Proceedings of the Second Workshop on Computational Approaches to Deception Detection,* 7–17, San Diego, California. Association for Computational Linguistics. http://aclweb.org/anthology/W/W16/W16-0800.pdf

Shane, T. (2020, June 30). *The psychology of misinformation: Why we're vulnerable.* First Draft. Retrieved from https://firstdraftnews.org:443/latest/the-psychology-of-misinformation-why-were-vulnerable/

Silverman, C., & Alexander, L. (2016, November 3). How teens in the balkans are duping trump supporters with fake news. *BuzzFeed News.* Retrieved from https://www.buzzfeednews.com/article/craigsilverman/how-macedonia-became-a-global-hub-for-pro-trump-misinfo

Southwell, B. G., Thorson, E. A., & Sheble, L. (2017, October 11). The persistence and peril of misinformation. *American Scientist.* Retrieved from https://www.americanscientist.org/article/the-persistence-and-peril-of-misinformation

Stencel, M., & Luther, J. (2020, October 13). Fact-checking count tops 300 for the first time. Duke Reporters' Lab. *Fact-Checking News Archives.* Retrieved from https://reporterslab.org/category/fact-checking/

Subedar, A. (2018, November 27). The godfather of fake news: Meet one of the world's most prolific writers of disinformation. *BBC.Co.Uk.* Retrieved from https://www.bbc.co.uk/news/resources/idt-sh/the_godfather_of_fake_news

Sydell, L. (2016, November 23). We tracked down a fake-news creator in the suburbs. Here's What We Learned. *NPR.Org.* Retrieved from https://www.npr.org/sections/alltechconsidered/2016/11/23/503146770/npr-finds-the-head-of-a-covert-fake-news-operation-in-the-suburbs

Taraborelli, D. (2008). How the web is changing the way we trust. *Current Issues in Computing and Philosophy, 12.*

Thurm, S. (2021, January 28). Facebook's oversight board has spoken. But it hasn't solved much. *Wired.* Retrieved from https://www.wired.com/story/facebook-oversight-board-has-spoken/

Twitchell, D. P., Nunamaker, J. F., & Burgoon, J. K. (2004). Using speech act profiling for deception detection. In *Intelligence and security informatics* (pp. 403–410). Retrieved from http://www.springerlink.com/content/ajeauy59kxjjttry

van der Linden, S., Leiserowitz, A., Rosenthal, S., & Maibach, E. (2017). Inoculating the public against misinformation about climate change. *Global Challenges, 1*(1). https://doi.org/10.1002/gch2.201600008

Vrij, A. (2000). *Detecting lies and deceit.* Wiley.

Vrij, A. (2004). Why professionals fail to catch liars and how they can improve. *Legal and Criminological Psychology, 9*(2), 159–181.

Vrij, A. (2008). *Detecting lies and deceit: Pitfalls and opportunities* (2nd ed.). Wiley. Retrieved from http://hdl.handle.net/2027/mdp.39015076166118

Walczyk, J. J., Runco, M. A., Tripp, S. M., & Smith, C. E. (2008). The creativity of lying: Divergent thinking and ideational correlates of the resolution of social dilemmas. *Creativity Research Journal, 20*(3), 328–342.

Wang, C. J., Ng, C. Y., & Brook, R. H. (2020). Response to COVID-19 in Taiwan: Big data analytics, new technology, and proactive testing. *JAMA, 323*(14), 1341. https://doi.org/10.1001/jama.2020.3151

Wang, W.-C., Brashier, N. M., Wing, E. A., Marsh, E. J., & Cabeza, R. (2016). On known unknowns: Fluency and the neural mechanisms of illusory truth. *Journal of Cognitive Neuroscience, 28*(5), 739–746. https://doi.org/10.1162/jocn_a_00923

Wiseman, R. (1995). The megalab truth test. *Nature, 373*, 391.

Wolf, M. (2018, August 25). Skim reading is the new normal. The effect on society is profound | Maryanne Wolf. *The Guardian*. Retrieved from http://www.theguardian.com/commentisfree/2018/aug/25/skim-reading-new-normal-maryanne-wolf

Zhou, L., Shi, Y. M., & Zhang, D. S. (2008). A statistical language modeling approach to online deception detection. *IEEE Transactions on Knowledge and Data Engineering, 20*(8), 1077–1081. https://doi.org/10.1109/Tkde.2007.190624

Zimmer, M. J., & Diply.com. (2021, February 10). 13+ trending pics that give us a unique view of the world. Diply.Com. Retrieved from https://diply.com/c/33089/trending-pics-give-unique-view-world

Chapter 3
Credibility Assessment Models and Trust Indicators in Social Sciences

Credibility is, after all, the most important thing a communicator has. A communicator in the news media who lacks credibility probably has no audience

(Severin & Tankard, 1992, p. 28)

The whole mechanism of society rests on confidence: it permeates all life, like the air we breathe: and its services are apt to be taken for granted and ignored, like those of fresh air, until attention is forcibly attracted by their failure

(Marshall, 1919, p. 113)

Trust, but verify (Russian proverb)

Abstract **Chapter 3** surveys credibility and trust research in information science, human–computer interaction, psychology, communication, and other social sciences. Several models explain the process of credibility assessment, dissecting it into stages and offering components for online content evaluation. Multiple predictive indicators have been considered for automated approaches in AI computing including source expertise and trustworthiness, surface features of the medium, and cognitive factors associated with information recipients. The overall credibility of a message (in a post, tweet, blog, or website article) is underpinned by the characteristics of its sources (such as their expertise and trustworthiness), and the characteristics of its specific content (such as plausibility, internal consistency, and quality), as seen through the eyes of the person who is receiving the message (with their cultural background, previous beliefs, or propensity to trust). The medium of delivery can also influence credibility judgements, be it a particular social media platform (like Twitter or Snapchat) or specific technology used for viewing the message (like iPhones). Research in psychology adds another layer associated with a cognitive effort on the part of the receivers of information. People's propensity to trust, thier perceptions of insitutional reputation, previous experiences with trust, and the context in which information is needed are among factors influencing people's decisions whether to trust the source and believe the message, or not. Thus, credibility assessments are layered (source, medium, message, receiver, context), and such judgements could be better supported by expert systems. Currently, surrogate measures like popularity, user engagement, or crowdsourced ratings systems stand in for credibility indicators or trust predictors in several social

V. L. Rubin, *Misinformation and Disinformation*,
https://doi.org/10.1007/978-3-030-95656-1_3

technologies. I identify current challenges and point out promising directions for AI to assist people with online information quality evaluations and layered credibility assessments or trust judgments.

Keywords Credibility · Believability · Credibility assessment · Credibility ratings · Credibility indicators · Trust · Dependability · Trust judgment · Trust Modeling · Trust predictors · Trust score · Information quality · Expertise · Trustworthiness · Cognitive authority · Reliability · Information Retrieval (IR) · Information seeking · Human–Computer Interaction · HCI · Library and Information Science · LIS · Communication · Psychology · Management information systems · MIS · Natural Language Processing · NLP · Computational methods · Artificial Intelligence · AI

3.1 Introduction

How do people decide whether to believe what they read online? Any online advice, news bit, or piece of vital information warrants some form of credibility check before one places confidence in it. Misplacing your trust in inadequate sources or otherwise attributing credibility inaccurately can easily leave you misinformed. People regularly make trust decisions about their web-based behavior that, for instance, "they are 'talking' to the right person (e.g. their bank), that their children are not accessing pornographic sites, etc. Perceived trust or credibility has a strong influence on people's willingness to engage with online activities such as shopping or banking, where sensitive information is involved" (Pickard et al., 2010, p. 307). As the range of our online activities and the decisions we make with the help of online resources (about health, politics, or investments) continues to increase, the stakes for our online literacy skills are higher than ever in the 2020s. So, how do we assess the information we encounter on the web? And how do we know who to place our confidence with?

Several disciplines (including information science, communication, psychology, education, and management sciences) offer insights into credibility, trust, and closely related concepts. Each strand of credibility research emphasizes the diverse aspects of credibility or isolates its constituent parts in slightly differently terms. The available research insights go beyond purely theoretical definitions and can be applied empirically to automate, at least partially, some aspects of online information evaluation. To use these theoretical concepts as potential indicators in predictive AI systems we have to consider their inventory, with its variety and overlaps, and select the most reliable and readily obtainable ones. (For a review of computational analyses with AI, see Chap. 7.)

Making trust judgments can also be explicitly taught as part of information literacy in terms of what factors to pay attention to. When digital media users realize that an online source lacks credibility, they can dismiss it as mis- or disinformative. Cutting the infodemic at its sources reduces the spread of rumors and other nonauthoritative information. In other words, "if a [digital media] user rejects new information as not credible, that information will not be learned, nor can it have any other impact" (Wathen & Burkell, 2002, p. 134). Making valid judgments to filter out information that is not credible is predicted to help interrupt the interaction of the causal factors of the infodemic: the hosts and the pathogens of mis- and disinformation (for a discussion of the (Rubin, 2019) infodemiological model and immediate digital literacy countermeasures; see Chap. 1, Sect. 1.2 and 1.3).

Evidence from credibility and trust research in multiple disciplines leads to the same conclusion: That "people rely on a wide variety of factors to decide whether to believe the information they obtain online" (Metzger & Flanagin, 2015, p. 448). "Some factors can be automatically evaluated by examining the given Web pages, for example, the presence or absence of an e-mail address in the Web page. Conversely, other factors, for example the objectivity of information on a Web page, can only be evaluated by humans" (Kakol et al., 2017, p. 1044).

Evaluating the credibility of the source, message, and medium independently, yet in connection with each other, was first strongly advocated by Tseng and Fogg (1999). Now, researchers in multiple disciplines generally agree with their layered analysis approach. "Taking the quintessential example of Wikipedia, credibility judgments can be made at the website level (is Wikipedia a credible source of information?), at the content level (is any specific entry within Wikipedia credible?), or regarding specific information author(s) (are specific contributors to Wikipedia credible?)" (Metzger & Flanagin, 2015, p. 447). The overall credibility of a message (in a post, tweet, blog, or website article) is underpinned by the characteristics of its sources (such as their expertise and trustworthiness), and the characteristics of its specific content (such as plausibility, internal consistency, and quality) as seen through the eyes of the person who is receiving the message (with their cultural background, previous beliefs, or propensity to trust). The medium of delivery can also influence credibility judgments, be it a particular social media platform (like Twitter or Snapchat) or specific technology used for viewing the message (like iPhones).

3.2 Credibility in Information Retrieval

In information science, credibility was not always an apparent criterion for information evaluation in an information search process. Initially, the task of *information retrieval* (IR) was formulated "as the problem of selecting texts from a database in response to some more-or-less well-specified query" (Belkin, 1993, p. 56). The classical IR was concerned with the representation, storage, organization, and accessing of information items for their efficient use (Salton & McGill, 1983). The IR systems' task was, first and foremost, to find methods to match a user's need, expressed as a query of its relevant text-based items (see Fig. 3.1). Early digital collections that contained representations of items were seen as analogous to library collections, and as such were seen to hold some base level of credibility. Representations of information (originally only texts) and queries and the techniques for the comparison of

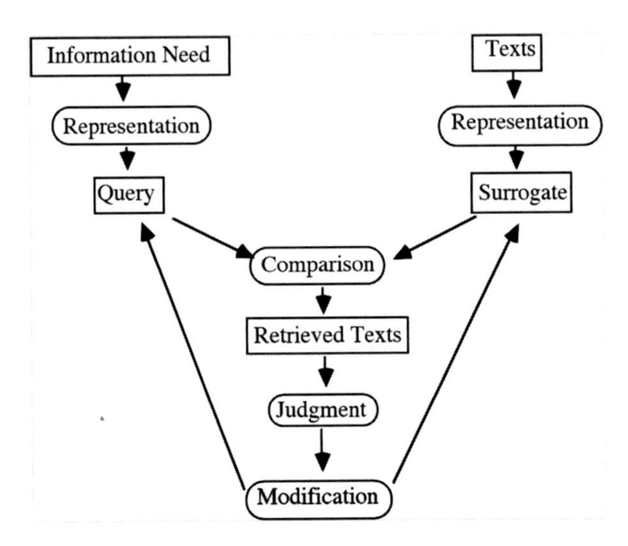

Fig. 3.1 A model of the standard view of information retrieval (Belkin, 1993)

these text and query representations were the dominant concerns. In this traditional view, the *topical relevance* of the information provided by IR systems allowed their users to express their information needs (i.e., what they wanted to find) and cut straight to the appropriate items. IR succeeded in its goals of reducing the vast amounts of web data to only the relevant information, giving rise to now ubiquitous search engines like Google Search, Microsoft Bing, Baidu, or Yandex.

In the mid-1990s, the emphasis shifted from the standard IR model to a newer view of the IR process as information seeking behavior. *Users' interactions* with information became central: How they seek, obtain, interpret, and evaluate information, and how they use it for decision-making and further actions. "People are not just passive recipients of messages, but rather active seekers of texts, and active constructors of meaning from these texts" (Belkin, 1993, p. 55). Users' perceptions, judgments, and cognitive processes were brought into the loop, undoubtedly influenced by studies in communication, psychology, and human–computer interaction.

3.3 Credibility Indicators in Social Sciences

Human–computer interaction (HCI), as a field at the crossroads of psychology, information, and computer science, examines the intricate relationship we have with our computers and digital technologies. In the late 1990s, HCI conceptualized computers as persuasive technologies that are able to change people's attitudes and behavior (Fogg, 1998), as opposed to mere simple computing mechanisms like calculators.

In HCI, *credibility* is synonymous with *believability*—"a perceived quality of a source, which may or may not result in associated trusting behaviors," while *trust* is roughly equivalent to *dependability*, and refers to a set of beliefs, dispositions, and behaviors associated with the acceptance of risk and vulnerability (Rieh & Danielson, 2007, p. 314). Early HCI works by Tseng and Fogg (1999) posited that to judge credibility, people simultaneously perceive and evaluate two distinct qualities of the source: *trustworthiness* (goodness or morality, described as well intentioned, truthful, or unbiased) and *expertise* (described as knowledgeable, reputable, or competent).

Around the turn of the twenty-first century, the *Stanford Web Credibility Research Project* conducted multiple studies on perceptions of web credibility, examining which elements boost, or conversely hurt, these perceptions (Rieh & Danielson, 2007). B. J. Fogg's Prominence-Interpretation Theory (2003) determined two necessary aspects in web credibility assessments, namely, the likelihood of an element being noticed when people evaluate credibility (*Prominence*) and the value or meaning assigned to the element based on the user's judgment of how the element affects the likelihood of the information being good or bad (*Interpretation*). "If one or the other does not happen, then there is no credibility assessment. The process of noticing prominent elements and interpreting them will typically happen more than once when a person evaluates a Web site, with new aspects of the site being noticed and interpreted until the person reaches satisfaction with an overall credibility assessment or reaches a constraint, such as running out of time" (Fogg et al., 2003, p. 11). Fogg (2003) originally outlined five factors which affect *prominence*: user involvement, information topic, task, experience level, and other individual differences, while three factors were found to affect *interpretation*: user assumptions, user skill and knowledge, and contextual factors such as the environment (Fogg, 2003).

With the goal of improving webpage design, Fogg et al. (2001) identified five elements that increase credibility perceptions (real-world feel, ease of use, expertise, trustworthiness, and tailoring), and two types that decrease them (commercial implications and amateurism, e.g., broken links, errors, font issues, rare updates). The follow-up study collected and analyzed study participants' comments and identified 18 most frequently noticed factors, when evaluating the credibility of websites in 10 contexts,[1] as determined by a website's intended function (Fogg et al., 2003). In their diminishing frequency of occurrence in the study participants' comments, these factors were design look (46.1%), information design/structure (28.5%), information focus (25.1%), company motive (15.5%), usefulness (14.8%) and accuracy (14.3%) of information, name recognition and reputation (14.1%), advertising (13.8%), bias of information (11.6%), tone of the writing (9%), identity of site sponsor (8.8%), functionality of site (8.6%), customer service (6.4%), performance on a test (3.6%), and affiliation (3.4%), and some other less noticed categories with less than 3% incidence that were not reported (Fogg et al., 2003). Notice that the

[1] Here are some representative examples: Amazon (e-commerce), E! Online (entertainment), E*Trade (finance), WebMD (health), CNN (news), American Red Cross (nonprofit), E-pinions (reviews), Yahoo! (Web searching), ESPN (sports), and Expedia (travel) (Fogg et al., 2003).

surface features (such as the design look and site structure) dominate the list, altogether accounting for three-quarters of the study participants' comments.

An alternative model of website credibility assessment was put forward by the *Media Effects Research Laboratory* at Penn State University. Four broad technological affordances—*Modality* (M), *Agency* (A), *Interactivity* (I), and *Navigability* (N)—were found to be present in most digital media, resulting in the use of certain heuristics that lead to user perceptions of digital content quality. Sundar (2008) stresses that the MAIN model explains how people perceive the credibility of digital media content "beyond what is explained by content characteristics" (p. 92). This model is consistent with psychology's assessment of credibility as a cognitive task, and its 26 qualities were listed as those forming the credibility assessment model with the most numerous constituent parts at the time of its publication (Sundar, 2008). These qualities, as shown in Fig. 3.2, later came to be understood as one of the more complete lists of credibility factors (Kakol et al., 2017).

These factors have subsequently been refined and augmented in a layered approach that sorted cues used in the credibility evaluation decisions by their relevance to these four layers: Information Source, Message, Medium, or Receiver (Metzger & Flanagin, 2015). "For example, author credentials constitute an author cue that information consumers might rely on to assess the relative credibility of information emanating from a source. Information currency, by contrast, is a message cue that has been shown to influence credibility perceptions. The presence or absence of website advertising is an example of a site or source cue affecting perceived credibility, and the degree of an individual's experience with a medium or a source is a receiver characteristic that has been shown to influence people's credibility evaluations" (Metzger & Flanagin, 2015, p. 448).

By the early 2000s, library and information science (LIS) studies were actively probing how people evaluate information and make credibility judgments in the context of the web 1.0, which lacked quality-control mechanisms. Rieh (2002) established that web users relied on a wide array of criteria of perceived information quality, including credibility. Credibility was theorized to consist of information accuracy, reliability, currency, comprehensiveness, and validity. Their data were comprehensive, including data obtained from search logs, verbal reports collected

List of credibility factors included in the Modality, Agency, Interactivity, Navigability (MAIN) model.			
Utility	Importance	Relevance	Believability
Popularity	Pedigree	Completeness	Level of detail
Variety	Clarity	Understandability	Appearance
Affect	Accessibility	Conciseness	Locatability
Representative quality	Consistency	Compatibility	Reliability
Trustworthiness	Uniqueness	Timeliness	Objectivity
Expertise	Benevolence		

Fig. 3.2 Sundar's (2008) MAIN Model rendered as credibility factors (by Kakol et al. (2017))

concurrently alongside search processes, and retrospective post-search interviews. Another criterion frequently mentioned by this study's participants was the authority of sources for online content—not just the authority of individual authors or experts, but also the host organizations, institutions, or books. This was termed *cognitive authority* to distinguish it from *administrative authority*, with someone exercising authority through their position (Rieh, 2002). Cognitive authority refers to "the extent to which users think that they can trust the information," or "influences that a user would recognize as proper because the information therein is thought to be credible and worthy of belief" (Rieh, 2002, p. 3). Prior literature in information science equated establishing cognitive authority in electronic resources to establishing credibility and influence over users. Olaisen (1990) states "what is generally known in information science theory—that although personal information sources may be the most trusted, they are not necessarily the most expert: we turn to other sources when expertise or competence is required" (cited in Wathen & Burkell, 2002, p. 137).

Rieh (2002) also distinguishes between two kinds of judgment: "*A predictive judgment* guides a decision about what kind of action the user is going to take given multiple choices (alternatives). As a result of this judgment, a new Web page is presented to the user, and when she/he looks at it, *an evaluative judgment* is made" (p. 146). The factors influencing each judgment of quality and authority were identified by the characteristics of information objects, the characteristics of sources, user knowledge, situation, ranking in search output, and general assumptions. Diverse facets in the judgments of information quality (good, accurate, current, useful, and important) and cognitive authority (trustworthy, credible, reliable, scholarly, official, authoritative) were thus derived empirically in this study (Rieh, 2002), which uncovered what people actually look for during their searches and how they explain their interactions with information in real time.

In their thorough (2002) literature review, Wathen and Burkell first consider traditional print media and summarize the various factors that have been found to impact source, message, and media characteristics, as well as factors related to receiver and context. Their review confirms that credibility is conceptualized as a multidimensional construct in the majority of the empirical evidence seen in psychology, consumer behavior, and communication studies such as Wilson and Sherrell's (1993) meta-analysis measuring factors influencing credibility perceptions.

Wathen and Burkell's (2002) factors influencing credibility in traditional print and interpersonal media show two main source-related components—expertise and trustworthiness, which are then complemented with credentials, likeability, goodwill, and dynamism. Wathen and Burkell's factors add detail to the layers of previous credibility determination models and elaborate on the nature of interactions within and between these layers. For instance, source effects interact with receiver-related factors through involvement. If the receiver of information is less involved with the topic, source characteristics (such as knowledge and beliefs) will be more influential. In addition, "source characteristics interact with message-related factors, such as discrepancy (from previous beliefs), incongruity in content, or timing of source identification within the message. For example, if a message is high in discrepancy, low in incongruity, and if the source is identified early in the message, then high-credibility sources will tend to have a greater impact on knowledge/attitudes/behaviors than low credibility sources (Wilson & Sherrell, 1993)" (cited in Wathen & Burkell, 2002, p. 136).

The factors for credibility interpretation from traditional media manifest themselves slightly differently in computer-based electronic media delivered via the Internet. "Surface aspects of the presentation take on particular significance, and the general attitude of the user toward computers has an impact on credibility" (Wathen & Burkell, 2002, p. 140). Rieh and Danielson (2007) found that academics with advanced degrees tended to make use of conventional indicators of credibility (such as an institution's name and an individual's affiliation) within the context of the web, but young adults reported primarily paying attention to surface features. "The appearance of the site's homepage (colors, graphics, etc.), the usability of the interface, how well the information is organized—in general, how well the site is designed. For Internet information, even a single spelling mistake can give the impression of 'amateurism' and lead the user to reject the site as not credible" (Wathen & Burkell, 2002, p. 140).

Their model shows two key factors influencing credibility in computer-based media: *cognitive qualities* such as influence, trustworthiness, competence, reliability, and relevance determine cognitive authority and features such as form, novelty, accessibility, and flexibility are the key *technical qualities* (Wathen & Burkell, 2002). "Combined, cognitive and technical features make up what he terms 'institutional quality' (Olaisen, 1990)—presumably, the greater the 'institutional quality,' the higher the judgment of cognitive authority, and thus the more credible and persuasive the information" (Wathen & Burkell, 2002, p. 137).

Wathen and Burkell work through a hypothetical scenario which helpfully describes how an information user proceeds in assessing information on a website. As the user enters a website, they make some immediate judgments about the surface characteristics of the site in an initial assessment, forming impressions about the appearance, interface design, download speed, interactivity, and site organization. Does the site look professional? Is the information easy to access? If the initial assessment meets enough of the user's criteria, the user moves to the next level of assessment; if not, the user is likely to leave the site and seek out other sources.

At the second level of evaluation, aspects of the source (like expertise, competence, trustworthiness, and credentials) are judged along with the content of the message (including its accuracy, currency, relevance, level of detail, examples, alternatives, etc.) If the site has the information the user is looking for, will they believe it? Does it apply to their context: How easy is it to tailor the information to their own situation? If both levels of assessment meet the user's criteria "for all (or enough?) of the dimensions, the user accepts the information as credible, and decides to evaluate the information content. If they do not, they will likely leave the site" (Wathen & Burkell, 2002, p. 142).

Other analytical frameworks identify similar factors used to determine the credibility of less-authoritative sources and suggest ways to automate such determinations. Rubin and Liddy (2006) proposed four categories that influence blog

1) Blogger's Expertise and Offline Identity Disclosure a) Name and geographic location b) Credentials c) Affiliations (personal and institutional) d) Blogrolls (i.e., hyperlinks to other sites) e) Stated competencies f) Mode of knowing 2) Blogger's Trustworthiness and Value System a) Biases b) Beliefs c) Opinions d) Honesty e) Preferences f) Habits g) Slogans	3) Information Quality a) Completeness b) Accuracy c) Appropriateness d) Timeliness e) Organization (by categories or chronology) f) Match to prior expectations g) Match to information need 4) Appeals and Triggers of a Personal Nature a) Aesthetic appeal (i.e., design layout, typography, and color schemes) b) Literary appeal (i.e., writing style and wittiness) c) Curiosity trigger d) Memory trigger (i.e., shared experiences) e) Personal connection (e.g., the source is an acquaintance or a competitor of the blog-reader)

Fig. 3.3 Rubin and Liddy's (2006) blog credibility assessment factors

credibility: (1) the blogger's expertise and the amount of offline identity disclosure; (2) the blogger's trustworthiness (or their overtly stated value system including beliefs, goals, and values); (3) information quality; and (4) appeals of a personal nature (Fig. 3.3). Some of these features can be extracted and systematized, using natural language statements, for instance, bloggers' overtly stated beliefs, goals, biases, opinions, preferences, ownership, and purchasing habits, relying on typical semantics such as "*do/don't love/like/care (much) for*," "*appealed to me (so much more)*," and "*(one of) my philosophy/(-ies) in life is*") (Rubin & Liddy, 2006).

Early psychological studies in communication and persuasion were foundational to the above-described seminal models in LIS. Even in 1950, psychologists and communication theorists (e.g., Hovland et al., 1953) saw credibility as a subjective perception on the part of the information receiver, and not as an objective property of a source, message, or medium. To extend this logic, the receiver's context, or the situation for which the information is needed, is also highly subjective. If "the credibility of the same source or piece of information may be judged differently by different people" (Metzger & Flanagin, 2015, p. 446), we should question the feasibility of ever achieving a complete automation of credibility judgements based on any set of intrinsic information qualities. Luckily, there is at least some evidence that using credibility assessment indicators can improve IR results. Weerkamp and de Rijke (2008) integrated several of Rubin and Liddy's (2006) indicators into their IR approach such as measures of post length, timeliness, spelling errors, or capitalization, and at blog-level, its spamminess, topical consistency, and the periodicity of its posts. The Weerkamp and de Rijke (2008) study ultimately showed that combining credibility indicators significantly improved retrieval effectiveness (Lukoianova & Rubin, 2014). This may imply that users tend to trust search results which have been filtered for credibility.

3.4 Trust Modeling

The concept of *trust* is also often used in everyday language and communication about making trustworthiness decisions (Lukoianova & Rubin, 2014). Ordinary language statements about trust seem to conceive of trust, at least partly, as a matter of behavior, rather than as an expectation or a reliance (Hardin, 2001), but trusting is neither an action nor a behavior. *Trust* is "an assured reliance on the character, ability, strength, or truth of someone or something" (Merriam-Webster Dictionary Online, 2021).

Many disciplines see trust as *a relationship* between an entity (*the trustor*) that relies on another entity (*the trustee*) based on a set of criteria. In social science, such entities are individuals who hold certain expectations of each other in social settings. Classic works in sociology define trust as the trustor's *subjective probability* (often treated as *a belief*) about whether the trustee(s) will perform a particular action that benefits the trustor (Cho et al., 2015). In psychology, trust is also "*a generalized expectancy* held by an individual that the word, promise, [or] oral or written statement of another individual or group can be relied on" (Rotter, 1980, p. 1).

Psychological studies of trust bring out the cognitive nature of trust assessments: They are decisions guided by prior trusting experiences and a general propensity to trust. A review of a series of psychological studies on the willingness to trust surmises that those prone to "believing others in the absence of clear-cut reasons to disbelieve" (i.e., high-trusting individuals) are better off because they are "less likely to lie and are possibly less likely to cheat or steal," and "less likely to be unhappy, conflicted, or maladjusted" (Rotter, 1980, p. 1). In addition, they were found to be "more likely to give others a second chance and to respect the rights of others," and they were "liked more and sought out as a friend more often, both by low-trusting and high-trusting others" (Rotter, 1980, p. 1). Yet, high trustors are not disadvantaged by their trusting nature: They do not show more gullibility (i.e., naïveté or foolishness) than low trustors. High trustors are "no less capable of determining who should be trusted" (or not), although high trustors "may be more likely to trust others" than low trustors in novel situations, and high trustors may be "fooled more often by crooks, but the low trustor is probably fooled equally often by distrusting honest people, thereby forfeiting the benefits that trusting others might bring" (Rotter, 1980, p. 7). Rotter concludes with a piece of advice to educators and parents for the "age of suspicion": "model and encourage a little more trust" within your "own smaller circles of influence" (Rotter, 1980, p. 6). This pre-web 1.0 advice seems counterintuitive in the age of misinformation. A proverb comes to mind: "Trust, but verify" (Доверяй, но проверяй *in Russian*), which stresses the need for vigilance, especially in matters of importance.[2]

---

[2]This Russian saying was apparently popularized in North America in the context of the Reagan–Gorbachev nuclear disarmament negations (Wikipedia., 2021a) at the time when the stakes were high and cross-Atlantic intergovernmental trust was low but sufficient to generate a probability of mutually beneficial actions.

Trust is also a core concern in the field of management information systems (MIS), which has extensively investigated the extent to which specific kinds of trustors (e.g., managers or other decision makers) rely on expert systems (decision support systems and other information systems) for their factual claims, general judgments, specific solutions, advice, or recommendations. MIS interprets entities as organizations, and may focus on both interpersonal and interorganizational trust in, for instance, virtual teams or technology acceptance in e-commerce, where *trust* refers to one partner's confidence, belief, and positive expectation that another partner will act in their best interests (Metzger & Flanagin, 2015). Humans' trust of automated operations, as researched in the fields of computing and engineering, has been modeled on the willingness of the trustor to take risks and be vulnerable to the ability, integrity, benevolence, and trustworthiness (or credibility) of the other entity they rely on (the trustee). *Distrust*, by contrast, is then "a negative expectation of a trustor regarding the behavior of the trustee, in a context that entails risk to the trustor" (Marsh & Dibben, 2003; Rousseau et al., 1998; cited in Rubin, 2009).

In economics, a decision to trust (or not) is seen as *an estimation of maximizing* a trustor's potential benefits (utility) *and minimizing* their risks. Such estimates are just as private and subjective as credibility judgments. "Trust has an emotional component and requires 'a leap of faith' (Möllering, 2006), and a willingness to tolerate uncertainty and accept vulnerability (Rousseau et al., 1998). Trust can serve as a gap-filler for explicit knowledge (Marsh & Dibben, 2003) in the absence of adequate information for rational decision-making" (cited in Rubin, 2009, p. 300). In computing and engineering, including the domains of AI, social network analysis, communication networks, cyber security, and related disciplines, such a lack of adequate information for rational choices is seen as the trustor's *imperfect knowledge* or *vulnerability*, and the goal of modeling trust is to reduce this inherent uncertainty in any potentially risky situation (Cho et al., 2015).

Using the terms *trust* and *credibility* almost interchangeably, a 2010 UK Joint Information Systems Committee–funded study posits a Trust Model for users' internal assessment of information. The model was driven by a trust-related literature review and was further verified in consultations with their study participants. The model also includes LIS and IR terminology such as information seeking and retrieval.

In this trust model, three types of identified factors—external cues, internal cues, and the user's cognitive state "interact in information seeking to lend trust in, and belief in the credibility of, the information found" (Pickard et al., 2010, p. 307). The user has to have trust in the methods for finding the information and then must decide whether to use the retrieved information for its purpose by considering perceived risks, usefulness, ease of use, and accessibility. The higher the perceived risk, the more effort is invested in searching and obtaining trustworthy information (Pickard et al., 2010). The users' *cognitive states* in the model are the need for closure, need for cognition, willingness to explore, motivation or disposition to believe, purpose, prior knowledge, time available, ability, past experience with site, author or author's institution, propensity to trust, trust in technology, risk propensity, faith in humanity, suspicion of humanity, and Internet anxiety (Pickard et al., 2010).

The trust model also contains *external factors* that impact user beliefs. These include the presence of a fee for content (students are unlikely to want to pay for information), presence of seals of approval, institutionally controlled credibility rating systems, preapproved database (e.g., JSTOR (short for "Journal Storage") or Education Resources Information Center (ERIC)) recommendations from others, digital signatures, rankings, offline credibility, presentation of the site or the provider, and ease of use of the site. Factors linked with *internal cues* are accuracy, freedom from errors, verifiability, authoritativeness (i.e., reputation of the source or qualifications), objectivity, currency, coverage (comprehensiveness and depth), presentation and format (quality of writing or structure), affiliations of source or site (traceability), source motivation, citations, and type of "object" (e.g., a journal, a blog, etc.) (Pickard et al., 2010).

By contrast with iterative LIS-based credibility assessment models, the trust model (Pickard et al., 2010) has a linear progression: users' cognitive state, internal and external cues are direct factors affecting trust/credibility of online information, and its consequent risk–benefit analyses (i.e., perceived risk, usefulness, ease of use, and accessibility) guide the users' intention to use the information (or not).

In sum, these seminal models of credibility assessment and trust should be used more extensively to inform current research into online quality assessment practices. The three models—Rieh's (2002), Wathen and Burkell's (2002), and Pickard et al.'s (2010)—share similarities: They each identify the sequential steps of an effortful (not intuitive) information assessment process, explaining how such judgments are made. In addition, the reviewed models provide a list of meaningful assessment criteria; none of the models found a single definitive criterium to predict the user's decision to believe any information online. A drawback of the models as a group is their lack of unified terminology which leads to synonymous terms used vaguely or in overlap with one another. (See also Cho et al. (2015), for a recent complete survey on modeling trust across disciplines.)

The complexity of existing criteria testifies to how difficult human credibility judgments are, in principle. Breaking the larger problem down into its component parts—namely, the credibility cues that instill trust—is a useful approach. The modeled criteria should be tested comparatively as a set for their predictive powers, prior to their adoption in AI systems. Social science models hold promise for solving part of the socio-technological problem of content credibility evaluation because they allow us to understand our humanity. That understanding can inform the construction of algorithms to evaluate any given webpage, or more specifically to examine any text or claim within the page, in an effort to assist humans with their decision-making.

3.5 Credibility Indicators in Journalistic Fact-Checking

Other expert communities that do credibility assessment professionally, such as journalists and fact-checkers, have also taken an interest in providing models of credibility assessments based on their practices. Major organizations are working on a set of common terminology for annotations of online news quality, to be applied consistently across fact-checking organizations. These terms could appear as markups on news content and be used to facilitate the eventual automation of fact-checking tasks. For example, the Credibility Coalition, borne out of collaborations among fact-checking initiatives in journalism,[3] developed its own credibility assessment system in which its various members within the coalition have agreed on credibility definitions and converged on its salient factors. This annotation scheme distinguishes eight article-level *content indicators* that can be ascertained from the title and text content, and eight external *context indicators* that require consulting the surrounding metadata (such as ads or layout) or external resources (Table 3.1), but do not include the page's publishers, authors, or multimedia content.

Zhang et al.'s (2018) indicators seem a bit repetitive, overemphasizing the role of sources, and their quotes or citations, but this is in line with the traditional investigative journalism practices of confirming what was said by authoritative sources (see Chap. 5 for investigative journalism "ways of knowing"). Their proposed indicator-based model aspires to keep potential AI applications in mind, as is evidenced by their use of schema.org's ClaimReview markup annotation,[4] and by their references to earlier works in computational fact-checking (e.g., Ciampaglia et al., 2015). However, many of their indicators still require human judgment and may be difficult to obtain by automated means.

Fact-checkers' attempts to develop quality-control instruments—to aid news readers and social media users—through standardized established criteria is reminiscent of the early efforts of the library community to catalogue information—to aid library patrons in the selection of high-quality materials. It is not surprising that journalists have been called up to collaborate with librarians and libraries, "their information-gathering cousins" to do more community outreach and explanations in order to repair the strained relationships between the news media and news readers and to rebuild trust in the eyes of the public (Beard, 2018). The next section will look at some legacy information-vetting practices inspired by librarianship.

[3] MisinfoCon: https://misinfocon.com; Climate Feedback process: https://climatefeedback.org/process/; Mozilla Festival (MozFest), London, Oct 2017: https://wiki.mozilla.org/Mozfest/2017 (accessed on 4 April 2022) the International Press Telecommunications Council Meeting (IPTC) Barcelona, Nov 2017: https://iptc.org (accessed on March 3, 2021).

[4] https://schema.org/ClaimReview

Table 3.1 Summary of content and context indicators of information quality at article level. (Condensed from Zhang et al. (2018))

Content indicators of credibility (determined by analyzing the title and text of the article)	Context indicators of credibility (determined by consulting external sources or examining the metadata surrounding the article)
Title representativeness: Is it off-topic, carrying little information, or overstating or understating claims?	*Originality*: Was the article an original piece of writing or duplicated elsewhere, and if so, was attribution given?
"Clickbait" title: Is it enticing its readers into clicking, and if so, in what format, e.g., listicle?	*Fact-checked*: Was the central claim of the article, if one exists, fact-checked by an organization vetted by a verified signatory of Poynter's international fact-checking network (IFCN) (accessed at https://www.poynter.org/international-fact-checking-network-fact-checkers-code-principles, on March 3, 2021)
Quotes from outside experts: Were experts quoted for journalistic rigor?	*Representative citations*: Were the first three sources accurately describing the original content?
Citation of organizations and studies: Were there one or more citations from a range of organizations or scientific studies?	*Reputation of citations*: What was the impact factor of the publication of any scientific study cited?
Calibration of confidence: Was there appropriate language used to show confidence in claims, acknowledging the authors' level of uncertainty (e.g., hedging, tentative, assertive language), and if so, where was it?	*Number of ads*: How many displayed ads, content recommendation engines (such as Taboola or Outbrain), and recommended sponsored content were there?
Logical fallacies: Was there any use of the straw man fallacy (presenting a counterargument as a more obviously wrong version of existing counterarguments), false dilemma fallacy (treating an issue as binary when it is not), slippery slope fallacy (assuming one small change will lead to a major change), appeal to fear fallacy (exaggerating the dangers of a situation), and the naturalistic fallacy (assuming that what is natural must be good)?	*Number of social calls*: How many calls to share (on social media, email,) or to join a mailing list were there?
Tone: Were there exaggerated claims or emotionally charged sections, especially for expressions of contempt, outrage, spite, or disgust?	*"Spammy" ads*: How does the "spamminess" of ads rate in terms of containing disturbing or titillating imagery, celebrities, or clickbait titles?
Inference: Was there correlation, singular causation, or general causation at play, and was there convincing evidence for the claims expressed?	*Placement of ads and social calls*: How aggressive was the placement of ads and social calls? Did they include pop-up windows, cover up article content, or distract through additional animation and audio?

3.6 Seals of Approval in Libraries and Health Informatics

Libraries have long produced detailed metadata records describing each item (book, image, artifact, etc.) in their collections for their catalogues. Libraries as institutions are known for providing access to reliable, credible, and truthful information from authoritative sources. You will not find a library whose mandate is to spread lies and to mislead the public. The profession still ponders whether, for instance, pseudo-sciences and conspiracy theories should deserve to be equally as visible as the critical literature debunking such theories. The debate between censorship versus free speech is alive and well in LIS scholarship. In contrast with the sporadic content moderation of the digital Wild West, libraries are largely successful information regulators due to their stringent policies and quality assurance practices that date back to turn of the twentieth-century ideas of "gate-keeping the universe of knowledge." In our universally connected pervasively digital twenty-first century, information professionals see themselves as "facilitators" of access, or perhaps, as "mediators" between information and the information needs of patrons. The gate-keeping role has been delegated to search engine developers and computer scientists, whose companies' moral compasses point more readily to profit, and not to the public good. When it comes to controlling the infodemic, LIS and library expertise has yet to be harnessed to its fullest potential. Technologists working on automation should collaborate with LIS scholars. Librarians should be more involved in the development of technologies that detect mis- or disinformation. Libraries, with the trust of the communities they serve, are well positioned to adopt such AI and to continue educating the public about media and information literacy in a proactive manner.

In the 1990s, library and information specialists collaborated with authoritative organizations in response to the proliferation of non-authoritative user-generated websites by creating standards and checklists to use in assessing online content with the intent of granting a certification of quality. One of the ideas was to display *a seal of approval* (also known at the time as a trust mark, certification, recommendation, or stamp of approval) that would directly demonstrate to users, with no extra effort on their part, that the site (and presumably its content) passed the granting agency's certification process.

Early ranking systems in health informatics, with or without the use of librarians as their accreditation proxies, also strived for quality control when faced with the variety of unwarranted medical and health advice online. The use of the seal of approval system assisted health-information consumers, in Burkell's words, "with the difficult task of sorting information 'wheat' from 'chaff'" (Burkell, 2004, p. 503). Seals of approvals in its graphic design and wording confirm one of the following: either a "self-reported voluntary compliance with a code of conduct (e.g., HONcode,[5] administered by the Health on the Net Foundation)," "a third-party verification for which participants pay a fee (e.g., URAC Health Web Site

[5] www.hon.ch (accessed on March 10, 2021).

Accreditation, administered by URAC),"[6] or "a collaborative combination of self-report, consumer report, and expert evaluation and [without] a fee (e.g., the MEDcertain project[7])" (Burkell, 2004, p. 495).

The practice of seals of approval took its inspiration from consumer reports in marketing and retail, as well as from other industries in which credibility is gained through certification processes and communicated by displaying visible badges or stamps. For example, *the Good Housekeeping Magazine* (Wikipedia., 2021b) regularly features recipes, diet, and health products and bestows their "Good Housekeeping Seal of Approval"[8] as a limited warranty to consumers that those products "have been assessed to perform as intended" (*Good* Housekeeping, 2014). Other verifiable seals of approval include, for example, eTrust, BBB, and Microsoft MVP (Shneiderman, 2014).

Now in the 2020s, seals of approval for online information are practically unheard of outside of niche health informatics and are certainly not widespread across online websites for a number of reasons. First, it is unclear whether users notice such external markers; second, whether users understand how seals or other systems work; and last, how well the meaning of these seals can be interpreted. A single seal is meant to be as clear and definitive as a verdict, but the processes by which the verdict was reached are hardly transparent, and never have been. There have been reports of mismatches between consumer expectations of core evaluated factors and the actual evaluation practice of accreditation bodies or their library proxies (e.g., Burkell, 2004). Plus, consultation with community members revealed that people distrusted the process; they distrusted how such certifications could be obtained by commercial organizations, their degree of integrity, and their policing procedures, wondering if "it could involve paying the subscription and getting the badge" (Pickard et al., 2010, p. 311). We now see similar scenarios of user distrust and discontent playing out in the process of giant tech companies moderating their content (e.g., by Facebook (Thurm, 2021)) and in the U.S. legislative attempts to increase the transparency of content moderation (McGill, 2020).

Prior research links the certification process itself (in the case of seals of approval) to the credibility of those providing the assessment, concluding that "you need to trust the ability/integrity/reputation of the information provider acting as third party that awards the certification such as a librarian, a manager of a digital repository, or a publisher before you trust the content of the information provided" (Pickard et al., 2010, p. 312). Trust, as it turns out, is indeed a precious commodity as the whole societal mechanism relies on it, to roughly paraphrase this chapter's epigraphs by Marshall (1919) and Severin and Tankard (1992). The public's perceptions of institutional quality amount to a reputation that was built through previously experienced trust.

[6] http://www.urac.org/ (accessed on March 10, 2021).

[7] http://www.medcertain.org/ (accessed on March 10, 2021).

[8] https://www.goodhousekeeping.com/institute/about-the-institute/a22148/about-good-housekeeping-seal/ (accessed on March 1, 2021).

3.7 Composite Reputation Ratings and Popularity Scores

Social technologies have begun to emphasize "public reputations based on long-term performance (e.g., eBay, Amazon), references from other users (e.g., likes, confirmations, badges, karma points), and visible histories of activities (e.g., Wikipedia edits, Amazon reviews)" (Shneiderman, 2014, p. 34).

These types of credibility assessments can be accomplished with or without the support of specialized expert systems. Such expert systems "attempt to learn quality ratings and predict the ratings of new content; however, we still do not have algorithms that automatically evaluate Web content credibility to an extent that is adequately accurate and helpful to Web users" (Kakol et al., 2017, p. 1044). In addition to individual assessments by nonexperts in the course of their daily information-related activities, assessments can also be crowdsourced, or requested from a group of (typically) volunteers. Nonexpert crowdsourcing is used to obtain credibility ratings such as Wikipedia's Article Feedback Tool (AFT),[9] the TweetCred[10] system for Twitter, or the Web of Trust (WOT)[11] system for evaluating Web portal credibility (Kakol et al., 2017).

Wikipedia's AFT was a short-lived experimental project that was discontinued in 2014 after it created awkward tensions in the Wikipedia editor community. The AFT was designed to "engage readers in the assessment of article quality" by asking for community feedback on individual pages using four quality indicators: trustworthiness, objectivity, completeness, and writing, on a four-star rating basis, and a self-reported assessment of the raters' knowledge: a relevant college/university degree, an expertise in profession, or a deep personal passion (Wikipedia., 2018). The wisdom of the crowd was put to the test and seems to have not survived as a viable solution for reasons that were never clearly articulated in the AFT Final Report (Wikipedia., 2020). It is possible to read between the lines to conclude that commenting became inflammatory, adversarial, or simply too difficult to manage. Many online newspapers in the mid-2010s, even such reputable legacy outlets as the *New York Times* (Etim, 2017), discovered similar controversies around free-range user feedback, and similarly chose to disable their comments.

An oft-cited example of crowdsourced reputation scores is the Web of Trust (WOT) Score[12] intended, like its predecessors, to help people make informed decisions about whether to trust a website or not. The score is computed by combining influence measures (including Alexa Rank[13] and Google PageRank[14]) and Twitter

[9] https://en.wikipedia.org/wiki/Wikipedia:Article_Feedback_Tool (accessed on March 10, 2021).

[10] t http://twitdigest.iiitd.edu.in/TweetCred/ (accessed on March 10, 2021).

[11] https://www.mywot.com/ (accessed on March 10, 2021).

[12] https://www.mywot.com/ (accessed on March 10, 2021).

[13] The Alexa Rank is a virtual ranking system set by Alexa.com, a subsidiary of Amazon, that audits and publishes the frequency of website visits as a geometric mean of reach and page views, averaged over a period of 3 months.

[14] Google PageRank is a link analysis algorithm that assigns a numerical weight to each element of a hyperlinked set of documents, such as the World Wide Web, with the purpose of measuring its relative importance within the set (https://www.domcop.com/openpagerank/)

Fig. 3.4 TweetCred user interface with feedback: Users can (**a**) agree ("thumbs up") or disagree ("thumbs down") with the rating on hover over tweets, and (**b**) provide their own credibility rating for a tweet (Gupta et al., 2014)

popularity, capitalizing on the ratings and reviews of a large community of users who are willing to share their comments and personal experiences. Again, popularity, user engagement, and crowdsourced consensus seem to be the top choices for the predictors of credibility.

TweetCred is another browser extension that uses 45 automated features together with human labels obtained using crowdsourcing converted into a single credibility rating (1/low to 7/high) for each tweet (Gupta et al., 2014). See an example of TweetCred crowdsourced credibility ratings user interface in Fig. 3.4.

The features used as TweetCred credibility factors relate to *tweet meta-data* (e.g., geocoordinates), *simple content* (e.g., number of characters or presence of smileys), *linguistic content* (e.g., swearwords or pronoun counts), as well as *author metadata* (e.g., followers), *network* (e.g., retweets) and *links* (*WOT* score or *YouTube* ratio of likes/dislikes). The automatically generated scores are reported to be fast—with 80% of them "computed and displayed within 6 seconds" and comparable to users' judgments—with "63% of users either agreed with [their]automatically-generated scores or disagreed by 1 or 2 points (on a scale from 1 to 7)" (Gupta et al., 2014, p. 3). However, these features admittedly take their lead from early studies of credibility and are focused on easily obtainable surface elements. In early HCI credibility research, surface elements were known to include, for instance, "spelling errors, willingness to provide contact information, professional appearance, rapid response, recognizable domain name, recency of content, and volume of information" (Shneiderman, 2014, p. 34).

Social popularity and *weblink structure* have also been used, perhaps misguidedly, as part of credibility assessments by Olteanu et al. (2013). Other credibility ratings have at times combined factors such as the *commonality* of the contents of articles among different news publishers, *numeric agreement or contradiction* (the match between different sources' reports of numerical values (e.g., 100 passengers), and *objectivity* (marked by the absence of subjective speculative phrases) (Nagura et al., 2006). Such strategies are creative but still rely heavily on consensus and commonalities in article content among different news publishers.

The question offers itself: What if the crowds are mistaken, the influence is manufactured, and the engagement is bloated or manipulated? It is apparent that more "liked" or "shared" content may communicate greater user engagement, but is such content likely to be the most credible? Viral content is known to thrive on speculation, sensationalism, and emotionality. A large-scale study of Twitter recently established that falsehoods diffused significantly faster and more broadly than the truth (Vosoughi et al., 2018). Thus, greater user engagement does not equal greater credibility. It is unfortunate that the richness of human experience, as described in many a model from the social sciences, is reduced to such abstracted technological affordances. AI should be looking for new, more sophisticated ways of computing credibility and trust predictors to surpass the current practical limitations.

3.8 Trust and Distrust Markers in Language

Instead of eliciting trust ratings, another solution is to mine how people speak about trust, or how they express distrust. Rubin's (2009) study models trust conceptually as a multipiece puzzle. The formulaic "A trusts B to do X in matters Y" suggests that each puzzle piece may change under different circumstances. For instance, in interpersonal relations, a trustor A (the trusting party) may trust a trustee B (the object of trust) in some things like house keys, but not in others like childcare. Similarly, when applied to information assessment online, A could trust B's Twitter feed on the latest fashion advice but not on health statistics. So, in principle, AI can process natural texts and analyze who says they trust whom, and on what matters specifically, or identify when trust or distrust can be implied from trustors' statements.

Rubin (2009) outlined indirect verbal indicators used in blogs to signal what elements gained trustors' trust or changed how they believed the trustee or the object of trust. Certain speech acts indicate trust: recommendations or referrals (e.g., "*he was the man to go to for*," "*who was known to be great at*"); stated praise, admiration, or thankfulness (e.g., "*helped me and he was amazing*," " *was the real deal*"); stated actions upon advice (e.g., "*took his advice and*," "*followed their advice*"); and successful leaps of faith or reliance on the trustee (e.g., "*had a good feeling about him*," "*had faith in him*," "*had confidence in her*," "*relied on their judgment to*") (Rubin, 2009).

Distrust, *doubt*, *disbelief*, and *suspicion* are often accompanied by uncertainty modifiers (such as *claimed*, *alleged*, *and supposedly*). Distrust is also palpable in three negative connotations: *anger* strongly expressed with *profanities* and *hostilities* (e.g., "a bunch of phonies," "absolutely stinks," "goddamn," and variations of "f*** idiots"); mildly expressed as *an intuition, suspicion,* or *apprehension* (e.g., "had a terrible gut feeling about"); or a strong *disapproval* (e.g., "*don't think you should*"), blame, criticism, or direct accusation (e.g., "*darn liars*") (Rubin, 2009). In addition, trust rhetoric was used as a linguistic façade, at times hypocritically or manipulatively, in a conscious attempt to invoke a socially desirable notion of trust, to control the reader, or to exploit the situation. The mere use of trust clichés can ironically cue distrust. Such trust rhetoric can be expressed through appeals for trust (e.g., "*Trust me!*" "*Believe me I'm an expert,*" or "*You have to trust me as his personal attorney that this is a risk free transaction…*"); ad-like overstatements, over-eager or overgeneralized statements (e.g., "*already gained trust from the customers*"); promises, reassurances, guarantees (e.g., "*fully committed to performing at my highest level at all times,*" "*I swear I'll never*"); loyalty and devotion pledges; and stated bets (Rubin, 2009).

Mining linguistic content, as a method, has the typical shortcomings of self-reports: we have to take people's word on it. However, if trust is seen as an estimation of expected reliance in people's minds, there is a good chance that what people say is meaningful, especially if their statements can be cross-referenced with their actions or behaviors. This approach is not simple to implement as it relies on the availability of self-reports and the ethical use of personal data, but it can underlie some factors involved in determining credibility computationally.

3.9 Contemporary Solutions to Credibility Assessments and Trust Ratings of Online Content

3.9.1 Digital Realities of Social Technologies in the 2020s

The information delivery medium has changed greatly since the 1990s and early 2000s. By 2015, "the popularity of social media platforms, such as Facebook and Twitter, has created an environment where information is pushed onto consumers either through sponsored messages or via 'shares' from friends" (Mitchell & Page, 2015). Nowadays, people receive personal messages from friends, family, and strangers, mixed in with global and national news, and viral content like cute kitten videos, all displayed in nearly identical formats. The market of online information is dominated by a few giant technology companies—including Facebook and Twitter—that have dictated homogeneous interfaces for any and all sources. "Unless the reader pays special attention, the article becomes decontextualized from its source and fact mixes freely with fiction" (Chen et al., 2015, p. 1).

More and more of our access to information and news is guided by social technologies (Vosoughi et al., 2018). The proportion of those who prefer to get their news online is growing. In 2016—28%, and in 2018—34% of American adults said they preferred to get news online, whether through websites, apps, or social media, with Facebook dominating as the most common social media site (about 4-in-10 Americans (43%) get news there) (Geiger, 2019). More broadly around the world, according to a recent news consumption online survey, commissioned by the Reuters Institute for the Study of Journalism across 40 countries, "the use of online and social media substantially increased in most countries, [… with] more than half of those surveyed (51%) [reporting the use of] some kind of open or closed online group to connect, share information, or take part in a local support network"

(Newman et al., 2021, p. 10). With such an extensive access to the news via social media newsfeeds, the formats of individual messages look homogenized, mixing in personal messages via online groups. While the format, its "look-and-feel" is specific to each platform or app, it is standardized enough that it can easily mask formerly clearly identifiable, surface credibility visual clues about the source of the message.

In early 2020s, digital environments the idea of the seal of approval method—with a single visible verdict—is rarely invoked in connection with social media labels but is undeniably similar. Labels like Twitter's "false news" or PolitiFact's "pants-on-fire" usually carry negative connotations and signal disapproval, distrust, and a warning, one step toward filtering out the information "chaff." If we can learn anything from past experiences, it is that the vagueness of a single seal of approval and the non-transparency of external assessments, makes people reluctant to adapt to such systems, even if they notice and understand what they mean.

If early trust and credibility research are completely overlooked, experimental works in computer science and the development community will have no foundational models to understand human information–related behaviors and cognitive assessment processes. Such a disconnect often results in rather simplistic ideas, with highly technical specialists working on computational solutions to problems while unaware of the previous lessons learned from HCI, LIS, communication, psychology, and so forth. A credibility label, seal, or score, signifying approval or disproval, may not be welcomed by contemporary social technology users, precisely because it lacks the credibility or transparency of the decisions that led to it being granted in the first place. By contrast, an approach that involves transparent explanations of the vetting process, supplemented by concise visual displays, may gain more traction with contemporary users and win their trust over time. In such approaches, theory should meet computational feasibility.

3.9.2 "Nutritional" Information Facts for Online Content as Explainable AI

Recently, a German research team put out an exciting proposal to display "Information Nutrition Labels" on each digital content item, mimicking the concise way nutrition labels are displayed on each food product (see Fig. 3.5) (Fuhr et al., 2018). The intent, just as with many other initiatives reviewed so far, is to help online information users make informed judgments about content. The difference in this approach is an explicitly stated intention to distance the developers from any ethical or moral judgement such as true or false, right or wrong, good or bad (Fuhr et al., 2018). (See also Chap. 4 on why philosophy matters.)

The Information Nutrition Labels contain nine components—*factuality, readability, virality, emotion, opinion, controversy, authority / credibility / trust* (as a single measure), *technicality*, and *topicality*. They are viewed as objective criteria

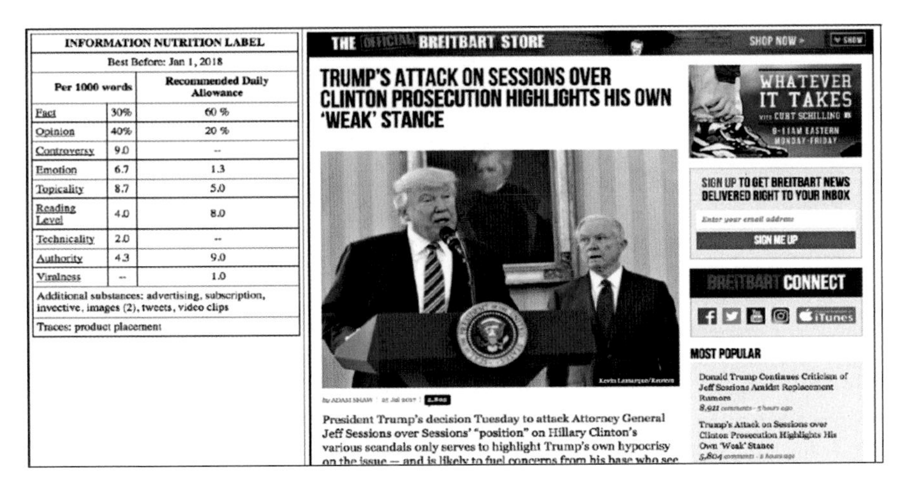

INFORMATION NUTRITION LABEL		
Best Before: Jan 1, 2018		
Per 1000 words		Recommended Daily Allowance
Fact	30%	60 %
Opinion	40%	20 %
Controversy	9.0	--
Emotion	6.7	1.3
Topicality	8.7	5.0
Reading Level	4.0	8.0
Technicality	2.0	--
Authority	4.3	9.0
Viralness	--	1.0
Additional substances: advertising, subscription, invective, images (2), tweets, video clips		
Traces: product placement		

Fig. 3.5 Fuhr et al.'s (2018) mockup of envisaged information nutrition label

that can be computed or measured with automated analyses, leaving the ultimate trust decision to the user.

Perhaps this is the way forward, but if we take into account the fiascos of previous attempts at external certifications, ranking systems, comments, labels, or seals of approval, we can easily predict that nutritional labels may have a similar uphill battle. Users have to notice these labels, spend time reading about the labels, and be able to interpret individual labels. They would have to rely on the trustworthiness and expertise of the developers and their computing mechanism that spits out these nutritional facts. The analogies are intuitive to an extent but require a steep learning curve to appreciate their value. Transparency in explanations should in principle win the users over, provided they are not overloaded cognitively. As we are cognitive misers online, we prefer to take shortcuts. (See also Chap. 2 on psychological factors that hinder our ability to evaluate information.)

It is disheartening to see that major technological developments are being undertaken on the heels of this proposal, yet the developers are seemingly unaware of prior trust and credibility research on how users make information quality judgments, how various cues fit together, and how users establish trust with technology. For further reading on trust in technology, see Tseng and Fogg (1999) explanations of *presumed credibility* (stereotypical assumptions, say about nutritional labels), *reputed credibility* (perceived by the virtue of credentials), *surface credibility* (an impression of the book by its cover), and *experienced credibility* (the most complex and reliable first-hand experience).

Several works have been built on the original Fuhr et al.'s (2018) proposal. For example, Gollub et al. (2018) reduced the number of Information Nutrition categories from nine to five and associated the five with visual labels which are allegedly intuitively easier to understand. To represent aspects of credibility in their Information Nutrition Dimensions (see Table 3.2, column 1) they use the physical qualities of

Table 3.2 Gollub et al.'s (2018) categorization of the Information Nutrition dimensions (column 1) into five categories (column 2)

Dimension	Category	Quantity	Range	Addressed User Question
Readability Technicality Verbosity*	I Effort	Time	0–120 min	Does time allow the reading?
Topicality Virality	II Kairos	Temperature	0–100 °C	Do others care?
Factuality Verifiability*	III Logos	Transparency	0–100%	How professional is the writing?
Emotion Opinion Controversy	IV Pathos	Sound pressure	0–120 dB	Is the article subjective?
Authority Credibility Trust	V Ethos	Credit rating	class A+... D	How reliable is the source?

time, temperature, transparency, and sound volume, as well as a credit rating system as in finance. These dimensions were then grouped into relatively well-known philosophical constructs according to Aristotle's modes of persuasion—logos, pathos, and ethos, roughly corresponding to logical argumentation, emotional sway, and established authority (see Table 3.2, column 2, five categories). The three are complemented by another classical rhetorical term "kairos" that translates approximately as the correct, critical, or opportune moment. Columns 3 and 4 (Table 3.2) show the related physical or financial quantities and their respective value ranges and column 5 states prototypical user questions for each of the five categories.

The Gollub et al. (2018) approach is highly creative and pays a lot of attention to visuals (see Fig. 3.6). Yet, it is superbly complex conceptually: It mixes physics, finance, philosophy, and nutrition analogies, most of which are fairly removed from the foundations of human cognition or credibility assessment.

An alternative implementation of Information Nutritional Labels can be seen in another browser plug-in, called NewsScan, which displays measurements in a single visualization with six icons/labels—source popularity, article popularity, ease of reading, sentiment, objectivity, and political bias (Kevin et al., 2018) (see Fig. 3.7). Light blue indicates low levels and dark blue indicates high levels of each label, with blue chosen for its association with trust, honesty, and security. When hovering over labels, users get short explanations about the labels and their scores. The developers put out a disclaimer with a hint toward their plug-in's role as an assistive technology: "To ensure that our tool only works as a guide and not a specific recommender, we do not interpret, for example, an easy-to-read article as being not worthwhile reading but just easy to understand. However, we believe some threshold label values about worthwhile and not worthwhile articles would indeed help readers in their decision making" (Kevin et al., 2018, p. 32). All is well enough, but what people typically associate with being "worthwhile reading" may entail a very different set of qualities such as expertise, trustworthiness, accuracy, and completeness, to name a few (consult any of the above models for a decent list). Also, notice another slippage to more easily computable aspects including the notorious crowdsourced popularity.

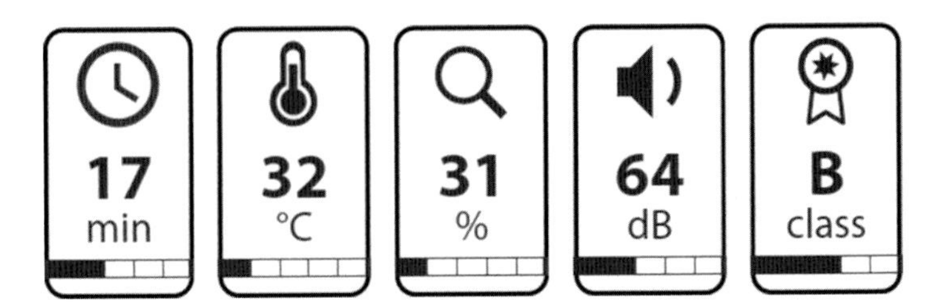

Fig. 3.6 Gollub et al.'s (2018) visual representation of the labels for time, temperature, transparency, volume, and credit ratings (for credibility), expressed as quantities to describe the nutrition facts of a document

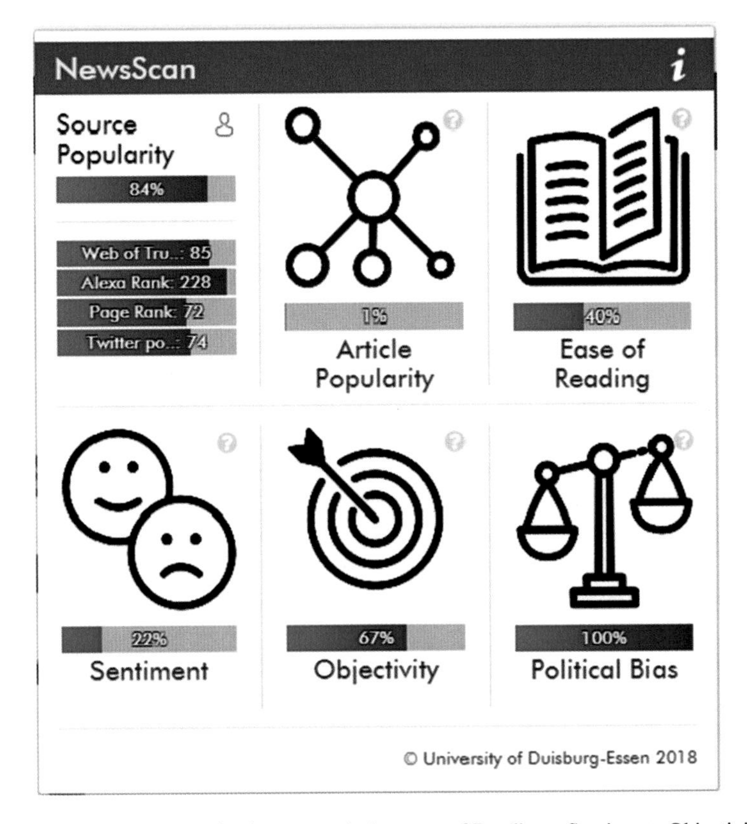

Fig. 3.7 The NewsScan plug-in shows popularity, ease of Readings, Sentiment, Objectivity, and Political Bias (Kevin et al., 2018)

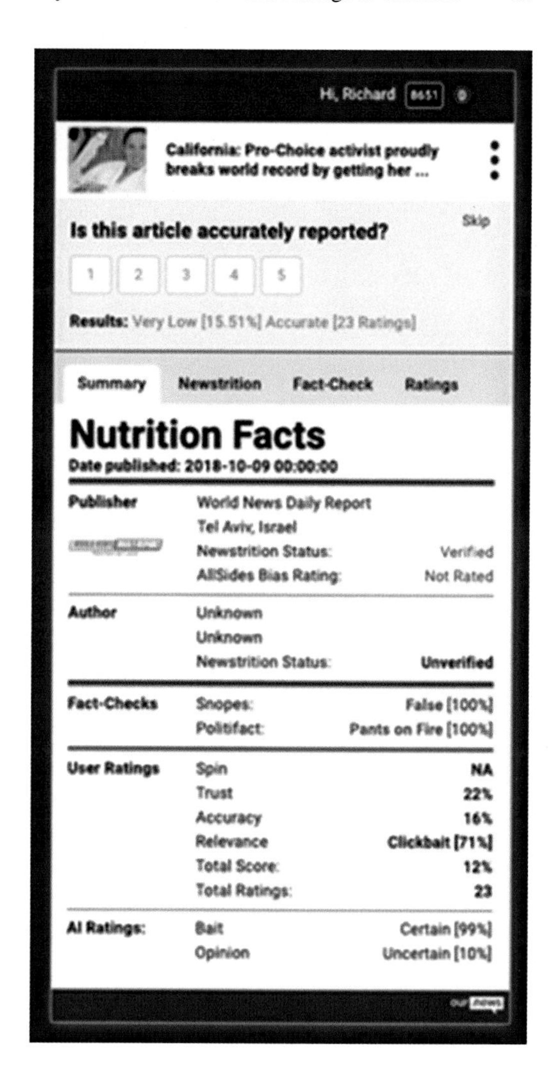

Fig. 3.8 *Our.News*, a nonpartisan online platform, developed the Newstrition Mobile App and browser plug-ins using Nutrition Labels for News (accessed at https://our.news/ on March 10, 2021)

Following this lead, start-up companies have produced proprietary applications experimenting with the presentation of News Nutritional Labels and combining them with other rating systems. One of them is *Our.News*,[15] a nonpartisan online platform that launched a Newstrition Mobile App[16] (Fig. 3.8.) and a Nutrition Labels for News browser plug-in version (Britzky, 2018)(Fig. 3.9). These applications allow users to read news anywhere they normally do (in a browser, through Facebook, Twitter, or any other news apps), and view a visual of a nutritional label, overlaid on top of the news article. According to a TechCrunch interview with *Our.*

[15] https://our.news/ (accessed on March 10, 2021).

[16] https://apps.apple.com/us/app/newstrition/id1460319017 (accessed on March 10, 2021).

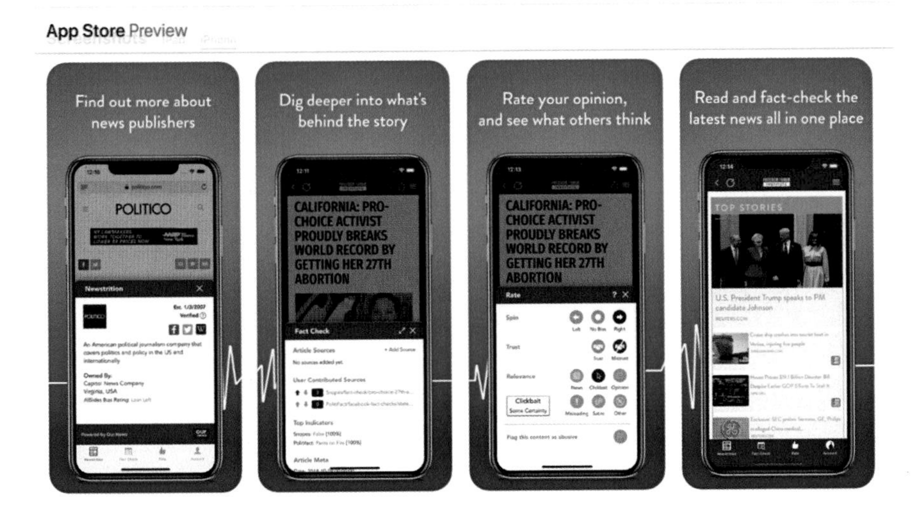

Fig. 3.9 The Newstrition mobile app allows users to both see detailed ratings and rate news themselves, according to the *Apple* App Store promotional imagery (accessed at https://apps.apple.com/us/app/newstrition/id1460319017#?platform=iphone on March 3, 2021)

News founder and CEO, Richard Zack, the label includes publisher descriptions; bias ratings; information about an article's sources, author, and editor; labels like "clickbait" or "satire"; as well as fact-checking information from sources like *PolitiFact*, *Snopes*, and FactCheck.org, and user ratings and reviews (Ha, 2020). *Our.News* has not readily disclosed further technical or computational details in any scientific literature to date, to the best of my knowledge.

Our.News is not unique in their solution. Other variants of information nutrition labels have recently flooded the market, which may indicate an uptick in their adoption and popularity. Another journalism and technology company, *NewsGuard*,[17] rates and ranks news websites for credibility and transparency (Mayhew, 2020) using strictly human expert judgments. Each of the nine criteria used for website assessments in the *NewsGuard*'s rating (2021) weighting scheme is worth a proportion of the total 100 points (see Fig. 3.10). For example, the first credibility-related criteria is about no repeated publishing of false content, worth 22 points in the weighting scheme, while two of the four transparency-related criteria—"discloses ownership and financing" and "clearly labels advertising" are worth 7.5 points each.

A cumulative score of 60 points or higher on a site receives a green rating, while a score lower than 60 points receives a red rating, going by the traffic light analogy. The categories indicate news sites that generally adhere to their standards of

[17] https://www.newsguardtech.com/ (accessed on March 10, 2021).

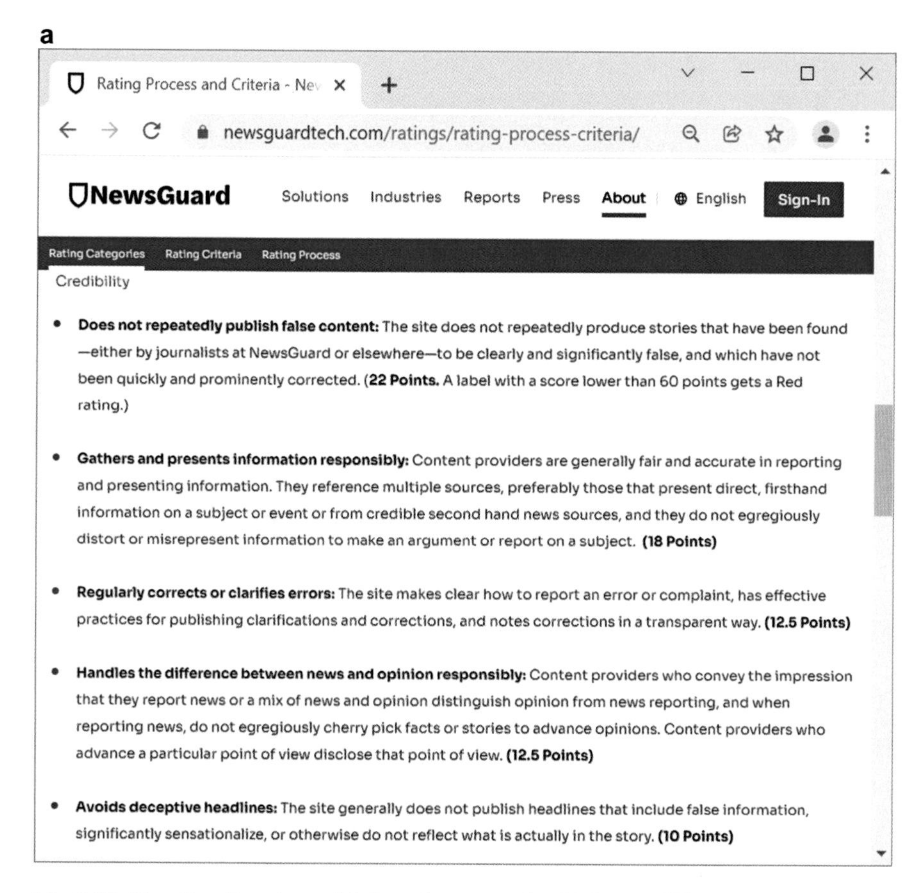

Fig. 3.10 *NewsGuard*'s rating weighting scheme contains (**a**) five credibility-related criteria and (**b**) four transparency-related criteria, totaling 100 points (accessed at https://www.newsguardtech.com/ratings/rating-process-criteria/ on December 17, 2021)

credibility and transparency (green) or fail to do so (red), with satire (orange) and platform (gray exclamation mark) in separate categories (Fig. 3.11).

Humorous satirical sites are not deemed "real" news site by *NewsGuard* and are thus simply described as satire, and marked in orange. The grey rating is for user-generated sites that hosts but do not vet users' content which may or may not be reliable. According to the company's explanations,[18] a trained analyst assesses the contents of a site against the criteria and fact-checks it, and their decision is reviewed by an experienced editor, both accountable for the rating.

[18] https://www.newsguardtech.com/ratings/rating-process-criteria/ (accessed on March 10, 2021).

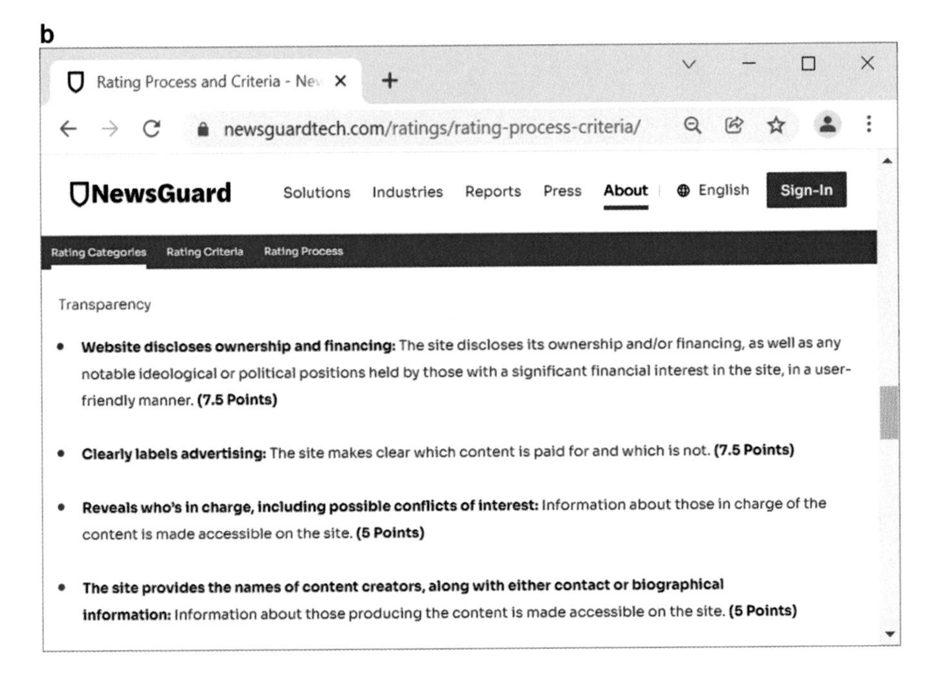

Fig. 3.10 (continued)

By the end of 2018, there were dozens of trust-related initiatives attempting to vet online information using a combination of journalistic fact-checking, crowd-sourcing, and some automation (Fischer, 2018). To name a few, The Trust Project[19] is one of the larger global initiatives to combat fake news; the *News Integrity Initiative*[20] includes *Facebook*, *Knight Foundation*, *Tow Foundation*, *AppNexus*, *Mozilla* and *Betaworks*; and *The Journalism Trust Initiative*[21] subsumes *Reporters Without Borders*, and *Agence France-Presse*, the *European Broadcasting Union*, and *the Global Editors Network.*

The intense cooperation between the fact-checking and computing communities, as evidenced by the founding of many start-ups and initiatives, must be a reflection of the dire need to solve the socio-technological problem of the proliferation of mis- and disinformation. The more clearly the problem of assessing credibility is formulated conceptually, modeled by tying it to online information quality, and partitioned into its other component trust parts, the easier it is to build on past research, refine the models, and offer practical solutions as a way forward. (See also Chap. 7 for review of computational methods around the problem of mis- and disinformation.)

[19] https://thetrustproject.org/

[20] https://www.journalism.cuny.edu/centers/tow-knight-center-entrepreneurial-journalism/news-integrity-initiative/

[21] https://ethicaljournalismnetwork.org/rsf-trust-initiative

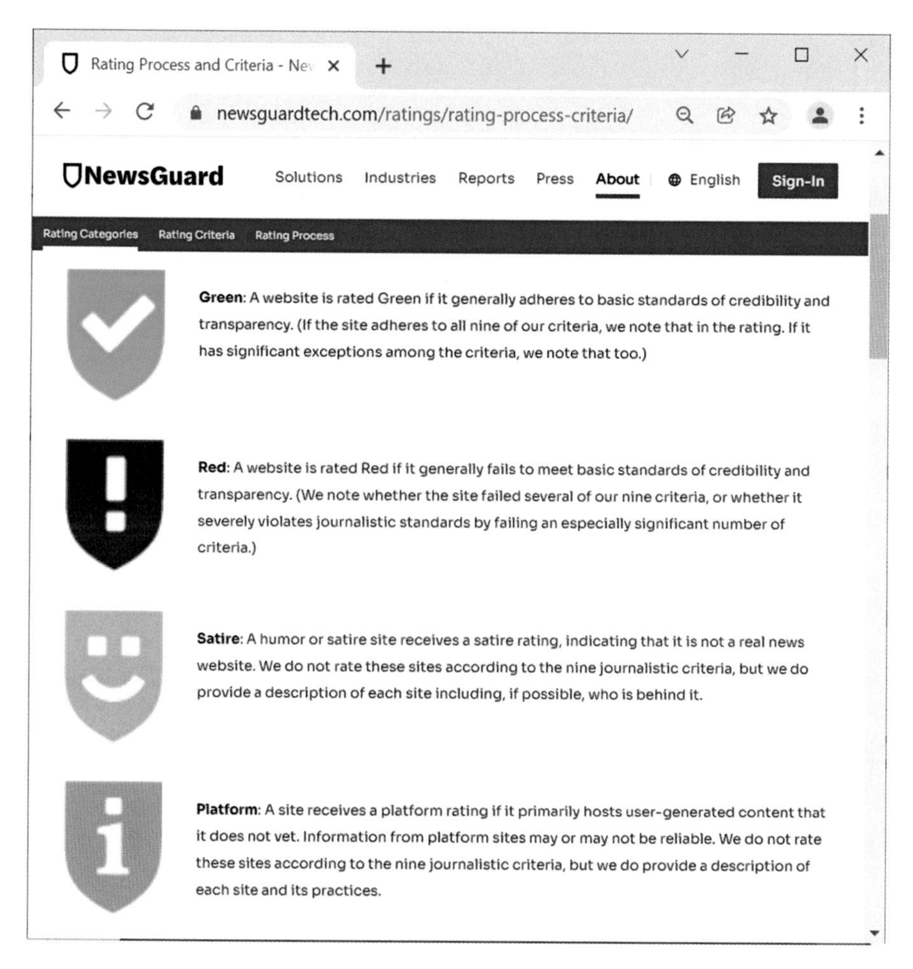

Fig. 3.11 *NewsGuard*'s rating categories indicate adherence of a news site to their standards of credibility and transparency (accessed at https://www.newsguardtech.com/ratings/rating-process-criteria/ on December 17, 2021)

3.10 Conclusions

Chapter 3 examined the concepts of credibility and trust, dissecting them into their component parts. Prior research in information science, human–computer interaction, psychology, communication, and other social sciences and associated computing fields all point to multiple components of credibility assessments for online information. Several theoretical models of information quality evaluation that include credibility assessments have been developed, other models explain people's decisions to trust (or not). Credibility assessments and trust judgments are at times used interchangeably in HCI, LIS, psychology, and communication studies, while MIS and economics literature distinguish the two. Each field that studies credibility and trust emphasizes its own definitions, and though some diverge on the exact steps in the evaluative procedure, or in the inventory of its subcomponents, all seem to agree that such assessments (of either trust or credibility) are accomplished in layers. The three largely agreed-upon layers of evaluation are (1) the credibility of the medium (such as broadly construed online environments or more specific search engines or platform newsfeeds), (2) the credibility of the source (such as specific websites, institutions, or authors), and (3) the credibility of the message (the actual statements or claims in the content). Each layer evokes a judgment to trust it or not; an estimation of the likelihood of something, or someone, being reliable.

Research in psychology adds another layer associated with the receiver of information, since credibility assessment requires a cognitive effort on the part of the receivers. This effort is on the top of their ability to pay attention and requires them to notice certain factors that can either boost or sink the overall information quality assessment. Specifically, prior experiences in trust behaviors (e.g., trust gained or lost by sources) and the personal propensity to trust are subject to individual variance and can impact the receivers' credibility judgments.

In several disciplines, credibility is generally equated with believability, and trusted sources are seen as dependable if they have cognitive authority. Cognitive authority is based on the receiver's perceptions of the source's trustworthiness and expertise. Credible people possess both trustworthiness and expertise, and it is wise to consider both elements when making a choice to trust someone or not. To assess how credible information is likely to be, we use a combination of factors: its consistency, completeness, accuracy, precision, and freedom from bias, to name a few. The results of search engine queries are additionally matched on relevance, while the content seen via social news feeds is only weakly filtered by one's social networks. The message (content itself) has to have these and other properties to be considered of high quality.

Due to information overload, people tend to be cognitive misers and spend the least amount of time and cognitive effort necessary. Early website developers in the 1990s–early 2000s knew to pay special attention to superficial signs of credibility such as professional look and feel. By the late 2010s–early 2020s, social technology news feeds became homogeneous in their presentation style, yet use imagery and snappy titles as ways to catch users' attention.

While social science research carefully elaborated on the subcomponents of credibility and trust, there are several stumbling blocks on the way to automating credibility predictions. First, the predictive powers of individual factors have rarely been experimentally compared and confirmed, so they cannot be readily plugged into algorithms. Expertise and trustworthiness top the lists of important factors in empirical studies of cognitive authority of the source. Second, while some subcomponents are readily obtainable (e.g., currency using the time stamp), others are harder to tease out or establish automatically (e.g., the logical accuracy of claims). And third, the ultimate decision about credibility and trust resides in the mind of the receivers of information. There may be ways to inform, educate, or even influence those judgments, assuming users are willing to pay attention or learn, but their ultimate judgments and decisions to act on information are private, highly subjective, and personal.

Efforts at computing credibility have been made by creating surrogate measures that stand in for credibility or trust in their approximation. Such efforts are usually limited by the feasibility of obtaining reliable predictors. For instance, popularity is easily computed by likes or other user engagement measures but may not be reliable. What is apparent is that there is a disconnect between the richness and depth of knowledge about human–information interactions and what AI can currently access empirically. Thus, further collaboration between disciplines is much needed.

The promise of AI is in revisiting the social sciences research on credibility and trust and in finding ways to extract and reconstruct more meaningful, truer-to-reality, and reliable indicators of information quality evaluation and credibility assessment. Such efforts should also support users' cognitive shortcuts in making trust judgments. Information nutritional labels, originating from journalistic fact-checking practices and picked up by software developers, may be approximating the idea of explainable AI. Such systems could be improved by deprioritizing the feasibility of surrogate measures (such as popularity and sentiment) and by establishing research agendas in search of reliable predictors outside of crowdsourced engagement ratings.

References

Beard, D. (2018, January 19). Tale of 2 polls: What do librarians have that journalists don't? *Poynter.* Retrieved from https://www.poynter.org/ethics-trust/2018/tale-of-2-polls-what-do-librarians-have-that-journalists-don't/

Belkin, N. J. (1993). Interaction with texts: Information retrieval as information-seeking behavior. *Information retrieval '93: Von Der Modellierung Zur Anwendung. Proceedings of the first conference of the Gesselschaft F_r Infomatik Fachgruppe information retrieval* (pp. 55–66).

Britzky, H. (2018, October 9). *Our.News* tool joins dozens of other sites fighting "fake" news. *Axios.* Retrieved from https://www.axios.com/newseum-partners-with-ournews-news-validation-tool-c12f8027-30a1-4ccf-9c72-a3d64f81ab87.html

Burkell, J. (2004). Health information seals of approval: What do they signify? *Information, Communication & Society, 7*(4), 491–509. https://doi.org/10.1080/1369118042000305610

Chen, Y., Conroy, N. J., & Rubin, V. L. (2015). News in an online world: the need for an automatic crap detector, Proceedings of the Association for Information Science and Technology, 52(1), 1–4. https://doi.org/10.1002/pra2.2015.145052010081

Cho, J.-H., Chan, K., & Adali, S. (2015). A survey on trust modeling. *ACM Computing Surveys, 48*(2), 1–40. https://doi.org/10.1145/2815595

Ciampaglia, G. L., Shiralkar, P., Rocha, L. M., Bollen, J., Menczer, F., & Flammini, A. (2015). Computational fact checking from knowledge networks. *PLoS One, 10*(6), e0128193. https://doi.org/10.1371/journal.pone.0128193

Etim, B. (2017, September 27). Why no comments? It's a matter of resources. *The New York Times.* Retrieved from https://www.nytimes.com/2017/09/27/reader-center/comments-moderation.html

Fischer, S. (2018, October 14). How pro-trust initiatives are taking over the Internet. *Axios.* Retrieved from https://www.axios.com/fake-news-initiatives-fact-checking-dfa6ab56-3295-4f1a-9b38-e61ca47e849f.html.

Fogg, B. J. (1998). Persuasive computers: Perspectives and research directions. *Proceedings of the SIGCHI Conference on Human Factors in Computing Systems (CHI '98).* ACM Press/Addison-Wesley Publishing Co. (pp. 225–232) doi: https://doi.org/10.1145/274644.274677.

Fogg, B. J. (2003). Prominence-interpretation theory: Explaining how people assess credibility online. *CHI '03 extended abstracts on human factors in computing systems* (pp. 722–723). doi: https://doi.org/10.1145/765891.765951.

Fogg, B. J., Swani, P., Treinen, M., Marshall, J., Laraki, O., Osipovich, A., Varma, C., Fang, N., Paul, J., Rangnekar, A., & Shon, J. (2001). What makes web sites credible?: A report on a large quantitative study. *Proceedings of the SIGCHI conference on human factors in computing systems—CHI '01* (pp. 61–68). doi: https://doi.org/10.1145/365024.365037.

Fogg, B. J., Soohoo, C., Danielson, D. R., Marable, L., Stanford, J., & Tauber, E. R. (2003). How do users evaluate the credibility of web sites? A study with over 2,500 participants. *Proceedings of the 2003 Conference on Designing for User Experiences* (pp. 1–15). doi: https://doi.org/10.1145/997078.997097.

Fuhr, N., Nejdl, W., Peters, I., Stein, B., Giachanou, A., Grefenstette, G., Gurevych, I., Hanselowski, A., Jarvelin, K., Jones, R., Liu, Y., & Mothe, J. (2018). An information nutritional label for online documents. *ACM SIGIR Forum, 51*, 46–66. https://doi.org/10.1145/3190580.3190588

Geiger, A. W. (2019, September 11). Key findings about the online news landscape in America. *Pew Research Center.* Retrieved from https://www.pewresearch.org/fact-tank/2019/09/11/key-findings-about-the-online-news-landscape-in-america/.

Gollub, T., Potthast, M., & Stein, B. (2018). Shaping the information nutrition label. In D. Albakour, D. Corney, J. Gonzalo, M. Martinez, B. Poblete, A. Vlachos (Eds.) *Proceedings of the NewsIR'18 workshop at ECIR*, 3.

Gupta, A., Kumaraguru, P., Castillo, C., & Meier, P. (2014). TweetCred: Real-time credibility assessment of content on twitter. In L. M. Aiello & D. McFarland (Eds.), *Social informatics: 6th international conference, SocInfo 2014, Barcelona, Spain, November 11–13, 2014. Proceedings* (pp. 228–243). Springer International Publishing. https://doi.org/10.1007/978-3-319-13734-6_16

Ha, A. (2020, February 7). Our.News fights misinformation with a 'nutrition label' for news stories. *TechCrunch*. Retrieved from https://social.techcrunch.com/2020/02/07/our-news/

Hardin, R. (2001). Conceptions and explanations of trust. In K. S. Cook (Ed.), *Trust in society* (pp. 3–39). Russell Sage Foundation.

Good Housekeeping. (2014, March 31). How the GH limited warranty seal protects you. *Good Housekeeping.* Retrieved from https://www.goodhousekeeping.com/product-reviews/history/about-good-housekeeping-seal.

Hovland, C. I., Janis, I. L., & Kelley, H. H. (1953). *Communication and persuasion: Psychological studies of opinion change*. Yale University Press.

Kakol, M., Nielek, R., & Wierzbicki, A. (2017). Understanding and predicting web content credibility using the content credibility corpus. *Information Processing & Management, 53*(5), 1043–1061. https://doi.org/10.1016/j.ipm.2017.04.003

Kevin, V., Högden, B., Schwenger, C., Şahan, A., Madan, N., Aggarwal, P., Bangaru, A., Muradov, F., & Aker, A. (2018). Information nutrition labels: A plugin for online news evaluation. *Proceedings of the first workshop on fact extraction and VERification (FEVER)*, 28–33. doi: https://doi.org/10.18653/v1/W18-5505.

Lukoianova, T., & Rubin, V. L. (2014). Veracity roadmap: Is big data objective, truthful and credible? *Advances in Classification Research Online, 24*(1), 4–15. https://doi.org/10.7152/acro.v24i1.14671

Marsh, S., & Dibben, M. R. (2003). The role of trust in information science and technology. *Annual Review of Information Science and Technology, 37*(1), 465–498.

Marshall, A. (1919). *Industry and trade*. Macmillan. Retrieved from http://socserv.mcmaster.ca/econ/ugcm/3ll3/marshall/Industry&Trade.pdf.

Mayhew, F. (2020, January 9). News website rating tool *NewsGuard* to start charging for service. *Press Gazette*. Retrieved from https://pressgazette.co.uk/news-website-rating-tool-newsguard-start-charging-service-paid-membership/.

McGill, M. H. (2020, June 24). Latest bill on tech shield takes aim at transparency on content calls. *Axios*. Retrieved from https://www.axios.com/latest-bill-on-tech-shield-pushes-for-transparency-in-moderation-deac8147-4bb1-45b6-be1a-ac417cc07349.html

Merriam-Webster Dictionary Online. (2021). *Trust*. Merriam-Webster Dictionary Online. Retrieved from https://www.merriam-webster.com/dictionary/trust.

Metzger, M. J., & Flanagin, A. J. (2015). Psychological approaches to credibility assessment online. In S. S. Sundar (Ed.), *The handbook of the psychology of communication technology* (1st ed., pp. 445–466). Wiley. https://doi.org/10.1002/9781118426456.ch20

Mitchell, A., & Page, D. (2015). *The state of the news media 2015* (p. 98).

Möllering, G. (2006). *Trust: Reason, routine, reflexivity*. Elsevier.

Nagura, R., Seki, Y., Kando, N., & Aono, M. (2006). A method of rating the credibility of news documents on the web. *Proceedings of the 29th annual international ACM SIGIR conference on research and development in information retrieval* (pp. 683–684). doi: https://doi.org/10.1145/1148170.1148316.

Newman, N., Fletcher, R., Schulz, A., Andı, S., & Nielsen, R. K. (2021). *Reuters institute digital news report 2020* (p. 112). Retrieved from https://reutersinstitute.politics.ox.ac.uk/sites/default/files/2020-06/DNR_2020_FINAL.pdf

NewsGuard Rating Process and Criteria. (2021, March 3). *NewsGuard*. Retrieved from https://www.newsguardtech.com/ratings/rating-process-criteria/.

Olaisen, J. (1990). Information quality factors and the cognitive authority of electronic information. In I. Wormell (Ed.), *Information quality: Definitions and dimensions* (pp. 91–121).

Olteanu, A., Peshterliev, S., Liu, X., & Aberer, K. (2013). Web credibility: Features exploration and credibility prediction. In P. Serdyukov, P. Braslavski, S. O. Kuznetsov, J. Kamps, S. Rüger, E. Agichtein, I. Segalovich, & E. Yilmaz (Eds.), *Advances in information retrieval* (pp. 557–568). Springer. https://doi.org/10.1007/978-3-642-36973-5_47

Pickard, A. J., Gannon-Leary, P., & Coventry, L. (2010). Trust in 'E': Users' trust in information resources in the web environment. In Quintela Varajão, E. J.; Cruz-Cunha, Manuela Maria; Putnik, Goran D.; Trigo, António (Eds.) ENTERprise information systems (pp. 305–314). doi: https://doi.org/10.1007/978-3-642-16419-4_31.

Rieh, S. Y. (2002). Judgment of information quality and cognitive authority in the web. *Journal of the American Society for Information Science and Technology, 3*(2), 145–161.

Rieh, S. Y., & Danielson, D. R. (2007). Credibility: A multidisciplinary framework. *Annual Review of Information Science and Technology, 41*(1), 307–364. https://doi.org/10.1002/aris.2007.1440410114

Rotter, J. B. (1980). Interpersonal trust, trustworthiness, and gullibility. *The American Psychologist, 35*(1), 1–7. https://doi.org/10.1037//0003-066X.35.1.1

Rousseau, D. M., Sitkin, S. B., Burt, R. S., & Camerer, C. (1998). Not so different after all: A cross-discipline view of trust. *The Academy of Management Review, 23*(3), 393–404.

Rubin, V. L. (2009). Trust incident account model: Preliminary indicators for trust rhetoric and trust or distrust in blogs. Proceedings of the third international association for the advancement of Artificial Intelligence conference on weblogs and social media, (ICWSM), 4. Retrieved from http://www.icwsm.org/2009/

Rubin, V. L. (2019). Disinformation and misinformation triangle: A conceptual model for "fake news" epidemic, causal factors and interventions. *Journal of Documentation*, ahead-of-print. https://doi.org/10.1108/JD-12-2018-0209

Rubin, V. L., & Liddy, E. (2006). Assessing credibility of weblogs. *AAAI Symposium on Computational Approaches to Analyzing Weblogs*, Stanford, CA. https://www.aaai.org/Papers/Symposia/Spring/2006/SS-06-03/SS06-03-038.pdf

Salton, G., & McGill, M. J. (1983). *Introduction to modern information retrieval*. McGraw-Hill. Retrieved from https://catalog.hathitrust.org/Record/000190320

Severin, W. J., & Tankard, J. W. (1992). *Communication theories: Origins, methods and uses in the mass media*. Longman.

Shneiderman B. (2014) Building Trusted Social Media Communities: A Research Roadmap for Promoting Credible Content. In: Bertino E., Matei S. (eds) *Roles, Trust, and Reputation in Social Media Knowledge Markets. Computational Social Sciences.* Springer, Cham. https://doi.org/10.1007/978-3-319-05467-4_2

Sundar, S. S. (2008). The MAIN model: A heuristic approach to understanding technology effects on credibility. In M. J. Metzger & A. J. Flanagin (Eds.), *Digital media, youth, and credibility* (pp. 73–100). The MIT Press. https://doi.org/10.1162/dmal.9780262562324.073

Thurm, S. (2021, January 28). *Facebook*'s Oversight Board has spoken. But it hasn't solved much. *Wired.* Retrieved from https://www.wired.com/story/facebook-oversight-board-has-spoken/

Tseng, H., & Fogg, B. J. (1999). Credibility and computing technology. *Communications of the ACM, 42*(5), 39–44.

Vosoughi, S., Roy, D., & Aral, S. (2018). The spread of true and false news online. *Science, 359*(6380), 1146–1151. https://doi.org/10.1126/science.aap9559

Wathen, C. N., & Burkell, J. (2002). Believe it or not: Factors influencing credibility on the web. *Journal of the American Society for Information Science and Technology, 53*(2), 134–144. https://doi.org/10.1002/asi.10016

Weerkamp, W., & de Rijke, M. (2008). Credibility improves topical blog post retrieval. *Proceedings of ACL-08: HLT.* Columbus, Ohio. Association for Computational Linguistics. (pp. 923–931).

Wikipedia. (2018). The Wikipedia article feedback tool (AFT). In *Wikipedia.* Retrieved from https://en.wikipedia.org/w/index.php?title=Wikipedia:Article_Feedback_Tool&oldid=861549818

Wikipedia. (2020). The *Wikipedia article feedback tool (AFT)/version 5/Report.* Retrieved from https://www.mediawiki.org/wiki/Article_feedback/Version_5/Report

Wikipedia. (2021a). Trust, but verify. In *Wikipedia.* Retrieved from https://en.wikipedia.org/w/index.php?title=Trust,_but_verify&oldid=1007750912

Wikipedia. (2021b). Good housekeeping. In *Wikipedia.* Retrieved from https://en.wikipedia.org/w/index.php?title=Good_Housekeeping&oldid=1004758938

Wilson, E. J., & Sherrell, D. L. (1993). Source effects in communication and persuasion research: A meta-analysis of effect size. *Journal of the Academy of Marketing Science, 21*(2), 101–112. https://doi.org/10.1007/BF02894421

Zhang, A. X., Robbins, M., Bice, E., Hawke, S., Karger, D., Mina, A. X., Ranganathan, A., Metz, S. E., Appling, S., Sehat, C. M., Gilmore, N., Adams, N. B., Vincent, E., & Lee, J. (2018). A structured response to misinformation: Defining and annotating credibility indicators in news articles. *Companion of The Web Conference 2018 on The Web Conference 2018—WWW '18* (pp. 603–612). doi: https://doi.org/10.1145/3184558.3188731.

Chapter 4
Philosophies of Truth

Philosophy is rightly called the knowledge of the truth.
For the end of theoretical knowledge is truth, that of practical knowledge being action.

Aristotle (1924 reprint, vol. 1, p. 214)

... though the philosopher may live remote from business,
the genius of philosophy, if carefully cultivated by several,
must gradually diffuse itself throughout the whole society,
and bestow [a spirit of accuracy and] a similar correctness on every art and calling.
The politician will acquire greater foresight and subtlety,
in the subdividing and balancing of power;
the lawyer more method and finer principles in his reasoning;
and the general public more regularity in his discipline,
and more caution in his plans and operations.

David Hume (1955) reprint, p. 19,

A Treatise of Human Nature (1739–1740))
Throughout life, no moral choice is more common
than that of whether to speak truthfully, equivocate, or lie—
whether to flatter, get out of trouble, retaliate, or gain some advantage.

Sissela Bok (2021, p. 1028)

Abstract **Chapter 4** establishes that truth can be seen from different philosophical perspectives, and our methods for connecting beliefs to reality and establishing facts matter for determining the resulting cumulative knowledge. We may not all agree on what truth is, but there is little doubt that truth matters. It is essential to us—as individuals and as a society—for the proper functioning of our legal, educational, and scientific institutions. Philosophical thought gives us a solid basis for understanding what constitutes facts, how to distinguish them from beliefs, and how to avoid logical fallacies. Schools of thought differ on their logical paths to establishing truth. Some theories suggest that truth exists in correspondence to facts; others indicate that it is in coherence with a set of prior beliefs or propositions. Alternatively, truth is seen as the ideal outcome of a rational inquiry. The concept of "verifiably justified true beliefs" is positioned here as most informative for rational decision-making and consequent actions. AI, inextricably connected to philosophy, needs to point its moral compass away from profit and to its truer north of moral

V. Rubin, *Misinformation and Disinformation*,
https://doi.org/10.1007/978-3-030-95656-1_4

virtues such as honesty and justice. Inspired by such philosophical considerations, I propose five concrete AI directions for truth-seeking.

Keywords Truth · Reality · Facts · Knowledge · Virtue · Morality · Ethics · Honesty · Metaphysics · Epistemology · Applied philosophy · Applied ethics · Logic · Logical fallacies · Deception justification test · Last resort test · Harms and benefits test · Publicity test · Truth judgments · Beliefs · Propositions · Assertions · Statements · Philosophical schools of thought · Inflationary theories · Deflationary theories · Epistemic theories · Correspondence theories · Coherence theories · Pragmatist theories · Verificationist theories · Redundancy theory · Performative theory · Realism · Idealism · Pragmatism · Perspectivism · Verificationism · Verifiability · Warranted assertability · Justified true beliefs · Correspondence · Consensus · Coherence · AI ethics · Moral compass · Codes of ethics · Codes of conduct · Ethical guidelines · NLP applications · Automated truth-seeking · Automated fact-checking · Political bias identification · Divergent opinion identification

4.1 Introduction

Honesty is universally understood to be a moral virtue and we all aspire to be truthful. Most of us would agree that truth is important to us as individuals and to society as a whole, but we may not all agree on what we consider "truth" in principle, how to establish and verify it, and how to relate it to the reality we live in.

The concepts of truth, reality, fact, and knowledge may be used interchangeably in a casual conversation, but they do not necessarily mean the same thing. If we want to minimize the spread of untruths online, we need to be precise in the meaning of these key concepts. For answers, we turn to philosophers who have debated the exact nature of truth and associated beliefs since the dawn of civilization.

No matter what school of thought, philosophers generally agree on the underlying premise that truth is discoverable and essential to us as individuals, and to society as a whole. Many institutions, such as our educational, legal, and scientific systems, function based on our ability to reveal, uncover, or discover the truth, and to use these discoveries to our benefit in schools, courts, medicine, technology, and codes of conduct.

Philosophy may appear largely inconsequential to some people, but our contemporary ways of thinking and reasoning are undoubtedly influenced by the thoughts of the luminaries of the past, passed down from generation to generation through cultural and societal knowledge and traditions. The great philosophers of the past are alike anyone in the present: In a perpetual state of truth discovery and learning, on a quest to figure out what is true and what is not, as it relates to our times and circumstances. Any time we encounter a new idea that has been made known to others, we can either accept its essence as true or reject it as untrue. Wittingly or not, we often make truth judgments about assertions[1] that surround us. If an assertion simply matches our reality and we know it, we are aligning with the correspondence theory school of thought, traced back to the ancient Greek classic philosophers like Plato and Aristotle and the European neoclassical luminaries like Bertrand Russell and Ludwig Wittgenstein. If we find a particular new bit of information to be true because it fits with our prior knowledge, we are adhering to some form of coherence theory as influenced by Baruch Spinoza, Immanuel Kant, and Georg Hegel. If we accept the truth because the consensus in our group is to accept it, we are in line with the perspectivist theory guided by the principles put forward by the early Greek Sophists and reaffirmed in its neoclassical forms by Friedrich Nietzsche. And, if the new bit of information does not conflict with your past experiences and is useful in the world, you are a pragmatist, even if you are unfamiliar with how pragmatism was defined and popularized by notable philosophers of the early 1900s such as William James, Charles Pierce, and John Dewey.

Let us consider these possibilities. I will give selective emphasis to philosophical ideas that may help both the human eye and artificial intelligence (AI)[2] to interpret a thought process vis-à-vis encountering online information and making a truth judgment about it. The aims here are fourfold.

First, philosophy is uniquely positioned to teach us about morality, ethics, and virtues. If honesty is a virtue, does it then mean it is always immoral to lie? Can any lie ever be justified? Philosophical writing is an excellent source for the history of thought on the justifiability of deception and mitigating circumstances. A specific justifiability test can help us make up our minds about the case of digital fakes in online information and their driving forces.[3]

Second, my canvassing of the philosophical definitions of truth, facts, and knowledge as a way of thinking about reality aims to rejuvenate discussions about ethics and morality in AI within the software development community. I propose how the discussed philosophical theories can be applied concretely as tasks for automated analysis of textual data. Directly observable phenomena—such as markers of factuality versus epistemic modality or uncertainty in beliefs—can be explicitly identified in stated assertions by automated means with natural language processing (NLP)[4] and machine learning (ML) techniques.

---

[1] In philosophy, ideas roughly expressed in a form of simple sentences, statements, or assertions—that can either be true or untrue—are called *propositions*. For example, "Ottawa is the capital of Canada," "smoking is bad for your health," "John bought a new car," and "snow is white" are four propositions.

[2] AI is essentially some form of computerized simulation of human intelligence (see Introduction to this book for more on AI and its link to language "understanding").

[3] See also Introduction to this book for causal factors of mis- and disinformation in the infodemiological triangle model, originally proposed in Rubin (2019).

[4] NLP is a subset of AI that enables computers to "understand," interpret, translate, and otherwise manipulate human language. (For a more formal definition and easy to understand explanations, see (Liddy, 2001).

Third, the human eye can be trained in the mechanics of verification, nudging the human mind toward the use of logical reasoning and rational thought instead of a reliance on emotions and intuition in snap judgements. Just as the art of constructing valid arguments can be taught in rhetoric classes, the art of identifying flawed reasoning can be part of media literacy education. Digital media users who have been alerted to the presence of fallacious reasoning can also make better credibility assessments (see Chap. 3), finding weaknesses in the arguments around controversial issues. Turning away from untruths already brings us closer to truths.

And last, but not least, understanding the philosophic debates and esoteric arguments in abstract allows us to take a mental inventory of how we make truth judgments about online content. Taking into account palpable influences from prior philosophical thought on how we reason and learn new facts, what else do we think interferes with our decisions to believe or disbelieve wtat we read, see, or hear? Awareness, self-monitoring, and conscious efforts can be powerful.

What follows is a synthesis of the key relevant ideas on truth. With the necessary simplifications, my quest is largely informed by meta-analytical and interpretative works written by philosophers (sometimes to the broader public), by selective readings of philosophical luminaries, and by public encyclopedias of philosophy. My goal is to obtain, as Aristotle puts it in his *Metaphysics*, "practical wisdom" for life and some guidance on truth. I start with the value of philosophy, its applicability to today's lives, and its link to AI.

4.2 Artificial Intelligence and Applied Ethics (Morality)

AI is inextricably linked to philosophy in several ways. It has been long accepted in philosophy, and probably more widely, that AI, as a newer digital technology, has the potential to impact the overall development of humanity (Müller, 2020). Recognizing this influence raises fundamental ethical questions such as "what we should do with these systems, what the systems themselves should do, what risks they involve, and how we can control these" (Müller, 2020). Discussions of AI ethics have revolved around AI and robotics (as in self-driving cars or autonomous drones) and more recently around the biases found in the data which AI uses to model its behavior, and the resultant need for algorithmic transparency (e.g., Rubin et al., 2020). In the case of the use of AI for curbing mis- and disinformation, we specifically focus on the instructions provided to the algorithms and their subsequent behavior in analyzing textual data (with NLP and ML). Such AI systems are capable of learning speech patterns, making autonomous predictions, and rendering verdicts about online content (e.g., "false," "opinion," "misleading," or "controversial"). These new technologies can directly enable automatic filtering, the de-ranking of misinformative search results (sorting and suppressing them from immediate display to the users), or the explicit labeling of undesirable, problematic, or hateful content, which in turn can route and reroute the flow of information online. Thus, this type of AI also influences and challenges the current norms and regulatory systems, which are both of particular interest to philosophy. For instance, if AI proceeds with straightforward disallowing of the content it deemed undesirable, this raises moral objections about freedom of speech, fairness, and justice for all. It is necessary to understand the trade-offs of the potential good and harms done by AI-enabled innovative technologies in the context of society's functioning, so that ethical concerns ("what ought to be done") are appropriately addressed by societal policies, regulations, and laws. In other words, just because a technology is capable of doing something, it does not necessarily mean that we should accept its capability to do so without questioning how beneficial (or harmful) this may be to the common good of society and the individuals within it.

Philosophy and AI have historically had a strong connection in that both have been concerned with deciphering shared concepts (e.g., action, goals, knowledge, belief, free will, and consciousness). "The AI point of view is that philosophical theories are useful to AI … and [they] provide a basis for designing systems with beliefs, reasoning, and plan," noted McCarthy[5] in his later writings (2006, p. 2). "AI takes what we may call the designer stance about these concepts; it asks what kinds of knowledge, belief, consciousness, etc. does a computer system need in order to behave intelligently and how to build them into a computer program. Philosophers have generally taken a more abstract view and asked what are knowledge, etc." (McCarthy, 2006, p. 3). Abstract notions which have been debated in philosophy for thousands of years are now finding their implementation in concrete decisions needed to guide some autonomous systems. For instance, the Trolley Problem—asking you to consider the value of human life (would you send one man to death by a runaway trolley to save five people?)—has traditionally been a moral dilemma in the realm of mental experiments, but is being applied literally to behaviors of self-driving vehicles that may hurt pedestrians (e.g., see Dierker, 2019).

More generally, AI attempts to construct computer systems that are capable of human-like tasks involving cognitive functions of the mind such as making decisions, applying reasoning, and learning from mistakes. AI goals are twofold. Technologically, they aim "to build useful tools, which can help humans in activities of various kinds, or perform the activities for them. The other [aim] is psychological: to help us understand human […] minds, or even intelligence in general" (Boden, 2005 p. 71). AI strives to mimic human cognitive functions, in essence modeling human intelligence or providing alternative ways to achieve similar goals such as producing a resemblance to a "sentient being" that is capable of conversing, "understanding," or which gives an appearance of "thinking." Philosophy puts forward theories, while AI uses and tests theories, describing theories of mind with the formalisms of logic and mathematics to model human intelligence.

[5] John McCarthy (mathematician and computer scientist), Marvin Minsky (cognitive and computer scientist), Nathaniel Rochester (invertor and developer of the first commercial computer at the IBM), and Claude Shannon (the father of information theory) are often credited with the proposal of a research agenda for "thinking machines." Their consequent conference with other scientists at Dartmouth College, in Hanover, New Hampshire in 1956 has also been widely recognized (Boden, 2005; Broussard, 2019; e.g., in Clocksin, 2003) as the inception of the field of AI.

The relationship between AI and philosophy is reciprocal. As AI makes ample use of philosophical concepts and logical symbolism, it is also turning to philosophy for ethical guidance to be used in practice. How can philosophy help?

Philosophy is useful in practice in several ways. Contrary to popular misconceptions, philosophy is not about esoteric arguments[6] that bear no relation to our everyday lives. Applied philosophy has been of interest since the times of the classical Greek philosophers like Aristotle who believed that "there was no point in studying ethics unless it would have some beneficial effect on the way one lived one's life," and this interest has been renewed in contemporary societies since roughly around the 1960s (Crisp, 2005, p. 244). Philosophy is most useful in practice when it is applied to better our lives. David Hume, the Scottish Enlightenment philosopher, believed that the genius of philosophy is its careful cultivation of ideas; and its practical value is its gradual diffusion throughout the whole of society, for greater foresight and subtlety of politicians, for finer reasoning of lawyers, and for more caution in the plans and operations of the general public (reprinted in Hume, 1955, p. 19). Thus, philosophy can be seen as an applied art with branches directly designated to our service.

Philosophy as a whole is generally concerned "with questions both of personal morality (what should I do?) and public morality (what is good society?)," but these issues are taken up fundamentally in the field of applied ethics to give attention to practical controversies with questions of morality arising in the family, workplace, and in the private and public spheres more broadly (Almond, 2005, p. 25). *Applied philosophy* and *applied ethics* are sometimes used as synonyms. "Applied philosophy is in fact broader, covering also such fields as law, education and art, and theoretical issues in artificial intelligence. These areas include philosophical problems—metaphysical[7] and epistemological[8]—that are not strictly ethical" (Almond, 2005, p. 24). Specifically for applied ethics, its central task is "to articulate what constitutes morality (which involves notions such as rightness and wrongness, guilt and shame) and [to] describe ethics (the systems of values and customs in the lives of particular groups of human beings)" (Crisp, 2005). In the public sphere, applied ethics may involve a range of issues for pluralistic societies such as assessing policies in the light of the impact of advances in new technologies, "assessing international obligations and duties to future generations in the light of

[6] I do admit that some philosophical writings can be unnecessarily off-putting.

[7] *Metaphysical* questions concern the existence and the nature of things that exist such as matter, mind, objects, and their properties, space and time, cause and effect, and possibility. At the very origin of metaphysics, "Aristotle provided two definitions of first philosophy: the study of 'being as such' (i.e., the nature of being, or what it is for a thing to be or to exist) and the study of 'the first causes of things' (i.e., their original or primary causes)" (Wolin, 2020).

[8] *Epistemological* questions are concerned with the nature, origin, and limits of human knowledge (Martinich, 2021). "Along with *metaphysics*, *logic*, and *ethics*, it [*epistemology*] is one of the four main branches of philosophy, and nearly every great philosopher has contributed to it" (Martinich, 2021).

environmental problems," or "aiming to provide guidelines for public policy" (Almond, 2005, p. 25).

In the context of the spread of mis- and disinformation, it is paramount to debate, protect, and regulate the ethical conduct of both human agents (e.g., digital users) and automated agents (AI systems that vet the information). The problems of providing truthful, accurate, high-quality information is of both personal and public ethical concern. Privately, there is a need to avoid the negative consequences of being misinformed when living our lives, while in the public arena, the problem of mis- and disinformation is strongly related to the requirement for ethical and moral guidelines for public policy and for laws of conduct to prevent the misdeeds of others. To place "ethical considerations at the heart of many areas of public debate" is to make them the objects of "analytic and morally sensitive reflection" (Almond, 2005, p. 29) above the commercial and political benefit of a few corporations or individuals, no matter how influential they may be.

4.2.1 Moral Virtues

In the philosophical account of ethics, the concept of a *virtue* is important. Virtue is considered in terms of which behaviors count as a virtue, how many virtues there are (including the question of unity—a single or many virtues), how virtues relate to our psychological identities, and how human virtues juxtapose with human vices. The historical variability of what is recognized as virtuous across different times and cultures is widely acknowledged. However, there are "constants in the psychology and circumstances of human beings that make certain virtues, in some version or other, ubiquitous: in every society people need (something like) courage, (something like) self-control with regard to anger and sexual desire, and some version of prudence" (William, 2005, p. 1048). The modern philosophical account of virtues is likely to agree with Aristotle in his view that vices are failings, while virtues are dispositions of character, acquired by ethical training, and displayed in action and in patterns of emotional reaction. Contemporary philosophers are also likely to amend Aristotle's position significantly, acknowledging more viciousness or evil in human nature, and admitting that virtues are not just rigid habits, but can be more flexible "under the application of practical reason" (William, 2005, p. 1046). Ethical theories, in principle, can center on *consequences* (good states of affairs); *right actions* as bearing ethical values; or alternatively on the virtues themselves that describe *a good, ethically admirable person* (William, 2005). The founder and general editor of the Internet Encyclopedia of Philosophy, contemporary philosopher James Fieser, sums it up nicely: "In short, we acquire what moral philosophers call *virtues*—positive character traits that regulate emotions and urges. Typical virtues include courage, temperance, justice, prudence, fortitude, liberality, and truthfulness. Vices, by contrast, are negative character traits that we develop in response to the same emotions and urges. Typical vices include cowardice, insensibility, injustice, and vanity. As a fully developed moral theory, *virtue theory* is the view that the foundation of morality is the development of good character traits, or virtues. A person is good if he or she has virtues and lacks vices" (Fieser, 2017).

There are intense theoretical debates about the possible unity of one single virtue, which Socrates called wisdom or knowledge, and the distinctions between various virtues such as justice, self-control, courage, generosity, honesty, etc. Can one possess certain virtues in the absence of others? Ethical theories address a number of other fascinating questions that would take us further away from the main discussion of the key virtue of interest here. Of those debates, of particular interest is the issue of justification of actions, or, whether it is always morally wrong to lie.

Intentional dishonesty online seems to attract little sympathy and is universally condemned, though perhaps not using terms as strong as vice or personal failing. Should it be? Sissela Bok, a Swedish-born American philosopher and ethicist, has written extensively on the justifiability of deception and lying since the late 1970s. Her book *Lying: Moral Choice in Public and Private Life* (Bok, 1978) contributed to the current revival of interest in the topic, as she pointed out that until then twentieth-century moral philosophers had been silent on the subject (Mahon, 2014).

More recently, Bok wrote in an article for *the Stanford Encyclopedia of Philosophy* that "all societies, as well as all major moral, religious, and legal traditions have condemned forms of deceit such as bearing false witness; but many have also held that deceit can be excusable or even mandated under certain circumstances, as, for instance, to deflect enemies in war or criminals bent on doing violence to innocent victims" (Bok, 2021, p. 1028). Common dilemmas familiar since antiquity have resulted in diverging opinions on whether it is justifiable for spouses to lie to one another about adultery or for physicians to lie to dying patients. New technologies enable digital quandaries for users facing similar practical moral choices: Whether a dating profile or a resume should be altered, a paper plagiarized, or an electronic signature forged. Spreading disinformation online—making inaccurate, misleading, or false claims for political or commercial gains—seems to cause public outrage and universal moral condemnation. While we all know that intentionally disinforming is morally wrong, we may still have to make a conscious effort to recognize such actions as a moral failing, or worse, as evidence of malicious intent.

In the case of doubt regarding whether a lie could be morally justified or mitigated, perhaps an explicit test—as articulated by Sissela Bok—should be applied. Her 1978 test for moral justification of lying includes three steps. "We must ask, first, whether there are alternative forms of action which will resolve the difficulty without the use of a lie" (Bok, 1978, p. 105). In other words, lies must be the "last resort," which is consistent with psychological literature stating that people prefer to achieve their goals with truthful means, if possible (see Chap. 3). Ethically speaking, "in any situation in which a lie is a possible choice, one must seek nonlying alternatives" (Mahon, 2014, p. 474). For Bok's second step, we must ask "what might be moral reasons brought forward to excuse the lie, and what reasons can be raised as counter-arguments" (Bok, 1978, p. 105). In other words, even within the last resort an all "harms and benefits" test should be applied. "Not merely the harm to the dupe of the lie but also the harm done to the liar in terms of effort to cover up the lie and damage to credibility, and the damage to the overall level of trust in communication in society. Her point was that the liar was biased to underestimate the risk of discovery and overestimate the benefits of lying and to ignore the difference between lies that are truly isolated occurrences and lies that tend to become institutional practices" (Mahon, 2014, p. 474). The third step is known as the "publicity test," which she elaborates on in much detail. In short, "we must ask what a public of reasonable persons might say about such lies" (Bok, 1978, p. 106). Bok saw this three-step test as a set of general principles and a logical continuation of the *veracity principle*[9] in which truth requires no justification, but the negative initial value of deception requires a burden of proof put upon the liar if it is to be morally justified. By this measure "although lies told to would-be murderers to protect innocents, lies told to protect one's life, and lies necessary to save groups against plague, invasion, or political or religious persecution can survive this test, few of the many lies told in society can survive this test" (Mahon, 2014, p. 475).

Most, if not all, intentionally disinformative messages circulating online would not find their moral justification with the public (assumed to be the humans reading the messages), once the motivating factors are properly disclosed and harms publicly debated. The problem, however, is that digital environments are increasingly mediated by algorithms. The users of social media platforms and other online services are constantly advised on what actions to take: What to purchase, how and when to exercise, who to contact, what to watch and read, and which routes to take when driving. In his 2011 TED Talk, Eli Pariser—author of the book *The Filter Bubble* and the founder of *Upworthy*, a socially bent aggregator oft-critiqued for its viral content—stated that we have largely passed "the torch of gatekeeping" from the humans to algorithms. He added that "algorithms don't yet have the kind of embedded ethics that the [traditional newspaper] editors [and librarians] did" (Pariser, 2011). Most AI algorithms that perform personalization, recommendation, filtering, and content promotion tasks are optimized for profit. They simply do not concern themselves with issues of moral rights and wrongs.

If *morality* is marking the differentiation between proper and improper intentions, decisions, and actions (how do we ought to live?), and *immorality* is an active opposition to good or right, then *amorality* is an unawareness or indifference to the questions of moral standards. Amorality is a disbelief in the importance of a particular set of moral principles (who cares?), it is an anything-goes attitude. Figure 4.1 illustrates the two contrasted concepts (of morality and immorality), setting the third option (of amorality) aside. The problem of today's AI systems is often not in their immorality but rather that much of their decisions are amoral or morally agnostic. To remedy such a state of affairs, AI developers should seek guidance on ethical principles and moral virtues. Though this view is getting louder in the AI community, it has not necessarily been voiced by the giant tech platforms. As Kyle Dent, writer for the *TechCrunch* points out, AI companies cannot be expected to police themselves; measures for accountability and transparency need to come from external regulators (Dent, 2019).

---

[9] See also Chap. 3 for psychological treatment of this principle.

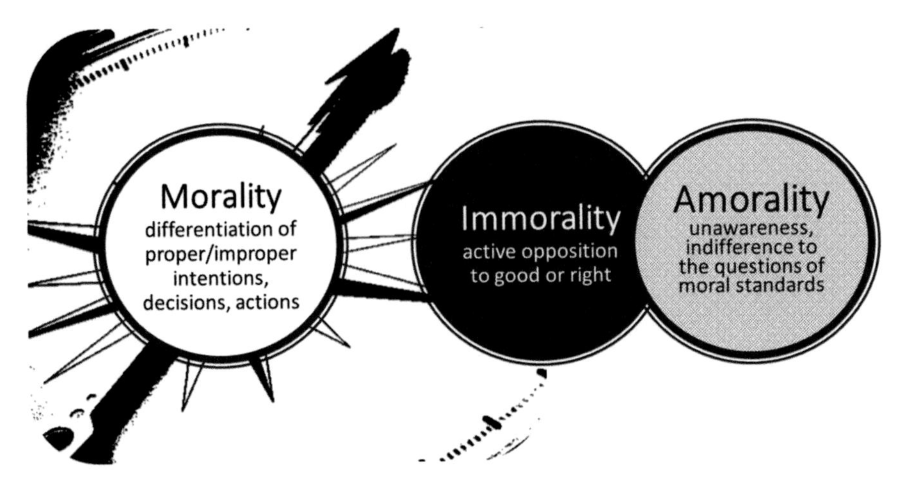

Fig. 4.1 Morality and immorality, often seen as the only binary choices, have a third counterpart—amorality. Both AI and human decisions require a moral compass pointing toward moral virtues

Bringing the principles of ethics and philosophical debate to the attention of AI software engineers is as important as educating digital users about the moral wrongs of disinformation and how to identify it.

In sum, in the absence of an explicit moral compass for an algorithmic mediation of online content, the discussion about the nature of moral virtues should be revitalized in AI and invigorated within the educational system. We can draw on hundreds of years of debates by philosophical luminaries and present their ideas in accessible formats for media literacy training for the general public and not just for students of philosophy.

The virtue of honesty is inextricably linked to the idea of determining what constitutes truth. The next section synthesizes how truth has been seen from various philosophical standpoints and how these arguments may inform our contemporary worldviews.

4.3 Theories of Truth

4.3.1 What Is Truth?

If honesty is a moral virtue that we all strive for, it is then essential to tell the truth. Telling the truth—being truthful—entails knowing the truth. How, then, does one find out what is true? The problem of truth is “in a way easy to state: what truths are, and what (if anything) makes them true”; it is one of the major central subjects discussed in philosophy for thousands of years and has given rise to “a great deal of controversy" (Glanzberg, 2021). How have philosophers defined truth and explained its relation to reality in the past?

"One virtually universal presupposition[10] [in epistemology[11]] is that knowledge is true belief, but not mere true belief. For example, lucky guesses or true beliefs resulting from wishful thinking are not knowledge" (Klein, 2005, p. 224). Epistemology thus asks the central question: What must be added to true beliefs to convert them into knowledge? And that is where several schools of thought differ on their views.

Dozens of philosophical schools of thought on truth can be boiled down to a few major perspectives on the application of one's philosophical stance to truth-seeking. I will discuss them in three broad categories—*inflationary*, *epistemic*, and *deflationary* theories of truth. Of the three groups, *inflationary theories* focus on the very nature of truth itself. Authoritative up-to-date reference materials in the field are used here to summarize a number of views on truth, starting with the prominent inflationary theories from the early part of the twentieth century: *the correspondence, coherence, and pragmatist theories of truth.* The following three subsections briefly detail how proponents of these theories reason. Each view undoubtedly warrants a more thorough investigation of the associated philosophical ideas (than the scope of this chapter) due to the importance of these ways of thinking in the modern society. Suffice it to say, negative pragmatist logic notably underlies much of scientific experimentation: hypotheses can never be proven, only disproven (see also Chap. 5, Sect. 5.4). The *epistemic* group includes *verificationists* and *perspectivists* who maintain that truth only exists in the human mind as constructs of belief and knowledge about the world. *Redundancy* and *performative* theories are in the third *deflationary* group, since they deny the importance of the notion of truth. Glanzberg (2018, 2021) recasts their main ideas in more modern "neoclassical" forms with "slogans" that capture each view. See Table 4.1 for a quick summary of the main positions held by the theories and their prominent theorists. A discussion of the lay understanding of truth and the applications of these theories in AI concludes this section, providing practical value.

4.3.2 *The Correspondence Theory*

"The basic idea of the *correspondence theory* is that what we believe or say is true, if it corresponds to the way things actually are—to the facts" (Glanzberg, 2021). If there is no such appropriate entity—a fact, event or state of affairs—to which it corresponds, the belief is false.

---

[10] Presupposition is something that is assumed in advance or taken for granted (Dictionary.com, 2021).

[11] Epistemology is one of the core areas in philosophy, concerned with "the nature, sources and limits of knowledge" (Klein, 2005, p. 224).

Table 4.1 Summary of key theories of truth with main premises, explanatory slogans, and representative luminaries. (These non-exhaustive lists are selectively assembled for ease of reference and are based on the synthesis of the philosophical literature reviewed for this chapter)

Type of theories	Neoclassical theories	Slogans (that reflect these views)	Explanations (of these slogans)	Prominent thinker examples
Inflationary theories (focus on the very nature of truth itself)	Correspondence theories (embody realism)	A belief is true if and only if it *corresponds to a fact*	Truth is a matter of whether an object in the world is paired up with a true proposition. Belief-to-world relation	Socrates, Plato, Aristotle. G. E. Moore, Bertrand Russell, Ludwig Wittgenstein
	Coherence theories (are associated with idealism)	A belief is true if and only if it is part of a *coherent* system of beliefs	Truth is a matter of how beliefs are related to each other. Belief-to-belief relation	Baruch Spinoza, Immanuel Kant, Georg Hegel, Harold Joachim, Brand Blanshard
	Pragmatist theories (incorporate aspects of the earlier theories)	A belief is true if it is does not conflict with experiences, is useful, is the result of inquiry. If a belief is disproved, it is assuredly false (Hocking's negative pragmatism)	True beliefs will remain settled at the end of prolonged inquiry. Truth is what is verifiable through experience	William James, Charles Sanders Peirce, John Dewey, William hocking
Epistemic theories (are forms of anti-realism)	Verificationist and assertability theories	A claim is correct just insofar as it is in principle *verifiable* with a verification procedure that in principle can be carried out to yield the answer	Truth is constrained by our abilities to verify. Truth *just is* verifiability	Michael Dummett, Crispin Wright
	Perspectivist theories	A proposition can only be true according to a specific perspective	Truth is established through specific perspective, ranging from the individual perspective (truth is only what is accepted by the individual)to the consensus perspective (truth is what is accepted by all members of the relevant community)	Early Greek Sophists, also used by Friedrich Nietzsche and Wilhelm Wundt
Deflationist theories (superfluous "is-true")	Redundancy theory	Stating something "is true" in a proposition adds no new information beyond the proposition itself	The concept of truth is superfluous	Frank Ramsey, Gottlob Frege
	Performative theory	The words "true" or "is true" in a proposition are only informative about the speaker's actions, beliefs, or intentions	Focuses on the speaker's beliefs, motivations, or intentions	Peter Strawson

What are facts here? "*Facts*, for the neo-classical correspondence theory, are entities in their own right. Facts are generally taken to be composed of particulars and properties and relations or universals, at least" (Glanzberg, 2021).

What are propositions then? *Propositions* are what are believed, they give the contents of beliefs and assertions. Propositions are the primary bearers of truth, and have structure roughly corresponding to the structure of sentences. For example, for a simple belief that "*Victoria writes*," the structure of the cast can be observed ⟨Victoria, writing⟩ to match the subject–predicate form of the *that*-clause that reports this belief, and it matches the structure of the belief itself.

The correspondence theory is one of the oldest and most generally accepted theories of truth which can be traced back to the definitions of truth proposed by Aristotle in *Metaphysics* and Plato in *Sophist* (Castro & Gramzow, 2014). "To say of what is that it is not, or of what is not that it is, is false, while to say of what is that it is, and of what is not that it is not, is true" (*Metaphysics* 1011b25); virtually identical formulations can be found in Plato (*Cratylus* 385b2, *Sophist* 263b)" (David, 2020).

Plato (429?–347 B.C.E.), considered "a major figure in the history of [Ancient Greek and Western] philosophy by any standard" (e.g., Kraut, 2017), held a correspondence position on truth and knowledge. He believed that "there are truths to be discovered; that knowledge is possible. Moreover, he held that truth is not [...] relative. Instead, it is objective; it is that which our reason, used rightly, apprehends" (Vaughn, 2014). Plato also distinguished between knowing and believing. "Since there are *objective* truths to be known, we may believe X, but belief alone does not guarantee we are correct. There are three necessary and sufficient conditions, according to Plato, for one to have knowledge: (1) the proposition must be *believed*; (2) the proposition must be *true*; and (3) the proposition must be supported by good *reasons*, which is to say, you must be justified in believing it. Thus, for Plato, knowledge is justified, true belief" (Vaughn, 2014).

This theory has been criticized because "it requires a concrete and comprehensive understanding of reality in which to check for correspondence, a so-called god's-eye view," nonetheless it now serves essentially "as a default definition of truth for most people," and its underlying logic "is very similar to the logic guiding modern scientific theory and its pursuit of truth" (Castro & Gramzow, 2014).

The neoclassical correspondence theory[12] is a more modern take using structured propositions woven together with the intuition that truth is "a content-to-world relation." This relation, in the most straightforward way, asks "for an object in the world to pair up with a true proposition," or more simply put, "what we say or think is true or false in virtue of the way the world turns out to be" (Glanzberg, 2021).

[12]It is a position held by the British "trinity of philosophers at Trinity College Cambridge"—G. E. Moore, Bertrand Russell, and Ludwig Wittgenstein (Baldwin, 2010), who together with the German philosopher, Gottlob Frege, are considered the founders of analytic philosophy.

4.3.3 The Coherence Theory

The coherence theory emphasizes that any individual belief or judgment we might make will only be partially true because the "whole complete truth" is a system of judgments. Expressed in a neoclassical slogan, it can be put as follows: "A belief is true if and only if it is part of a coherent system of beliefs" (Glanzberg, 2021).

One typical critique of this view is that there is a possibility of any given statement and its contradicting statement to be simultaneously true if they are part of two different belief systems. For example, two popularly held beliefs among the US adults, according to a recent survey, are "stricter environmental laws are worth the costs" (commonly held among Democrats), and stricter environmental laws are not worth the costs because they "cost jobs and hurt the economy" (more commonly held by Republicans) (Pew Research Center, U.S. Politics and Policy et al., 2019, p. 98). Each assertion coheres with its respective political parties' systems of belief, but they cannot both be true, because they are mutually contradicting.

Another common criticism is that coherence theory does not have a mechanism to deal with nonsensical statements—such as alien abductions claims and sightings of extraterrestrial life forms. These assertions have to be accepted as true, according to this theory, as long as they are coherent with each other.

To contrast with the correspondence theory, in which the world provides a suitable object to mirror a proposition, coherence theorists idealistically believe that truth is a matter of how beliefs are related to each other in an appropriate system of beliefs (Glanzberg, 2021). Coherence theorists are typically associated with *idealism*, a view that "see[s] little (if any) room between a system of beliefs and the world it is about" (Glanzberg, 2021). By contrast, the correspondence theory embodies *realism*, a view in which "the world exists objectively, independently of the ways we think about it or describe it; and our thoughts and claims are about that world" (Glanzberg, 2021). Realists hold that our claims are objectively true or false, depending on how the world those claims are about turns out to be, and assumes that the world represented in our thoughts or language is an objective world (Glanzberg, 2021). Another important mark of realism is the principle of *bivalence*: Every truth-bearer[13] is either true or false, and "a realist should see there being a fact of the matter one way or the other about whether any given claim is correct" (Glanzberg, 2021).

[13] All theories of truth start with the concept of *truth-bearers*—beliefs, propositions, sentences, or utterances—understood to be meaningful, that is, able to say something about the nature of the world or what the world is like.

A truth-maker—a fact, event, or state of affairs—is a term given to any entity which makes the truth-bearer true. A classic example is a statement "snow is white" is a truth-bearer, and then the fact that snow is actually white is the truth-maker. Philosophers disagree on the candidates for truth-bearers and truth-makers. (See also the brief discussion of propositions and their structure above.)

4.3.4 Pragmatist Theories

Pragmatist theories are a more recent approach to defining truth which were popularized by American philosophers, Charles Pierce, William James, and John Dewey in the early 1900s (Castro & Gramzow, 2014). Pragmatists believe that something is true if it does not conflict with past experiences, if is cognitively useful in the world, and if it is the end result of self-correcting systematic processes of scientific, philosophical, or cultural inquiry or debate. Some weaknesses of the theory have been pointed out including the idea, among others, that what is cognitively useful to one individual in their particular time, place, and culture may differ from what is useful to others under different circumstances. "The crux of Peirce's pragmatism is that for any statement to be meaningful, it must have practical bearings" (Atkin, 2021).

Pragmatists do not always contradict with previous theories. For instance, they allow for a correspondence "insofar as the scientific method of inquiry is answerable to some independent world," "have an affinity with coherence theories, insofar as we expect the end of inquiry to be a coherent system of beliefs," as well as "maintain an important verificationist idea: truth is what is verifiable" (Glanzberg, 2021).

Peirce's "pragmatist principle" guides scientific laboratory experimentation by listing hypotheses with their expected experiential consequences, if they were to be true. "If an object is fragile, and we were to drop it, we would probably see it break. Peirce applied his principle to explain truth in terms of the eventual agreement of responsible inquirers: a proposition is true if it would be accepted eventually by anyone who inquired into it" (Hookway, 2005, p. 774).

An alternate view of pragmatism was put forth in the early twentieth century by a negative pragmatist, William Hocking: "that which can be proven using the pragmatism criteria still may be either true or false, but something that is disproved is assuredly false. This basic logic underlies much contemporary thinking about scientific experimentation—hypotheses can never be proven, only disproven" (Castro & Gramzow, 2014). "Peirce saw the pragmatic account of meaning as a method for clearing up metaphysics and aiding scientific inquiry. This has led many to take Peirce's early statement of pragmatism as a forerunner of the verificationist account of meaning championed by logical positivists" (Atkin, 2021). (See also Chap. 5 for further details on the scientific method of inquiry.)

4.3.5 Epistemic (Anti-Realist) Theories

The *epistemic theories* define the concept of truth using epistemic ideas such as belief and knowledge. In this anti-realist view, truth is not some innate property of objects but it exists only as a construct of humans in relation with the world. Thus, truth cannot be independent of human thought. Instead, truth is constrained by our abilities to verify, and therefore it is constrained by our epistemic situation. "A claim is correct just insofar as it is in principle *verifiable*, i.e., there is a verification procedure we could in principle carry out which would yield the answer that the claim in question was verified" (Glanzberg, 2021).

Verificationism is traced back to "the original empiricists, such as John Locke, and came into its own during the late 1600s to 1800s with philosophers such as David Hume and Auguste Comte" (Castro & Gramzow, 2014). Some verification theories, like those of Michael Dummett's sort, do not support bivalence (see above). "Any statement that reaches beyond what we can in principle verify or refute (verify its negation) will be a counter-example to bivalence. Take, for instance, the claim that there is some substance, say uranium, present in some region of the universe too distant to be inspected by us within the expected lifespan of the universe. Insofar as this really would be in principle unverifiable, we have no reason to maintain it is true or false according to the verificationist theory of truth" (Glanzberg, 2021).

Verificationists have been criticized for the "inability to deal with vague terms or terms that can have multiple or shifting definitions. For example, the truth of a statement that a given plant is a tree can be difficult to discern given that the definition of a tree is somewhat vague" (Castro & Gramzow, 2014).

A similar tenet guides *perspectivist theory*—a proposition can only be true according to a specific perspective, in which truth can be established. "These range from the individual perspective, in which truth is only that which is accepted by the individual, to the consensus perspective, in which what is true is defined by the acceptance of all members of the relevant community. Forms of perspectivism have existed since the early Greek Sophists and have been used in the philosophy of Friedrich Nietzsche and the psychological theories of Wilhelm Wundt. Critics of this perspective argue that some facts are true no matter what perspective is taken and that their results are nothing more than opinions about truths instead of truths themselves" (Castro & Gramzow, 2014).

4.3.6 Deflationary (or Deflationist) Theories

Deflationary theories deny inflationary assumptions and deflate the importance of the nature and notion of truth, with the view that the predicates "*true*" or "*is true*" are superfluous linguistic constructs (Castro & Gramzow, 2014). "Deflationist ideas appear quite early on, including a well-known argument against correspondence in [Gottlob] Frege (1918–19)" (Glanzberg, 2021).

The redundancy theory, popularized by philosophers and logicians in the 1910s and 1920s, including Frank Ramsey and Gottlob Frege, posits that the concept of truth is superfluous and that saying a proposition "is true" adds no new information beyond the proposition itself (Castro & Gramzow, 2014). Basically, this view sees no difference in saying "*it is true that Victoria is writing*" and simply saying that "*Victoria is writing.*" It posits that "there is no property of truth at all, and appearances of the expression 'true' in our sentences are redundant, having no effect on what we express" (Glanzberg, 2021). Critics of the theory point to the use of the predicate "is true" in open sentences such as "everything he says is true," in which "is true" is not redundant but is necessary for the meaning of the sentence.

The performative theory of truth moves away from defining truth and focuses on the speaker's beliefs, motivations, or intentions. The theory was conceived in the 1950s by Peter Strawson and holds that there is no separate property of truth in a proposition, but that the words "*true*" or "*is true*" are still informative about the intentions or beliefs of the speaker (Castro & Gramzow, 2014). That is, to say that "*it is true that John retired*" is still informative since it reveals the speakers' awareness and endorsement of the idea. In other words, simply appending "*it is true*" to any statement only discloses the speaker's beliefs, motivation, or intentions. The theory's critics point out that it shifts its focus away from actually defining truth. Furthermore, not all intentions can have "*is true*" added to them, for example, the statement "*please shut the door is true*" is so awkward that it distracts from the ability to detect whether the speaker endorses shutting the door or not (Castro & Gramzow, 2014).

To sum it up, I have canvassed several prominent varieties of theories of truth and highlighted their main premises and representative thinkers in each of the three broad groups—*inflationary*, *deflationary*, and *epistemic*. Table 4.1 offers a quick visual summary. By nature or by education, each human being may be drawn to the logic of one school of thought or another. At the same time, ambiguities and information overload force us to develop coping techniques that combine truth-seeking approaches for us to proceed more smoothly in our decision-making (see also Chap. 2). The next section considers briefly how these influential ideas may take form in ordinary people's everyday thinking and in their conceptions of truth.

4.4 Practical Applications of Philosophies of Truth

4.4.1 Ordinary Life Decisions

What is truth to a layperson who is proceeding ordinarily in everyday life and is possibly unaware of the intricacies of philosophical debates? How do we typically know when something is true? I invite the readers to consider these questions by reflecting on their own experiences. For most of us, correspondence theory—the oldest one of them all—sounds reasonable. Philosophers observe that we may have a general feeling that if the information we gain through our senses reflects what is really going on in the world, it is true, and if it does not, then it is false (Castro & Gramzow, 2014). Some things we have to presume true, since it is impossible to verify everything through our senses and experiences, so we have adapted to live in presumed realities.

Most of us use a correspondence theory approach when we have some information to consider. If the information matches our reality, then it must be true. If we have prior knowledge of reality, this experience will guide our decision and we may judge the truth automatically. Without prior experience—when the situation is ambiguous or novel—explicit judgments of truth are more common. We may resort to a coherence approach if the information can be smoothly integrated with older information. Alternatively, in the absence of experience, we may rely on tradition, authority, or the consensus of a group we belong to. This follows the practice of the perspectivist approach: we may rely on the *consensus* within the group we trust for an accepted answer to the truth judgment.

Arguing that we are all "hardwired to be duped," Levine (2019) asks: "how many tweets, posts, e-mails, etc. do you receive and how many [do] you question in terms of their honesty" (p. ix)? He quickly responds that nearly all of our communication is uncritically accepted as "honest," as it is our virtually universal human tendency to live our lives in a truth-default mindset. He says that even when we suspect deception, we often pay attention to the wrong cues (see Chap. 2), such as assessing the sincerity of the communicator's demeanor, which can be misleading. A much surer bet is to question the consistency of whatever communication we receive in terms of its correspondence (alignment of known facts and evidence) or coherence (logical consistency within communication) (Levine, 2019, p. 259). Checking for correspondence to reality, looking for established facts, and verifying a communication's internal logical consistency are also well-known professional lie-catchers' and fact-checkers' techniques (see Chap. 5). Understanding the logic behind philosophical arguments, knowing what makes something true, and knowing how these mechanisms of inference work enhances our reasoning ability, critical thinking, and evaluative skills. Coping mechanisms that allow us to unburden ourselves from overly taxing cognitive tasks may play a big role (see psychological biases and tendencies in Chap. 2; see credibility assessment factors in Chap. 3).

When we come across ambiguities and cannot fall back on the correspondence approach, we are likely to opt to accept propositions with fewer conflicts with our prior knowledge, therefore making our mental models more coherent (i.e., the coherence approach). Also, if people who we trust and identify with culturally already agree that a piece of information is true, it is easier for us to process the information by agreeing with the consensus (the perspectivist approach) (Castro & Gramzow, 2014).

4.4.2 On Truth, Reality, Fact, and Knowledge

Most people probably use these terms—facts, truths, knowledge—interchangeably if they do not realize the specificity of their use in philosophy. In laypersons' terms, *truth* is what a person presumes or believes to be a fact, but it is not identical with the fact, since the same reality is viewed differently by different people based on their beliefs. Facts are representations of the reality as it is, and many of them can be verified by anyone at any time without exceptions. Those who claim to know the truth may disagree with each other if they do not know the same facts or do not share a belief in facts with one another.

Beliefs may or may not correspond with *reality*, and when they do, they are true. To figure out whether a given belief is true or false, we use empirical evidence and logic, but whether we can prove a belief or not, is different from its being true. When we have sufficient evidence that a belief is true and *it is in fact true*, then we *know* it to be true.

Knowledge is something we learn from a source. Other sources may have different beliefs about the same fact. When we know something to be true and find our belief about this truth to be justified, it becomes part of our knowledge. What people believe to be true changes from time to time, while the facts do not change. Thus, knowledge is flexible because it changes as new knowledge is acquired. Rational people, then, may at times revise their belief system if they discover internal inconsistency or if their beliefs do not correspond with reality (see also Chap. 5 on professional practices to discern facts).

4.4.3 Contemporary Debates of "Post-Truth" and "Truth Decay"

The question of what constitutes truth and how we know when something is true have long been debated by philosophers, and today's public discourse is no less heated. Social media, opinion columns, and broadcast interviews went through a stage of declaring an era of *post-truth* and *Oxford Dictionaries* made it the Word of the Year in 2016[14] . "In some ways, we have become *too* concerned with truth, to the point where we can no longer agree on it"—wrote a columnist for the *Guardian* in 2019 who went on to blame technology for "encouraging us to believe we can all have first-hand access to the 'real' facts—and now we can't stop fighting about it" (Davies, 2019). Books and collections of essays elaborate on the meaning of the term (e.g., Condello & Andina, 2019). Other philosophers "shocked by the idea of post-truth" have called on other scientists and philosophers to "speak up when scientific findings are ignored by those in power or treated as mere matters of faith," imploring them to remind society "of the importance of the social mission of science—to provide the best information possible as the basis for public policy", and to "publicly affirm the intellectual virtues that they so effectively model: critical thinking, sustained inquiry and revision of beliefs on the basis of evidence" (Higgins, 2016). The social science community at the service of policy makers has produced reports that refocus attention on an alternative metaphor, that of "truth decay." This emphasizes "the importance of facts and fact-based analysis" in the face of disagreements, the increasing influence of opinions, and the decline of trust (Kavanagh & Rich, 2018). The issue of how we establish facts and build up societal knowledge remains a hot topic (see Chap. 5 for scientific, journalistic and jurisprudence practices).

[14] https://languages.oup.com/word-of-the-year/2016/

4.5 Application in AI and NLP

How can these philosophical theories of truth be of any practical use in the context of text analysis using AI? In see several ways. First, perhaps the most evident, is that what is accessible to text-processing algorithms is only the wording of the way we talk, post, tweet, and otherwise make our thoughts knowns. Digital statements, assertions, or sentences, termed as *truth-bearers* in philosophy, can be checked for correspondence between them, or more formally, between their structured propositions. Since the reality outside of our linguistic expression (or what we say) is not accessible for verification, pairing up propositions with actual facts (what happens beyond utterances) is for the most part impossible. What is however accessible for verification is a wealth of assertions that has been made publicly available and can be traced, aggregated, and examined. These assertions may concern our daily lives, but more importantly, they may concern public proclamations and attempts to influence society at large. The publicly available assertions are legally accessible as data for mining[15] with AI and can be analyzed more thoroughly for digital fingerprints revealing their alignment with moral virtues, for their logical reasoning, and the internal coherence between their prior stated intentions and current actions (e.g., comparing campaign promises to subsequent law enactments). Data mining's moral compass can be creatively incentivized to turn away from constantly pointing to profit, and reset to focus on the moral virtues such as honesty, generosity, kindness, etc. It is an aspiration that will find its critics, but as any other moral choice, it is good to conceive of, as a goal in principle, no matter how unrealistic or naïve it may sound in light of an overbearing ad revenue model as the main business model of online media.

Checking for consistencies is supported by at least two types of inflationary theories of truth—correspondence and coherence. To be more precise, AI systems can check for (a) an internal coherence within a certain group (a perspective on an issue, an existing consensus); (b) a coherence between one specific individual and another (within a group seeking diverging, dissenting, or opposing opinions); or (c) a coherence within the statements from the same person over time or across locations. These three tasks, along with many others, are essential components of what reporters may do to fact-check a story by hand (see investigative journalism practices in Chap. 5). Automated fact-checking is in its dawn, though some previous work has been done on the identification of political bias and consensus (reviewed in Chap. 7).

A second way in which AI may be put to use is in aiding lay digital media users who are known for having hard time differentiating facts from opinions in news (Pew Research Center's Journalism Project, 2018). Philosophy tells us opinions can be dangerous since they are roughly beliefs that may be untrue or unjustified. AI can help humans to distinguish between facts and opinions based on their phrasing.

---

[15] Though privacy concerns often arise, and mining procedures may be considered morally questionable in cases when digital users do not realize how extensively intrusive data-mining algorithms are.

Such assistive technologies can be used to more explicitly label online content. There is already promising work in NLP on *stated beliefs, opinions, epistemic modality, negation, certainty,* and *factuality* (Elsayed et al., 2019; Morante & Sporleder, 2012; Rubin, 2010; Rubin et al., 2005; e.g., Sauri & Pustejovsky, 2009; Saurí & Pustejovsky, 2012) (see Chap. 7 for review).

A third promising direction stemming directly from philosophical logical reasoning is *the identification of logical fallacies* in arguments. A scan for several specific informal logic fallacies can be done on long-form narratives such as speeches that attempt to persuade the audience. Dowden (2021) divides fallacies into three categories according to the epistemological or logical factors that cause the error: "(1) the reasoning is invalid but is presented as if it were a valid argument, or else it is inductively much weaker than it is presented as being, (2) the argument has an unjustified premise, or (3) some relevant evidence has been ignored or suppressed." Other typologies include, for example, fallacies that occur due to reliance on an irrelevant reason (e.g., Ad Hominem, Appeal to Pity, and Affirming the Consequent), due to ambiguity (Accent, Amphiboly, and Equivocation), or fallacies of illegitimate presumption (Begging the Question, False Dilemma, No True Scotsman, Complex Question, and Suppressed Evidence). For a human being to identify such or other fallacies may be tedious and cognitively taxing, but AI can learn the prototypical templates and highlight fallacious writing as a new type of assistive technology. Multiple resources in philosophy and rhetoric have compiled extensive lists of logical fallacies—upward of one hundred[16] or two hundred[17] specific types—with representative case scenarios. Library guides have also been put forward with video materials and ample examples.[18]

Fourth, if both algorithms and humans are indeed in great need of a moral compass, this should be considered as the next problem to solve in AI. Principles have to be overtly coded into algorithmic decisions and informed by open public consultations and debates. With the bulk of the responsibilities for ethical training shared between the elementary education system and the family, teachers and parents often seek advice on how to teach kids, for instance, kindness and compassion to oneself and others. A common piece of advice is to explicitly notice and praise kids' acts of kindness, for instance when someone is attempting to console a hurt friend. To some this idea may appear childish, but perhaps AI systems could respond well to similar point-and-naming approaches. Algorithms could learn from statements labeled with their encoded moral virtues, provided such statements were discoverable in social media, and recognized in speech patterns using the symbolic logic of conditional statements and machine-derived observations. When we think of labeling mis- and disinformation, negative connotations come to mind—labels communicating that a missive was attempted to be verified but failed the test of factuality. Creating a

[16] See http://utminers.utep.edu/omwilliamson/ENGL1311/fallacies.htm

[17] See https://iep.utm.edu/fallacy/#H6

[18] See, for example, https://guides.lib.uiowa.edu/c.php?g=849536&p=6077643 from the University of Iowa Libraries, that highlight five common fallacies: the ad hominem fallacy, post hoc *ergo propter hoc*, straw man, slippery slope, and false dichotomy.

database of reaffirming labels to annotate content is an opposing approach. The source of such content can come from narratives that are freely occurring and publicly available in digital media.

Another way toward instilling ethical virtues is to create codes of ethics, for example, within a profession to govern the professional actions of its members. The practice of creating codes of ethics can help all parties involved clearly see the rights and wrongs. Though it may seem unnecessary in the case of telling the truth and being honest, it creates a precedence for public scrutiny and a clear standard against which to verify the acceptability of a statement.

There have been precedents in such efforts coming from some governments, journalism, industry, and social platforms. Some have established ethical boards and committees, signed up for voluntary codes of ethics or agreements, and some countries have even enacted regulations through fines. For instance, to deal with the resurgence of false news in the months leading up to the Indian 2019 General Elections, "social media platforms (including Facebook, WhatsApp, Twitter, Google, ShareChat, TikTok, etc.) and the Internet and Mobile Association of India (IAMAI), an industry body, agreed to a Voluntary Code of Ethics for the General Election 2019" in early 2019 (Ahmad, 2019, p. 107). Similar measures were put in place for platforms operating in India (including the five mentioned above) and the Internet and Mobile Association of India (Ahmad, 2019, p. 10). In Argentina, concerned about their own October 2019 Presidential Election, "press associations, digital platforms, and political parties signed an agreement on Digital Ethics with the Cámara Nacional Electoral (CNE) (National Court on Elections) aimed at fighting misinformation during political campaigns and election periods in social networks, through cooperation with the Argentinean authorities to protect the accuracy of information within their purview" (Ahmad, 2019, p. 12). In Sweden, journalists have been "bound by industry ethical guidelines (Ethical Rules for Press, TV, and Radio) and have a general duty to correct information that is false," with "the maximum fine of SEK 32,000 (or about US$3,500) to be paid by the publisher" in case of violations of the ethical rules (Ahmad, 2019, p. 148). The effectiveness of these measures has been widely questioned (e.g., Lomas, 2019). Their success hinges on independent researchers and monitors being able to gain access to data, and many giant tech companies are not ready to share their coffers. In the wake of the CEOs of Facebook, Google, and Twitter testifying before the US Congress in 2021, some experts voiced their position loudly in the media; arguing that these platforms need to allow access to data for the public to see how "large numbers of users react to malicious content" and "what happens when countermeasures are introduced" (e.g., Newmark & Bateman, 2021). The code of ethics documents themselves and the cases that they generate are potential data for AI mining and further scrutiny, should they ever be made public.

And finally, my fifth point, AI systems could play a significant role in the greater dissemination of knowledge by promoting well-known facts, that is true justified beliefs, that have met a high standard of scientific consensus. AI could explain the rationale or justification behind facts, increasing the volume and ways in which true scientific knowledge is mobilized from experts to the population.

4.6 Conclusions

Chapter 4 untangles the commonly confused concepts of truth, reality, facts, and knowledge. Philosophy gives us a solid basis for understanding what constitutes truth according to a few prominent schools of thought. Philosophical worldviews—such as realism, idealism, pragmatism, or perspectivism—have seeped into todays' consciousness. These worldviews mirror how the people ascribing to those points of view relate truth to reality. Being able to name these outlooks on life, among others, and studying these views and their pros and cons allows us to expand the horizons of our personal understanding of the mechanisms of logical thinking and to cultivate a more disciplined mind. The spread of mis- and disinformation has only heightened our awareness of how dysfunctional the system of user-generated content has become and how hard it is to establish truth when nonsense and pseudo-knowledge is actively propagated by agents of disinformation.

In practice, when you have some information to consider, realists would use the correspondence theory approach, especially given prior knowledge or experience that can facilitate truth judgments. In novel and ambiguous situations, more effortful analytical thinking is required. The new information can be smoothly integrated with some older information, in accordance with the idealistic coherence theory approach. Alternatively, with the perspectivist approach, tradition, authority, or the consensus (within the group you belong to or trust) may offer you an accepted answer. The logic of one school of thought or another may appeal to you and seem more natural. The choice of explicit truth-seeking approaches—both coherence and perspectivist—is likely driven by a cognitive mechanism known as processing fluency, or the ease and speed of our brains' processing and memory recall (see Chap. 2 for an overview of psychological aspects of information processing). The way in which novel information is presented may create ambiguities, and its sheer volume may also force you to take further shortcuts. This is where it pays off to actively pay attention instead. If identifying truth is important, it is best advised to manage explicit decisions about truth judgments with the disciplined mind that knows how to apply logic, think through ideas critically, and avoid logical fallacies.

In the age of mis- and disinformation, philosophical ideas on truth are valuable in many ways for the private and public spheres. These ideas (definitions, stepwise procedures, and general principles) can help individuals separate truths from non-truths and facts from beliefs in the current digital media landscape. A rational person clearly understands that only verifiably justified true beliefs are the building blocks of knowledge based on which it is possible to make well-informed decisions and actions. The logical flaws within each theory have been thoroughly laid out by philosophical debates, and they can help lay people avoid mistakes in reasoning if they are explicitly taught as part of media literacy training. For the public sphere, applied ethics can guide policies and ethical codes of conduct, thereby revitalizing the emphasis on the virtue of honesty, among others, and providing a moral compass for the currently amoral (or morally agnostic) AI algorithms.

AI works toward automating the identification of logical fallacies in the inferences and reasoning of explicit public statements and the process of checking arguments' internal coherence and correspondence to previously verified dated facts. In addition to automatically spotting negative phenomena stemming from errors in judgment or the vice of dishonesty (e.g., false news), more attention should be paid to the explicit cultivation and labeling of positive virtuous instances (e.g., mined from the codes of ethics and adjudicated cases). I have elaborated on five ways to harness philosophical ideas to move AI systems forward on curbing mis- and disinformation. Some work on these fronts (e.g., automated fact-checking) has already begun (see Chap. 7 for review) and some are novel ideas, to the best of my knowledge.

References

Ahmad, T. (2019, September). *Government responses to disinformation on social media platforms* [Web page]. Retrieved from https://www.loc.gov/law/help/social-media-disinformation/canada.php

Almond, B. (2005). Applied ethics. In E. Craig (Ed.), *The shorter Routledge encyclopedia of philosophy* (2nd ed.). Routledge.

Aristotle. (1924). *Aristotle's metaphysics*. The Clarendon Press.

Atkin, A. (2021). *Peirce, Charles Sanders: pragmatism | internet encyclopedia of philosophy*. Retrieved from https://iep.utm.edu/peircepr/

Baldwin, T. (2010). George Edward Moore. In E. N. Zalta (Ed.), *The Stanford encyclopedia of philosophy (Summer 2010)*. Metaphysics Research Lab, Stanford University. Retrieved from https://plato.stanford.edu/archives/sum2010/entries/moore/

Boden, M. A. (2005). Artificial intelligence. In E. Craig (Ed.), *The shorter Routledge encyclopedia of philosophy* (2nd ed.). Routledge.

Bok, S. (1978). *Lying: moral choice in public and private life* (1st ed.). Pantheon Books.

Bok, S. (2021). Truthfulness. In E. N. Zalta (Ed.), *The Stanford encyclopedia of philosophy (Summer 2021)*. Metaphysics Research Lab, Stanford University. Retrieved from https://plato.stanford.edu/archives/sum2021/entriesruth/

Broussard, M. (2019). *Artificial unintelligence: how computers misunderstand the world*. MIT Press. Retrieved from https://mitpress.mit.edu/books/artificial-unintelligence

Castro, J. R., & Gramzow, R. H. (2014). Truth. In T. Levine (Ed.), *Encyclopedia of deception* (Vol. 1, pp. 904–906). SAGE Publications. https://doi.org/10.4135/9781483306902

Clocksin, W. F. (2003). Artificial intelligence and the future. *Philosophical Transactions of the Royal Society of London Series A: Mathematical, Physical, and Engineering Sciences, 361*(1809), 1721–1748. https://doi.org/10.1098/rsta.2003.1232

Condello, A., & Andina, T. (Eds.). (2019). *Post-truth, philosophy and law*. Routledge.

Crisp, R. (2005). Ethics. In E. Craig (Ed.), *The shorter Routledge encyclopedia of philosophy* (2nd ed.). Routledge.

David, M. (2020). The correspondence theory of truth. In E. N. Zalta (Ed.), *The Stanford encyclopedia of philosophy (Winter 2020)*. Metaphysics Research Lab, Stanford University. Retrieved from https://plato.stanford.edu/archives/win2020/entriesruth-correspondence/

Davies, W. (2019, September 19). Why can't we agree on what's true any more? *The Guardian*. Retrieved from http://www.theguardian.com/media/2019/sep/19/why-cant-we-agree-on-whats-true-anymore

Dent, K. (2019, August 25). The risks of amoral AI. *TechCrunch*. Retrieved from https://social.techcrunch.com/2019/08/25/the-risks-of-amoral-a-i/
Dictionary.com. (2021). Definition of presupposition. In *Www.dictionary.com*. Retrieved from https://www.dictionary.com/browse/presupposition.
Dierker, B. R. (2019, March 9). *The trolley problem and self-driving cars | Benjamin R. Dierker*. Retrieved from https://fee.org/articles/the-trolley-problem-and-self-driving-cars/
Dowden, B. (2021). *Fallacies | internet encyclopedia of philosophy*. Retrieved from https://iep.utm.edu/fallacy/
Elsayed, T., Nakov, P., Barrón-Cedeño, A., Hasanain, M., Suwaileh, R., Da San Martino, G., & Atanasova, P. (2019). Overview of the CLEF-2019 CheckThat! lab: Automatic identification and verification of claims. In F. Crestani, M. Braschler, J. Savoy, A. Rauber, H. Müller, D. E. Losada, G. H. Bürki, L. Cappellato, & N. Ferro (Eds.), *Experimental IR meets multilinguality, multimodality, and interaction* (pp. 301–321). Springer International Publishing. https://doi.org/10.1007/978-3-030-28577-7_25
Fieser, J. (2017). Virtues. In *Moral issues that divide us*. Retrieved from https://www.utm.edu/staff/jfieser/class/300/virtues.htm.
Glanzberg, M. (2018). *The Oxford handbook of truth* (1st ed.). Oxford University Press.
Glanzberg, M. (2021). Truth. In E. N. Zalta (Ed.), *The Stanford encyclopedia of philosophy (Summer 2021)*. Metaphysics Research Lab, Stanford University. Retrieved from https://plato.stanford.edu/archives/sum2021/entriesruth/
Higgins, K. (2016). Post-truth: a guide for the perplexed. *Nature, 540*(7631). https://doi.org/10.1038/540009a
Hookway, C. (2005). Peirce, Charles Sanders (1839–1914). In E. Craig (Ed.), *The shorter Routledge encyclopedia of philosophy* (2nd ed.). Routledge.
Hume, D. (1955). An inquiry concerning human understanding: with a supplement. An abstract of A treatise of human nature. Bobbs-Merrill.
Kavanagh, J., & Rich, M. D. (2018). *Truth decay: an initial exploration of the diminishing role of facts and analysis in American public life*. Retrieved from https://www.rand.org/pubs/research_reports/RR2314.html.
Klein, P. D. (2005). Epistemology. In E. Craig (Ed.), *The shorter Routledge encyclopedia of philosophy* (2nd ed.). Routledge.
Kraut, R. (2017). Plato. In E. N. Zalta (Ed.), *The Stanford encyclopedia of philosophy (Fall 2017)*. Metaphysics Research Lab, Stanford University. Retrieved from https://plato.stanford.edu/archives/fall2017/entries/plato/
Levine, T. R. (2019). *Duped: Truth-default theory and the social science of lying and deception*. University of Alabama Press. http://ebookcentral.proquest.com/lib/west/detail.action?docID=5964203
Liddy, E. (2001). Natural language processing. In M. A. Drake (Ed.), *Encyclopedia of library and information science* (2nd ed.). Marcel Dekker. Retrieved from http://surface.syr.edu/cnlp/11
Lomas, N. (2019, January 19). Facebook, Google and twitter told to do more to fight fake news ahead of European elections. *TechCrunch*. Retrieved from https://social.techcrunch.com/2019/01/29/facebook-google-and-twitter-told-to-do-more-to-fight-fake-news-ahead-of-european-elections/
Mahon, J. E. (2014). History of deception: 1950 to the present. In T. R. Levine (Ed.), *Encyclopedia of deceptions*. SAGE.
Martinich, A. P. (2021). Epistemology: definition, history, types, examples, philosophers, & facts. In *Encyclopedia Britannica*. Retrieved from https://www.britannica.com/topic/epistemology.
McCarthy, J. (2006). *The philosophy of AI and the AI of philosophy*. Retrieved from http://jmc.stanford.edu/articles/aiphil2.html
Morante, R., & Sporleder, C. (2012). Modality and negation: An introduction to the special issue. *Computational Linguistics, 38*(2), 223–260. https://doi.org/10.1162/COLI_a_00095
Müller, V. C. (2020). Ethics of artificial intelligence and robotics. In E. N. Zalta (Ed.), *The Stanford encyclopedia of philosophy (Winter 2020)*. Metaphysics Research Lab, Stanford University. Retrieved from https://plato.stanford.edu/archives/win2020/entries/ethics-ai/

Newmark, J., & Bateman, C. (2021, March 24). *Social media disinformation discussions are going in circles. Here's how to change that* slate magazine. Retrieved from https://slate.com/technology/2021/03/online-disinformation-congressional-hearing-amazon-google-twitter-ceos.html

Pariser, E. (2011). *TED talk transcript of "Beware online 'filter bubbles.'"* Retrieved from https://www.ted.com/talks/eli_pariser_beware_online_filter_bubbles/transcript

Pew Research Center U.S. Politics and Policy, Doherty, C., Kiley, J., & Asheer, N. (2019, December 17). *In a politically polarized era, sharp divides in both partisan coalitions* [Pew Research Center U.S. Politics and Policy]. Retrieved from https://www.pewresearch.org/politics/2019/12/17/7-domestic-policy-taxes-environment-health-care/

Pew Research Center's Journalism Project. (2018, June 18). Distinguishing between factual and opinion statements in the news. *Pew Research Center's Journalism Project*. Retrieved from https://www.journalism.org/2018/06/18/distinguishing-between-factual-and-opinion-statements-in-the-news/

Rubin, V. L. (2010). Epistemic modality: from uncertainty to certainty in the context of information seeking as interactions with texts. *Information Processing & Management, 46*(5), 533–540. https://doi.org/10.1016/j.ipm.2010.02.006

Rubin, V. L. (2019). Disinformation and misinformation triangle: a conceptual model for "fake news" epidemic, causal factors and interventions. *Journal of Documentation*, Journal of Documentation, 75 (5), 1013-1034. https://doi.org/10.1108/JD-12-2018-0209

Rubin, V. L., Burkell, J., Cornwell, S., Asubiaro, T., Chen, Y., Pawlick-Potts, D., & Brogly, C. (2020, October 22). *AI opaqueness: What makes AI systems more transparent?* (Panel). In *the CAIS/ACSI2020 Proceedings of the 48th Annual Conference of the Canadian Association for Information Science Conference/l'Association canadienne des sciences de l'information. An Open Virtual Conference: "Diverging Trajectories in Information Science."* Retrieved from https://www.cais2020.ca/talk/ai-opaqueness/CAIS2020-panel2-Rubin.pdf

Rubin, V. L., Liddy, E. D., & Kando, N. (2005). Certainty identification in texts: categorization model and manual tagging results. In J. G. Shanahan, Y. Qu, & J. Wiebe (Eds.), *Computing attitude and affect in text: theory and applications* (pp. 61–76). Springer-Verlag.

Sauri, R., & Pustejovsky, J. (2009). FactBank: a corpus annotated with event factuality. *Language Resources and Evaluation, 43*(3), 227–268. https://doi.org/10.1007/s10579-009-9089-9

Saurí, R., & Pustejovsky, J. (2012). Are you sure that this happened? Assessing the factuality degree of events in text. *Computational Linguistics, 38*(2), 261–299. https://doi.org/10.1162/COLI_a_00096

Vaughn, L. (2014). *Living philosophy: a historical introduction to philosophical ideas*. Oxford University Press. Retrieved from https://global.oup.com/us/companion.websites/9780190628703/sr/ch4/summary/

William, B. (2005). Virtues and Vices. In E. Craig (Ed.), *The shorter Routledge encyclopedia of philosophy* (2nd ed.). Routledge.

Wolin, R. (2020). Metaphysics: definition, problems, theories, history, & criticism. In *Encyclopedia Britannica*. Retrieved from https://www.britannica.com/topic/metaphysics

Part II
Applied Professional Practices and Artificial Intelligence

Chapter 5
Investigation in Law Enforcement, Journalism, and Sciences

One of the most salient features of our culture is that there is so much bullshit. Everyone knows this. Each of us contributes his share. But we tend to take the situation for granted. Most people are rather confident of their ability to recognize bullshit and to avoid being taken in by it. So the phenomenon has not aroused much deliberate concern, or attracted much sustained inquiry.

(Frankfurt, 2005)

Abstract **Chapter 5** focuses on empirical knowledge as it is applied in three sample professions using stepwise procedures to establish facts, detect lies, or discern truth. Law enforcement, scientific inquiry, and investigative reporting each use well-established traditions for truth-seeking, systematic ways of collecting strong supportive evidence, and conducting thorough inquiries to reach conclusions. Following best practices leads to establishing facts and advancing reliable knowledge. When experts are not well trained, diligent, or honest, mistakes and missteps may happen, resulting in cases of wrongful convictions, scientific dishonesty, or journalistic fraud.

I considered the pros and cons of several methodological approaches in each field, and their potential insights for informing the creation of AI-enabled algorithms to detect disinformation in digital online content. In law enforcement, several checklist-like techniques for interrogation and analysis of truthfulness and credibility are amenable to automation. Journalism offers aspirational norms and standards in "the discipline of verification." Fact-checking statements (who in fact said what, names, dates, and locations) seems feasible with AI, if there is already a digital trace. Practicing journalists are sometimes reluctant to further verify substantive assertions. Accomplishing this with AI may also prove to be difficult. The success of the enterprise of scientific inquiries is credited to the rigorous scientific method which places high value on systematicity, transparency, and replicability. None of those values are concrete enough to emulate with AI at a generic level. These principles, with more procedural elaboration from applied fields, can help us to take further concrete steps with NLP, ML, and other AIs to combat disinformation.

V. L. Rubin, *Misinformation and Disinformation*,
https://doi.org/10.1007/978-3-030-95656-1_5

Keywords Law enforcement · Forensic credibility assessment · Statement analysis · Statement validity assessment · SVA · Content-based criteria analysis · CBCA · Reality monitoring · RM · Memory characteristic questionnaire · MCQ · Polygraph · Lie detector · Assessment criteria checklist · Story reconstructability · Police interrogations · Interviews · Investigative reporting · Journalism · Skeptical ways of knowing · Discipline of verification · Completeness · Source credibility · Proof of evidence · 5-W questions (in reporting) · Healthy skepticism · Journalistic codes of ethics · Fact-checking · Yellow journalism · Newsworthiness · Sensationalism · Scandal-mongering · Exaggeration · Fabrication · Scientific inquiry · Scientific method · Empirical methodology · Empirical evidence · Hypothesis testing · Scientific principles · Falsifiability · Reproducibility · Verifiability · Confidence in scientific knowledge · Transparency · Replicability · Generalizability · Systematicity · Validity

5.1 Introduction

The first half of this book (Part I) serves cumulatively as a theoretical knowledge base for our further discussion of its applications in practice. Chapter 1 provides a framework for the problem of mis- and disinformation and some background knowledge in this area of research. A review of behavioral research from psychology and communication studies has covered deception and deceptive behaviors as the underlying phenomena of mis- and disinformation (Chap. 2). The understanding of how we assess the credibility of online content and build trust has come primarily from information sciences and human–computer interaction research (Chap. 3). A review of applied ethics on deception and philosophical theories on truth has led us to different methods for establishing facts and distinguishing beliefs from reality (Chap. 4).

The second half of this book (Part II) starts from the applied practices of several professional expert groups that discern truth from deception (in this Chap. 5). Experts in the fields of lie detection, truth seeking, and the discovery of facts rely primarily on their stepwise procedures and methods (done by hand) to discern lies from truths and establish facts. Artificial intelligence (AI) applications can perhaps mimic and augment some of these experts' efforts and be applied in the context of disinformation detection.

Alternatively, other professional groups—such as marketers and advertisers, set on winning their potential customers' eyeballs and purses—may manipulate perceptions of reality, regardless of what is known to be true, bending reality to their advantage, and at times ignoring the boundaries of moral ethics. Certain questionable techniques for propaganda and mass persuasion to influence the public's minds—for personal, commercial, or political gains—are well documented and deserve a review of their origins and the means by which they achieve their goals (Chap. 6).

Chapter 5 opens the applied half of this book with a review of the common practices and procedures in use by three groups of experts operating in jurisprudence, journalism, and science. In the criminal justice system and law enforcement, detectives and other practitioners are confronted with the need to figure out which of the suspects, witnesses, or victims are telling the truth, which testimonies are to be considered credible, and which should carry weight in investigations. The second expert group—investigative journalists, reporters, and fact-checkers—have a similar objective to make sure that their sources, informants, and eyewitnesses render their side of the story honestly and truthfully, so that their news readership is provided with a complete and contextually accurate picture, to the best of their storytelling abilities. The third expert group—the community of scientists as a whole, be they in natural sciences, social sciences, or humanities—communicate amply with colleagues in their respective specialized fields. Some scientists are proactive in communicating their findings to the general public as part of their community outreach. Efforts in combatting disinformation—such as professors Carl Bergstrom and Jevin West's "calling out BS" curriculum at the University of Washington—have received much-deserved attention. Their contributions have been compared to the renowned works of other notable popularizers of science who similarly encouraged critical thinking: Darrell Huff's (1993) "How to Lie With Statistics," Carl Sagan's "Baloney Detection Kit" in his (Sagan, 1996) "The Demon-Haunted World: Science as a Candle in the Dark," and Harry Frankfurt's (2005) "On Bullshit" (profiled by the *Washington Post* in Guarino, 2019).

Many a government granting agency has put in place "knowledge mobilization" mandates which instruct scientists to provide access to their publications via open access academic forums and other popularized science outlets (e.g., Social Sciences and Humanities Research Council of Canada (SSHRC), 2012). Canadian researchers are also encouraged to speak to the general public and the media, and to use social media venues as their megaphones (Government of Canada Tri-Agency: CIHR, NSERC, and SSHRC, 2015). To add credibility to what the scientific community is communicating to the public, it is important that the public understand the general process of scientific inquiry as it pertains to establishing incremental facts and truths in growing our scientific understanding of the shared reality we live in.

Each of these expert groups—investigators, journalists, and scientists—already use technologies[1] that provide assistance to their decision-making. Some of these procedures and common practices are good candidates for mimicking human expert judgment with artificial intelligence (AI) algorithms. The purpose of this chapter is to investigate these expert best practices more deeply, looking for insights to inform the creation of automated procedures for deception detection, fact-checking, and other systematic ways of establishing valid facts. (See Chap. 7 for a review of how associated AI systems have been implemented to-date.)

[1] By "technologies" here I do not mean the current most popular everyday use of the term (referring to information communication technologies, or ITCs, like cellphones or laptops) but I rather mean "technology" in its second (original) sense: "a manner of accomplishing a task" (Editors of Merriam-Webster's Dictionary, 2021), referring to stepwise procedures in a practical field.

Concluding the chapter with a comparison of the best practices' pros and cons across jurisprudence, journalistic, and scientific procedures, I provide a brief assessment of the feasibility of automating these procedures with AI for the applied purpose of revealing online disinformation. (For an elaboration on the task and problem statement, see Chap. 1 on the spread of mis- and disinformation in digital environments.) A word of caution is threaded throughout this chapter's sections: Human expert practices can go wrong, resulting in malpractices; this could cause compounding errors in AI implementations and a variety of potential negative consequences. Considerations of abuse and misapplications should certainly be taken into account when striving for appropriate, transparent, and effective uses of AI to fight against the infodemic.

5.2 Lie Detection in Law Enforcement and Justice System

"So-called 'lie detection'
involves inferring deception through analysis of physiological responses
to a structured, but unstandardized, series of questions. [...]
For now, although the idea of a lie detector may be comforting,
the most practical advice is to remain skeptical about any conclusion wrung from a polygraph."
(American Psychological Association, 2004)

Let us talk about various law enforcement techniques used to evaluate the truthfulness of testimonies in the North American justice systems. Police, border patrol, and other law enforcement officers verify the information they get by interrogating suspects, victims, and witnesses by matching it to prior known facts and by checking testimonies for consistencies. These experts are trained to collect and verify evidence. Specialists in credibility assessment and lie detection look for semantic and stylistic patterns in gathered data, specifically what is being said and how, and they do so under specific guidelines. Well-known systematic approaches and procedures are somewhat reminiscent of other credibility assessments such as those performed in library and information science (LIS) and health informatics (see Chap. 3). Both detectives and librarians aim to establish credible sources and use composite evidence to progress from concrete checklists of predefined parameters to more abstract notions (e.g., clarity or fullness of details).

5.2.1 *Polygraph (Lie Detector)*

A polygraph, better known as a lie detector, has been essential in police work since the 1920s. Polygraphs operate on the premise that liars exhibit subtle but measurable changes in respiration, heartbeat rate, blood pressure, and galvanic skin response or moisture in the fingertips.

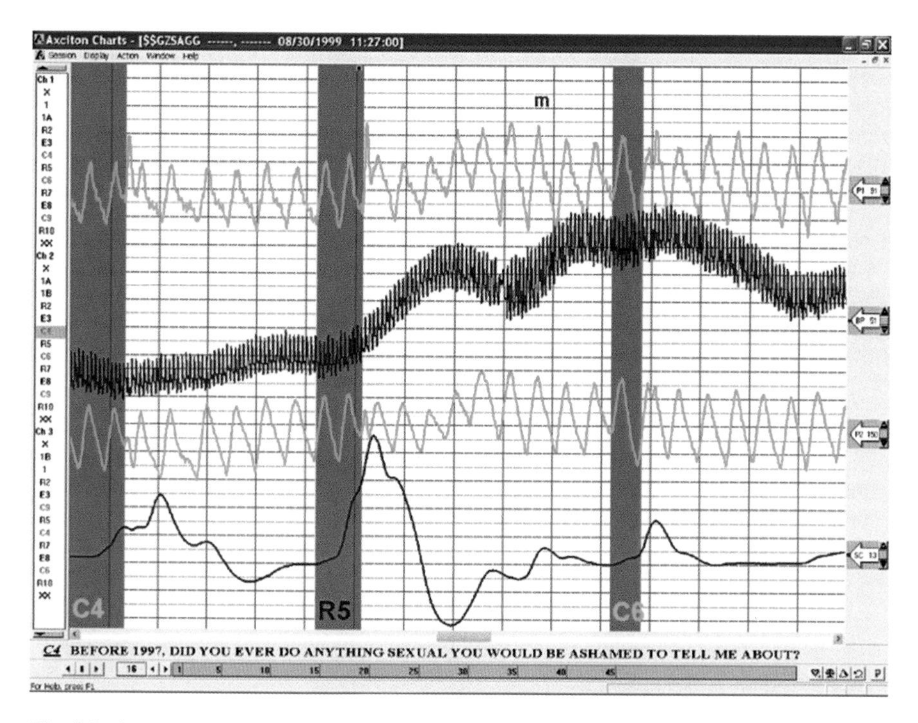

Fig. 5.1 A screenshot of a computerized polygraph reading of physiological parameters—chest breathing (top green line), blood pressure (red line), stomach breathing (bottom green line), and electrical conductivity of the sweat gland (blue line) (Axciton Systems, Inc., 2021). A demonstration image published by Axciton Systems, Inc., the specialized manufacturer (accessed https://www.axciton.com/demos.htm on April 16, 2021)

Polygraph systems are nowadays computerized: A specialized polygraph software is typically provided by companies (such as Axciton Systems, Inc.,[2] or Lafayette Instrument[3]) along with the equipment or hardware such as sensors and pads. A polygraph collects physiological data from at least three systems in the body (American Polygraph Association, 2021), commonly using physiological data sensors such as blood pressure cuffs, electrodermal sensors, and respiration sensors positioned on the chest and abdomen of the examinee (Synnott et al., 2015).

Digital outputs from the sensors and other instruments are connected to the specialized polygraph software, and measurements of physiological changes during a polygraph examination are directly recorded, stored, and analyzed by specialized software. For example, see a demonstration screenshot in Fig. 5.1 which shows the fluctuations of the cardiovascular, respiratory, and electrodermal activities.

[2] See https://www.axciton.com/about.htm (accessed on April 16, 2021).

[3] See https://lafayettepolygraph.com/ (accessed on April 16, 2021).

"The [polygraph] test itself is usually embedded in a longer interview that may last for as long as 4 h. At certain points in the interview the polygrapher will ask a series of yes/no questions. Some of these are intended to elicit physiological states that provide baseline measurements, others are intended to elicit departures from the baseline that indicate an emotionally aroused state. Aroused states presumably encode a flight instinct indicative of deception" (Fitzpatrick et al., 2015, p. 9). Thus, a conclusion reached by the polygraph expert is not final or definitive, but is rather an inference of the presence of deception based on the verbal interview and the physiological responses to the procedure. It is not than surprising that the American Psychological Association in their advisory column on Law and Psychology in real-world applications remains skeptical about any conclusions, as they put it, "wrung from a polygraph" (, 2004).

Nonetheless, polygraph examiners go through special training to be certified in the proper interrogation procedures and be able to work cases and interpret polygraph test results. Trained experts combine polygraph measurements with additional verbal analyses to decide if there is an attempt to deceive. Lesser known physiological analyzers include those that measure voice-stress, thermal images of facial skin temperature, and brain wave activity.

None of these devices can "read the mind," as some people sometimes mistakenly believe, but a refusal to undergo a polygraph test may raise suspicions of guilty consciousness. Since the validity of polygraph tests have been questioned by psychologists, their use remains controversial. For instance, polygraph readings are not admissible in most North American or Western European courts, though they are admissible in Japan and India. Evaluations of the accuracy of polygraph tests conclude that they are "well above chance, though well below perfection": Accuracies "range from 74% to 89% for guilty examinees, with 1% to 13% false-negatives, and 59% to 83% for innocent examinees, with a false-positive ratio varying from 10% to 23%" (Synnott et al., 2015, p. 70). Despite skepticism about polygraph exams, "the absence of other viable alternatives makes the polygraph a widely used technique for detecting deception" (Fitzpatrick et al., 2015, p. 10). The polygraph is but one tool, and a controversial one, as far as the justice system toolbox has to offer.

5.2.2 *Verbal Content Analyses*

While polygraph tests are widely recognizable in the general public as lie detection technology, they are not of particular use for detecting lies in online contexts. As is self-evident, the covert nature of disinformation, and the lack of knowledge about its sources and origins precludes us from being able to gain direct physical access to the agents of disinformation, let alone gain their consent to be wired up with sensors and undergo a polygraph test. So, for the purpose of automating lie detection, we should be focusing on the verbal content of disinformation that is digitally available for AI analyses, rather than looking at physiological nonverbal signs of deceiving.

Verbal content analyses—what is being said and how—are commonly used in the justice system when experts assess verbal statements. Law enforcement officers and other experts in the field cross-check the information they find during direct interviews with after-the-fact analyses of the statements. Transcripts of interviews, testimonies, or confessions become part of the evidence in legal investigations and court proceedings. This chapter aims to shed light on how these various verbal content analyses are conducted in the justice system: by what logic, in what order, and with what principles in mind. What is done "manually," using human analytical skills, can inform AI-based deception detection algorithms that rely on natural language processing (NLP) and machine learning (ML) techniques (see Chap. 1, Sect. 1.5 for background on AI and NLP; and Chap. 7, Sect. 7.1–7.2 for a primer on key computational concepts, principles, and methodological goals). Here I review the three main manual assessment methodologies: Statement Validity Assessment (SVA), Content-Based Criteria Analysis (CBCA), and Reality Monitoring (RM) techniques. I discuss each method in turn, considering both their pointed questions and sequencing, and their usefulness and limitations in potentially informing the creation of AI-enabled algorithms to detect disinformation in digital online content.

5.2.3 Statement Validity Assessment (SVA) and the Undeutsch Hypothesis

The scientific study of deceptive language and statement analysis for credibility dates back to the work of legal practitioners and psychologists in 1950s' Europe. Child witnesses in sexual abuse cases in West Germany were psychologically assessed for their credibility in accordance with a list of criteria originally proposed by Udo Undeutsch (1954). Undeutsch maintained that it is "not the veracity of the reporting person but the truthfulness of the statement that matters, and there are certain relatively exact, definable, descriptive criteria that form a key tool for the determination of the truthfulness of statements" (Undeutsch, 1954). What came to be known as the *Undeutsch hypothesis* stated that the memory of a real-life self-experienced event differs in content and quality from a fabricated or imagined event (Undeutsch, 1967). So, it is assumed that language reveals information about the veracity of statements regardless of the facts mentioned in its content. Attributes of truthfully described, experienced events are supposed to include anchoring of the account, concreteness, wealth of detail, originality, internal consistency, and mentions of specific details. Undeutsch elaborated on how his criteria could be manifested in deceptive language, for instance, in reports of subjective experiences, mentions of unexpected complications, spontaneous corrections, self-deprecating interspersions, and references to details that exceed the child's capability to understand (Fitzpatrick et al., 2015). Undeutsch assumed that experts have the ability to identify such language cues in testimonies, and moreover, they can teach others how to follow such procedures (Adams, 1996).

During the 1960s and 1970s, this Undeutsch hypothesis evolved through the integration of other proposed criteria and was effectively constituted as a new police interrogation[4] technique—Statement Validity Analysis (SVA). SVA is a diagnostic procedure and a set of techniques to assess the truthfulness of a statement. Its proponents see it as an instrument for systematic credibility assessment that has with time extended from its use in assessing child witnesses in sexual assault cases to being applied to adult witnesses in other contexts. With SVA, trained experts generate and test their hypotheses about the source of a given statement: does it describe a personal experience of the witness or does it have another source? The latter implies that statements were made-up using fantasy, imagination, etc. "SVA includes methods of collecting data which are relevant with regard to the hypotheses in question, techniques of analysing these data, and guidelines for drawing conclusions regarding the initial hypotheses" (Köhnken, 2004, p. 43).

Procedurally, SVA has four stages: "(1) a background case-file analysis in which hypotheses about the source of the statement are generated, (2) a semi-structured interview, (3) an analysis of the content of the interview, and (4) an evaluation of stage 3 based on a validity checklist" (Fitzpatrick et al., 2015, p. 34). A more fine-grained account of the SVA components is narrated in Fig. 5.2.

5.2.4 *Content-Based Criteria Analysis (CBCA)*

The Criteria-Based Content Analysis (CBCA) instrument integrates the Undeutsch criteria into a categorical system of cues. CBCA is a content analytical technique often used as the third stage of SVA. "The rationale of CBCA is that a true statement differs in content quality from a fabricated account because (a) a truth teller can draw on an episodic autobiographical representation containing a multitude of details, whereas a liar has to relate to scripts containing only general details of an event; and (b) a liar is busier with strategic self-presentation than a truth teller" (Volbert & Steller, 2014, p. 208). CBCA is used to distinguish between event-based and intentionally fabricated statements and it is now widely accepted as a method of credibility assessment in many European courts (Griesel et al., 2013). CBCA has 19 criteria (Fig. 5.3) judged by trained experts on a three-point scale (0—the criterion is absent; 1—present; and 2—strongly present).

Niehaus (2008) extended Steller and Köhnken's (1989) CBCA checklist with more details, as the original criteria were deemed non-exhaustive and they were simply meant to be examples. Her compilation distinguishes characteristics pointing to the differences in the cognitive processes of liars and truth tellers from those due to characteristics referring to aspects of strategic self-presentation (see Fig. 5.4).

[4] In my limited personal experience of interacting with Canadian lie detection officers, they prefer to use the term "interviews" in the 2020s rather than the term "interrogations" as was used in some dated professional literature. Those terms appear to be equivalents when referring to law enforcement questioning procedures, and the change is a nod to more politically correct times.

Analysis of the case file

Generation of hypotheses about the source of the statement

Decision on assessment methods which are appropriate to test the hypotheses

Examination of the witness
- ➢ history-taking interview with witness or parents
- ➢ application of personality questionnaires and cognitive abilities tests
- ➢ interview of the witness about the event in question

Content analysis of the statement (or relevant parts of it) using CBCA criteria

Evaluation of the CBCA results taking into account
- ➢ cognitive abilities of the witness (e.g., verbal fluency, creativity, general intelligence)
- ➢ particular knowledge and experiences of the witness (e.g., previous sexual experiences in a case of sexual abuse, access to porno videos, literature, conversation with others about similar events)
- ➢ case characteristics (e.g., time between the event in question and the interview, complexity of the event)

Analysis of consistency of repeated statements

Analysis of the origin of the statement and its further development
- ➢ circumstances of initial disclosure
- ➢ responses of others to first disclosure
- ➢ previous interviews

Evaluation of hypotheses in light of the results of the various assessments methods; hypotheses, which are not compatible with the diagnostic results are rejected; if all hypotheses for other sources than own experience are rejected, the veracity of the statement is assumed.

Fig. 5.2 Components of statement validity analysis (SVA) (Köhnken, 2004)

Either version of the CBCA criteria has limitations, and just as polygraph methodology, both SVA and CBCA have their critics, especially around the subjective nature of the procedures and their lack of inter-reliability among different examiners following the same procedures. Notably, Aldert Vrij, a leading expert in the deception detection research in psychology, conducted empirical evaluations of SVA and CBCA. In (Vrij, 2005), Vrij concludes that SVA evaluations are not accurate enough to be admitted as expert scientific evidence in criminal courts due to high error rates (around 30% in laboratory studies) and the disputed nature of the method in the relevant scientific community. By contrast, for police investigations, he sees SVA/CBCA as a valuable tool, "for example, in the initial stage of investigation for forming rough indications of the veracity of various statements" (Vrij, 2005, p. 34).

Experts in legal psychology emphasize that CBCA must be conducted in the context of the broader assessment of each specific legal case. The more comprehensively involved procedure of SVA (that subsumes CBCA) "is not a quick-to-apply check-list tool," they say, and it should be seen as "a *comprehensive ideographic*

General characteristics
1. Logical structure (statement is coherent and logically consistent);
2. Unstructured production (information is presented in non-chronological order);
3. Quantity of details (statement is rich in details).

Specific contents
4. Contextual embedding (events are placed in time and location);
5. Descriptions of interactions (statement has information that links the alleged perpetrator and witness);
6. Reproduction of conversation (specific dialogue, not summaries of what people said);
7. Reporting of unexpected complications during the incident.

Peculiarities of content
8. Unusual details (tattoos, stutters, individual quirks);
9. Superfluous details (details that are non-essential to the allegation);
10. Accurately reported details misunderstood (mentioning of details outside a person's scope of understanding);
11. Related external associations;
12. Accounts of subjective mental state (description of a change in a subject's feelings during the incident);
13. Attribution of perpetrator's mental state (witness describes perpetrator's feelings).

Motivation related contents
14. Spontaneous corrections;
15. Admitting lack of memory;
16. Raising doubts about one's own testimony;
17. Self-deprecation;
18. Pardoning the perpetrator.

Offence-specific elements
19. Details characteristic of the offence.

Fig. 5.3 Statement validity analysis (SVA) relies on the assessment of the validity assessment of 19 checklist criteria. (Adapted from Steller and Köhnken (1989) with explanatory comments in parenthesis by Fitzpatrick (2015, p. 35))

Level	Autobiographic episodic memory versus script information	Strategic self-presentation
Single characteristics	Characteristics of episodic autobiographical memory, e.g.: – Contextual embedding – Spatial information – Temporal information – Description of interactions – Reproduction of conversations – Peripheral details – Sensory impressions – Emotions and feelings – Own thoughts – Personal implications – Perpetrator's mental state Script-deviant/script-irrelevant details, e.g.: – Unexpected complications – Unusual details – Related external associations Details not comprehended – Accurately reported details not comprehended	Indications of memory-related shortcomings, e.g.: – Spontaneous corrections – Admitting lack of memory – Efforts to remember – Reality controls Questioning credibility, e.g.: – Raising doubts about one's own testimony – Raising doubts about one's own person Other problematic contents, e.g.: – Self-deprecation – Pardoning the perpetrator
Statement as a whole	– Reconstructability of the event – Vividness of event – Quantity of details – Unstructured production – Spontaneous supplementing	

Fig. 5.4 Niehaus's (2008) modified system of Steller and Köhnken's (1989) CBCA content characteristics (Volbert & Steller, 2014). (Used with permission from *European Psychologist* 2014; Vol. 19(3):207–220. ©2014 Hogrefe Publishing www.hogrefe.com doi: 10.1027/1016-9040/a000200)

approach enabling psychological experts to perform what is, in many cases, a crucial assessment of the credibility of a specific testimony" (Volbert & Steller, 2014, p. 216). In other words, it requires both much care during the analysis and training prior to applying it.

5.2.5 Reality Monitoring (RM)

Another technique in police work that may be seen as an alternative to both polygraph and SVA/CBCA is Reality Monitoring (RM). Reality monitoring in itself is a kind of meta-memory processes "by which people reflect on varying degrees of understanding about the nature of their own memories" (Johnson & Raye, 1981, p. 70). The technique of RM in applied psychology and in the justice system nowadays refers to the process of deciding whether the memory has an internal or external source. Originally, the model was developed in 1981 by Marcia Johnson and Carol Raye, two psychologists who, following the lead of philosophers like David Hume, were concerned with separating the real (remembered external experiences) from the imagined (internal thoughts). RM provides potential explanations for why the CBCA criteria work: spelling out why the language of made-up testimonies differs from the authentic memories.

Johnson and Raye (1981) essentially held that the origins of people's memories differ. "People remember information from two basic sources: that derived from external sources (obtained through perceptual processes) and that generated by internal processes such as reasoning, imagination, and thought" (Johnson & Raye, 1981, p. 67). Our sensitivity to the origins of information is reflected by differences between memory traces as we recount our experiences. Thus, the assumption in this line of thinking is again that remembering perceived events is different from self-generated events (such as thinking and imagining) because "they are different operations of the mind" that possibly have "characteristically different amounts of various types of information" (Johnson & Raye, 1981, p. 82). These characteristics, also called attributes, are central to RM. External memories have more spatial and temporal contextual attributes coded in the representation of the event, more sensory attributes, and more semantic details (both more of them and more specific ones), while internal memories include more information about cognitive operations (Johnson & Raye, 1981).

Johnson, Raye, and their colleagues later developed a 39-item Memory Characteristic Questionnaire (MCQ) and asked their study respondents to self-report their assessments of their own specific autobiographic memory (e.g., a visit to a dentist, a library, or a social event) as opposed to an imagined event (a dream, fantasy, or unfulfilled intentions) (Johnson et al., 1988). The resulting differences between perceived and imagined events were encouraging and provided the basis for making RM decisions, as originally hypothesized by Johnson and Raye (1981). For example, for "recent events, perceived events were given higher ratings than imagined events on the following characteristics: visual detail, sound, smell, taste, realism, location, setting, spatial arrangement of objects and people, and temporal questions (time, year, season, day, and hour). Perceived events were also more positive in tone and included more supporting memories from before and after the target event. Recent imagined events were more complex than were recent perceived events, seemed at the time to have more implications, were more intense, and were thought about more often" (Johnson et al., 1988, p. 373).

5.2.6 Integration of CBCA and RM Approaches

Sporer (1997) used a factor analysis on both sets of criteria, CBCA and RM MCQ, and extracted eight abstract subscales of their common underlying dimensions: (1) *clarity*, (2) *sensory experiences*, (3) *spatial information*, (4) *time information*, (5) *emotions and feelings*, (6) *reconstructability of the story* (despite complexity of the action, presumed and factual consequences, and certainty/doubts about the memory), (7) *realism*, and (8) *cognitive operations.* In these RM subscales, the first seven subscales are expected to occur more in the truthful statements, and the eighth (*cognitive operations*) are expected to occur more in the invented statements (Sporer, 1997, p. 373). The eight RM subscales "can lead to an integrative cognitive-social theory of the detection of deception that unifies both types of approaches. For example, in both approaches, logical consistency, realism and the possibility of reconstructing a story line seem to be essential global characteristics. A quantitative richness in detail, with contextual embedding in space and time, appears also important. At a more specific level, internal processes like feelings, emotions and cognitive operations can be identified" (Sporer, 1997, p. 392).

Fitzpatrick et al. (2015) report that researchers in deception find RM particularly appealing because, while SVA/CBCA provide a set of heuristic tools, RM's cognitive model has an empirically proven theoretical basis that explains why truthful statements differ from invented ones. The integration of the two approaches (CBCA and RM) is logical due to some overlaps in their criteria, as well as due to their complementary natures: Careful examination of the lists can provide fuller and more precise criteria.

5.2.7 Suitability for Automation

Both techniques (CBCA and RM) are prime candidates for automation with AI/NLP algorithms to distinguish truthful narratives online from deceptive ones, and the experts in the field of deception detection generally agree that in spite of their somewhat subjective nature, they are "worth considering in attempts" to automate the process (Fitzpatrick et al., 2015, p. 37).

The CBCA lists of criteria have been of particular interest to those automating the process of identifying deception (see a review of the trailblazing works in Chap. 7). Several listed criteria fall neatly within the types of language analytics already conducted with NLP techniques for other purposes of language "understanding." For instance, temporal and spatial information is easy to inventory due to the formulaic ways in which people speak about time and space. Much has been accomplished with various methods for identifying emotions and feelings, as they constitute the goals of sentiment analysis, a burgeoning subfield of NLP (or Liu, 2012; e.g., Pang & Lee, 2008). (See also Chap. 1 for more background.)

The principle danger in adopting and adapting checklist-like procedures from police interrogations and content analyses lies in oversimplifying and decontextualizing these methods. It may not be sufficient to simply convert the SVA/CBCA approach into straightforward AI-based analytics of digital narratives. Police investigators have access to the broader context in their case files, while in suspected online disinformation cases, dis-informants are typically anonymous, and there is certainly little chance to interview them. Thus, any AI prediction based on checklist-like interrogation procedures will likely lack the required depth of contextual facts, and so it should be used merely as an assistance to a human making the ultimate judgement call.

5.2.8 A Word of Caution

Historically, among the adverse consequences of unjust actions of the justice system and law enforcement professionals, there have been errors in judgment, and errors with the incorrect uses of the procedures and instruments at the disposal of the experts. Cases of overzealous and prejudiced police interrogations may result in false confessions or elicitations of false memories, in turn leading to false arrests, wrongful convictions, and the unjust incarceration of innocent persons. There have been efforts in the nonprofit U.S. sector, for example, by *The National Association for Civilian Oversight of Law Enforcement* (NACOLE) and *the Innocence Project* to "prevent wrongful convictions, improve police practices," and increase "public trust in police, police legitimacy, and the relationship between police and the communities they serve" (National Association for Civilian Oversight of Law Enforcement, 2021).

While commonly practiced in police fieldwork and admitted in some countries' courts as evidence, the status of some of the above-discussed investigative techniques have been challenged for their lack of solid scientific evaluation, lack of proof of the consistency of judgments across examiners, and the lack of applicability of these methods across contexts (i.e., generalizability). Applications to areas outside of the analysis of written testimonies from interviewed suspects, witnesses, and victims in legal cases are very rare thus far.

In using these procedures (heuristics, theoretical models, empirical checklists, criteria, etc.), as the basis for automating the identification of disinformation (inaccurate, false or misleading information spread with the intent to deceive), we need to be aware of the potential for errors and the subjective nature of these analyses. There is a great need for further R&D in how to best, if at all, adopt and adapt these methods to massive scale information processing technologies. As with any other composite tools, AI-based techniques are likely to have an element of built-in bias (e.g., coming from its training data). Prior to accusing anyone of lying based on AI-driven recommendations, humans should review and validate AI conclusions. In the case of news readers or social media users encountering potential disinformation online, the ultimate decision is certainly in their own mind, but having access to the list of systematic criteria that AI used and some explanations — will ultimately help humans weigh the facts of each case like a pro.

5.3 Investigative Reporting and Fact-Checking in Journalism

> *"'Journalistic truth' is more than mere accuracy. It is a sorting-out process that takes place between the initial story and the interaction among the public, newsmakers, and journalists.*
>
> *The first principle of journalism—its disinterested pursuit of truth—is ultimately what sets journalism aside from all other forms of communication....*
>
> *[W]hat journalism is after[is]a practical or functional form of truth. It is not in the absolute or philosophical sense. It is not the truth of a chemical equation. Journalism can—and must pursue the truths by which we can operate on a day-to-day basis."*
>
> (Kovach & Rosenstiel, 2007, pp. 42–43)

Turning attention to a different kind of inquiry, journalistic investigations, let us review traditional reporting practices. Professionals in the justice system and in journalism appear to have similar goals—to establish the facts and find out the truth—but they go about it with slightly different methods and hardly comparable standards for internal or external oversight. Also, the fruits of police and journalistic investigations are of distinctly different flavors: both in terms of the formats of resulting reports and the legal consequences. In this section, we drill down specifically into how journalists (are taught and, in practice) verify and fact-check information for their investigative reporting. It may help to start with agreeing on what journalism is, according to the journalists who define this practice.

5.3.1 Journalistic Goals and Aspirations in Pursuit of Facts

"Unlike doctors or lawyers with high degrees of government and professional regulation, anyone can call themselves a journalist and claim to practice journalism regardless of their education or training" (Jahng et al., 2021, p. 3). Even though many journalists self-identify as professionals, journalism is not a registered profession, and it is typically defined as an activity, at least in the North American tradition. According to the API, "journalism is the activity of gathering, assessing, creating, and presenting news and information" and the product of these activities (American Press Institute, 2021). These journalistic activities are guided by distinguishable norms and ethical virtues, often spelled out in codes of ethics which are typically not legally enforced (see Schmuhl, 1984). In an effort to increase standardization and professionalization, journalists have various internal and external mechanisms for (institutional) self-criticism and self-evaluation such as the work of ombudspeople, editors, and factcheckers, and the practices of retraction and correction. Various boards, councils, review boards, and other organizations create, oversee, and commit to journalistic standards but only exert some internal control over the practices of their members. Internal control is largely done by encouraging voluntary adherence to standards, commitments, tradition, and best practices.

In principle, journalists strive for accuracy, completeness, fairness, unbiased balance, accountability, and transparency in their journalistic work (Schmuhl, 1984). The ethical underpinnings of the investigative reporting process are laid out in many a textbook and practical guide. "Journalists seek to tell us things which are true. However, our ability to discover the whole truth about any matter is severely circumscribed by our own intelligence, perspicacity, time, and resources. This is no less true for the reporter. In the western tradition, a number of practices have evolved to provide structural safeguards to the pursuit of truthfulness in the reporting of facts" (Sanders, 2003, ch. 4, p. 3). The ways of the tradecraft including its norms, standards, and expectations are traditionally passed down in newsrooms, on the job, or taught explicitly in professional training programs. Adhering to the traditional ideals is highly encouraged.

The technological changes in the use, production, and dissemination of online news has brought about economic hardships in the trade due to the decoupling of newsrooms from their ad revenues. Extreme shortages in funding have put the industry in the unprecedented situation in which society expects reporters to continue their crucial role as the watchdog of democratic society, while many news outlets have had to close down. Those who survived had to adapt to new digital production realities, and inevitably have made severe cuts in their essential newsroom positions that typically provide the quality-control positions such as editors and fact-checkers. (See also Chap. 1 for the current media landscape.) Online news readership has alternative sources to turn to, including self-proclaimed satirists, citizen reporters, bloggers, or tik-tokkers, who are hardly bound by any journalistic norms or ethics but still self-identify as journalists.

Kovach and Rosenstiel emphasize that good journalists sort out facts and "journalistic truths" in the context of the day-to-day functioning of their society.

> "To understand this sorting out process it is important to remember that journalism exists in a societal context. Out of necessity, citizens and societies depend on accurate and reliable accounts of events. They develop procedures and processes to arrive at what might be called 'functional truths.' Police track down and arrest suspects based on facts. Judges hold trials. Juries render verdicts of guilty or innocent. Industries are regulated, taxes are collected, laws are made. We teach our children rules, history, physics, and biology. All of these truths—even the laws of science—are subject to revisions, but we operate by them in the meantime because they are necessary and they work" (Kovach & Rosenstiel, 2007, pp. 42–43).

Reporters are also storytellers: They tell day-to-day stories based on what they know at the time. Their storytelling should be guided by ethical principles and should serve their community. Telling a story inevitably adds a layer of interpretation to position the news in the existing cultural context. Storytelling is an important role that historically has been performed by much respected individuals such as elders, priests, bards, or poets, but now the story-telling role has typically been relegated to anyone in the media (Sanders, 2003).

Even though reporters interpret reality when they tell their stories and construct their narratives, their practice is aimed at truthfulness and impartiality. Kovach and Rosenstiel, in their classic journalism textbook (2001, 2007, 2014) proclaim journalism to be "the discipline of verification," setting the essence of journalistic communication apart from other kinds of activities (e.g., entertainment, propaganda, fiction, or art). They call on journalists to "sift out the rumor, the innuendo, the insignificant, and the spin and concentrate on what is true and important about a story" (Kovach & Rosenstiel, 2007, p. 61). Bill Kovach and Tom Rosenstiel's call and their other instructional writings resonate throughout the community, as they are two of the most recognized and influential thinkers on the future of news in the U.S. Their 10 elements of journalism—its essential principles and practices—are recited in journalism programs like commandments:

1. Obligation to tell the TRUTH.
2. LOYALTY to citizens.
3. VERIFICATION of information.
4. Practitioners' INDEPENDENCE from those they cover.
5. MONITOR those in power.
6. Provide a forum for PUBLIC CRITICISM and COMPROMISE.
7. Keep news INTERESTING and RELEVANT.
8. Keep news COMPREHENSIVE and PROPORTIONAL.
9. Maintain personal sense of ETHICS and RESPONSIBILITY.
10. CITIZENS, too, have rights and responsibilities (adapted from Kovach and Rosenstiel (2007).

These 10 elements are written into news literacy websites (such as the one by the *Committee of Concerned Journalists*, a far-reaching consortium of reporters, editors, producers, publishers, and academics) as explanations to the general public of what it is that journalists do. For example, see the 10 elements almost verbatim with explanations on the online *API's Journalism Essentials Guide* (Dean, 2021).

These elements are consistent with the long-standing tradition of verification in which any newcomer in the trade is advised by its veterans to "corroborate and verify even the most likely assertions they encountered: 'Kid, if your mother says she loves you, check it out'" (Kovach & Rosenstiel, 2010, p. 36).

Yet, the practice of sifting out the rumors and sorting out the facts leaves quite a bit of room for interpretation and creativity for story-tellers, writers, and interpreters of reality. The journalistic method in search of 'journalistic truth' is neither as strictly prescribed as the methodological procedures of the justice system and law enforcement (see Sect. 5.2), nor as rigorous as in the scientific method of inquiry (see Sect. 5.4). So, how do journalists establish facts?

5.3.2 *Skeptical Way of Knowing in the Best Traditions of Investigative Reporting*

Journalists are typically taught that for good quality reporting any news story should cover its five W-s: the "Who," "What," "When," "Where," and "Why" of the recounted event. These five W-s are considered the basic line of questioning for information-gathering and problem-solving, with another H-question (for How) sometimes added as well. Kovach and Rosenstiel reinforced and reframed this mantra of five W-s for investigative reporters in their book, *Blur* (2010). The book is aimed at anyone who more generally wants to know what is true in the age of information overload. These veteran journalists present a systematic process for analyzing the content and media by asking reporters to ask themselves the following six questions:

1. What kinds of content am I encountering?
2. Is the information complete; and if not, what am I missing?
3. Who or what are the sources, and why should I believe them?
4. What evidence is presented, and how was it tested or vetted?
5. What might be an alternative explanation or understanding?
6. Am I learning what I need to? (Kovach & Rosenstiel, 2010).

They suggest that in the face of any novel information, the first step is first and foremost to orient yourself, get to "know the neighborhood." This is accomplished by asking what kinds of content you are looking at, for example, raw information, news, propaganda, advertising, publicity, or entertainment (Kovach & Rosenstiel, 2010, p. 34). The rest of the questions put a high value on completeness (Question 2), source credibility (Question 3), and proof of evidence (Question 4) that would allow news readers to put the events into a context that gives meaning. This is a kind of storytelling that is well researched and substantiated.

These questions constitute a slight departure from the simpler five W-s mantra toward a more systematic content analysis. The issue of the source credibility (Question 3) was particularly dire during the Trump 2016–2020 Administration in the U.S. As Luke Burns, a reporter for the *New York Times* elegantly put it, that in addition to the five "W"-s, reporters also needed to start asking questions such as "Seriously?" and "Have you no shame?" (Burns, 2017). While postproduction fact-checking may be seen as the ultimate quality assurance procedure in journalism, the verification process overall is rooted in the conscientious questioning of sources and a healthy dose of skepticism about the information at hand from the start of the journalistic investigation. Unsurprisingly, these six concerns—although framed in journalism as essential to good-quality reporting—closely resonate with credibility assessment methods in LIS and health informatics (see Chap. 3) and justice system investigations (see Sect. 5.2. above.)

5.3.3 *Fact-Checking within Investigative Journalism and as a Newly Emerged Genre*

Fact-checking as an activity is portrayed as central to the practice of (investigative) news reporting. What journalism textbooks prescribe and what reporters actually do, however, may not necessarily be one and the same. "While a concerted quest for accuracy is seen by many journalists as central to their professional identity, informal rules of practice for achieving news accuracy are elusive and highly nuanced" (Shapiro et al., 2013, p. 657). Meta-studies of journalistic practices attempt to establish what in fact they do for their fact-checking procedures, reconstructing in detail the process of verification. For example, from self-reported evidence obtained in interviews with 28 Canadian journalists, it was surmised that "there is considerable diversity in verification strategies, at times mirroring social scientific methods (source triangulation, analysis of primary data sources or official documents, semi-participant observation), and different degrees of reflexivity or critical awareness of journalists' own blind spots and limitations" (Shapiro et al., 2013, p. 657). Specifically, proper names, numbers, locations, and some other concrete details were verified with greater care than some other types of factual statements. Perhaps, broader and more substantive factual assertions may go unchecked due to either a lack of expertise, time, resources, or the willingness to verify them. Shapiro et al.'s (2013) findings also attested vagueness in the verification criteria, though adherence to the spirit of verification was frequently mentioned and highly praised.

It is customary for journalists to rely on their sources, be they knowledge experts, event participants, victims, or eye-witnesses. If reporters are reluctant to make a judgment on whether their sources' substantive assertions stand up to scrutiny, they simply practice what has been dubbed by Tuchman (1972) "a strategic ritual" which they use defensively against accusations of subjectivity.[5] This reliance on direct quotations "defers responsibility for accuracy and truthfulness to the source, and takes refuge in what is often (especially in the U.S.) disparagingly referred to as 'he said, she said journalism'" (Birks, 2019).

---

[5] Tuchman (1972) notes that other professions and occupations (such as doctors and lawyers) also often appeal to the notion of *objectivity* that is understood operationally to "equate objectivity with the ability to remain sufficiently impersonal to follow routine procedures appropriate to a specific case" (p. 677). From the ritualistic behavior point of view, "the formal attributes of news stories and of newspapers would appear to entail strategic rituals justifying a claim to objectivity. They enable a newsman to say, pointing to his evidence, 'I am objective because I have used quotation marks'" (p. 677). For social scientists, by contrast, objectivity roughly means *replicability*: Research procedures "are described with such explicitness that others employing them on the same problem will come to the same conclusions. In effect, then, this is a notion of objectivity as technical routinization and rests, at bottom, on the codification of the research procedures that were employed" (p. 677). "Objectivity is often considered to be an ideal for scientific inquiry, a good reason for valuing scientific knowledge, and the basis of the authority of science in society" (Reiss & Sprenger, 2020). See further discussion of replicability and transparency of the scientific method in Sect. 5.4.

The point of "the discipline of verification" is not limited to confirming that the source in fact stated what the news report purports them to have said, if it did, it would then fall short of separating facts from fiction. In investigative journalism, substantial claims should be verified through double sourcing, sources' credibility and agenda should be checked, and any problems of controversy and disagreement in sources' assertions should be corroborated (Birks, 2019). What constitutes corroboration of "pivotal facts" is quite ambiguous, as journalists decide how much discrepancy to tolerate between (presumed human) sources' accounts which in turn may be flawed in some way; across news genres, journalists for the most part are dealing with human sources rather than documents, statistics, or direct evidence (Birks, 2019).

A recently emerged separate genre of "fact-checking" journalism is a curious trend that emerged from outside the industry. In this genre, verification procedures are conducted post-factum after a news release or may be applied to corroborate or challenge politicians' claims, statistics, or other "factual assertions of interest to the general public." "Fact-checking" was the inevitable result of economic cuts and reorganizations on the newsroom vetting structure, perhaps in part due to the availability of no-longer-employed qualified freelance journalists, now "fact-checkers." The number of fact-checking organizations has grown worldwide in no small part due to the increased demand from audiences—netizens who are actively engaged in the news feedback loop, and who expect greater attention to be paid to the truth, accuracy, and precision of digital news. According to the database of fact-checking organizations maintained by *the Reporters' Lab* at Duke University, there were 304 initiatives in 84 countries counted in October 2020 (Stencel & Luther, 2020).

One of these organizations, The International Fact-Checking Network (IFCN), is run out of the Poynter Institute for Media Studies: A nonprofit journalism school and research organization located in St. Petersburg, Florida, that also owns the *Tampa Bay Times* newspaper and operates *PolitiFact* (Wikipedia Contributors, 2021). The IFCN promotes nonpartisan and transparent fact-checking as an instrument of accountability in journalism and condemns "unsourced or biased fact-checking [that] can increase distrust in the media and experts while polluting public understanding" (The International Fact-Checking Network (IFCN), 2016). There are five aspirational ethical commitments laid out in the IFCN Fact-Checkers' Code of Principles, namely, the commitment to (1) nonpartisanship and fairness; (2) transparency of sources; (3) transparency of funding and organization; (4) transparency of methodology; and (5) open and honest corrections (see also their website,[6] for further decoding of these principles). Originally launched in 2016, this Code provides guidance to journalistic or fact-checking "organizations that regularly publish nonpartisan reports on the accuracy of statements by public figures, major institutions, and other widely circulated claims of interest to society. It is the result of consultations among fact-checkers from around the world and offers conscientious practitioners principles to aspire to in their everyday work" (IFCN 2016).

[6] See https://www.poynter.org/ifcn-fact-checkers-code-of-principles/ (Accessed on May 14, 2021).

Within the industry, there is an ongoing debate about the newly emerged genre of fact-checking. "Journalists and editors are consciously reluctant to incorporate fact-checking into their general practice, not for practical, resource-related reasons, but because accusing a politicians of lying looks too much like bias" (Birks, 2019). (See also Chap. 2 for Sect. 2.6.1 on content moderation and Chap. 3 for Sect. 3.5 on credibility indicators in fact-checking).

5.3.4 Newsworthiness, Sensationalism, and "Yellow Journalism"

Ideally, once journalists have their verified information from their vetted credible sources, there is still a story to tell. News—as a constructed story—is guided by both the reporters' knowledge of the truth at the time and by the story's newsworthiness. When the value of newsworthiness supersedes the publication's focus on truthfulness and verifiable facts, news stories can be presented provocatively, without any regard for the journalistic ethic of adhering to verified facts. The misinformative tactics of distortion, fabrication, and exaggeration make for sensationalized news and are primarily concerned with increased print circulation or its equivalent popularity metric in digital media.

Sensationalism is an American term from the times of early print media news circulation in the 1890s. It describes the exaggeration and dramatization of news stories in order to capture the attention of the reader by appealing to their emotions and by deceiving them (Editors of Wikipedia, 2021). Newspapers at the end of the 1890s were vital sources of information and news publishers had the power to mold and shape public knowledge and opinion. William Randolph Hearst, the owner of the *San Francisco Examiner*, in a competition for readers with Joseph Pulitzer's newspaper, was the first to devote over a third of his newspaper to stories of crime and corruption which he framed as dramatic moral dilemmas (Editors of Encyclopedia Britannica, 2021). The origin of the then-coined term *yellow journalism*—referring to the circulation of news with no legitimate value or which has little or nothing to do with verified facts—is typically associated with Hearst's name, as documented in historical cartoons (see Fig. 5.5) (Huynh & Balcetis, 2014). The deceptive tactics of sensationalism and fabrication in yellow journalism were then (and still now) used to increase circulation numbers without much regard for the truth.

The name comes from the *Yellow Kid* comic strip character (see Fig. 5.6a), created and drawn by Richard F. Outcault for purely entertainment purposes, which was published in one of the first Sunday supplement comic strips, *Hogan's Alley*, in 1890s American newspapers (PBS Online Documentary Film, 1999).

The circulation war between Joseph Pulitzer's *New York World* and William Randolph Hearst's *New York Morning Journal* in the late 1890s, was also marked by scandal-mongering, exaggeration, and other sensationalizing tricks such as the use of misleading visuals and headlines (Editors of New World Encyclopedia, 2021;

Fig. 5.5. "The Yellow Press," by L. M. Glackens, portrays William Randolph Hearst as a jester distributing sensational stories. (Accessed at http://www.newworldencyclopedia.org/entry/William_Randolph_Hearst on February 20, 2020)

Fig. 5.6 Yellow (Kid) Journalism
(**a**) Richard F. Outcault's Yellow Kid Comic Character. (Accessed at https://en.wikipedia.org/wiki/The_Yellow_Kid on May 20, 2020). (**b**) Pulitzer and Hearst shown as Yellow Kids Competing for the Ownership of the 1898 Spanish-American News. (Photograph by Leon Barritt; U.S. Library of Congress Prints and Photographs Online Catalogue, accessed at https://www.loc.gov/pictures/item/95508199/ on May 20, 2020)

Editors of Wikipedia, 2021). The "Yellow Journalism" cartoon shows Pulitzer and Hearst, both attired as Yellow Kid comics characters, competitively claiming ownership of the Spanish-American war of 1898 (Fig. 5.6b).

During the civil uprising in Cuba in 1896, the Spanish government sent General Valeriano Weyler to quash the rebellion; it was overrun with an overwhelming number of Spanish troops and Cuban people were detained in concentration camps, devoid of basic necessities and resulting in numerous deaths. While this was an egregious situation, Hearst's and Pulitzer's newspapers (the *New York Morning Journal* and the *New York World*, respectively) were competing with overly dramatized sensational accounts of the conditions in Cuba, accompanied by unrealistic drawings. Frederick Remington, an artist for Hearst's *Morning Journal* who went to Cuba to illustrate events, recounted a telegraph to Hearst: "… there is no trouble here. There will be no war." Hearst is accused of replying with, "Please remain. You furnish the pictures and I'll furnish the war" (Huynh & Balcetis, 2014, p. 437). The Spanish-American War is often referred to as the first "media war" (PBS Online Documentary Film, 1999). These acts of yellow journalism, credited with pushing the United States into war with Spain, are among the most notoriously deceptive tactics in the history of journalism and have impacted modern day journalistic ethics (Huynh & Balcetis, 2014).

5.3.5 News Literacy for all

Debunking unethical journalism in the age of information overload is proving to be difficult. Even if newsrooms issue corrections, they are largely ineffective due to the speed of news propagation on social media and the fact that readers are more likely to retain frequently repeated information, even if it is false. Corrections fail in many ways: They neither spread as "fervently" as the initial falsehoods, nor do they reach the same audiences (Silverman, 2015, p. 48).

The system used for safeguarding traditional newsrooms against errors was in additional layers of verification by the fact-checkers, copy editors, and editors. Now it has either been eroded or outsourced due to the budgetary constraints in news production, the increasing demands for popularity, and the rush to be first to break the news. Due to the current weakening of verification mechanisms within news-making organizations and the time pressures in news production, the function of fact-checking is often outsourced to external organizations such as Snopes.com or Politifact.com. To the best of my knowledge, such organizations employ human fact-checkers and content moderators, but also sometimes turn to AI as a means of verification (see also Chap. 3 for credibility assessment practices).

The unfortunate situation is that the onus of verification may ultimately rest with news readers. Given the digital media landscape (see Chap. 1), every news reader and/or social media user is well advised to ask the same kinds of questions as any ethical journalist should. The process of vetting sources, verifying assertions, and interrogating the validity of the reported facts, images, and attributed quotations has now become an essential part of news literacy (see Chap. 2, Sect. 2.6 for further discussion of literacy education as an intervention for being misinformed).

5.3.6 *Implications for AI*

Were journalistic verification or fact-checking procedures to be emulated with AI, more concrete procedural details would need to be worked out, since these practices are highly varied beyond verifying spellings, locations, names, and other super-specific details. General principles are laid out at the aspirational level, and by and large they are adhered to by conscientious practitioners, but little is offered in the industry by way of concrete step-by-step guidance. Such instruction is essential for algorithmic implementations and the next steps to implement journalistic AI would be to cross-check the known steps journalists take in their trade to ensure verification and apply modern NLP technologies. Procedurally, journalism leaves much room for creativity and story construction, perhaps as an art form like any other writing, but jounralistic truths are written with an obligation to adhere to the facts.

Fact-checking organizations that verify already existing factual claims and assertions (in the news or elsewhere posted digitally) have worked out agreed-upon procedures for verification and indicators of credibility. These sets of concrete questions are moving toward standardization and are now accepted internationally in some collaborating fact-checkers' networks (see Chap. 3 for Zhang et al.'s (2018) content and context indicators of information quality at the article level). If these "journalistic secrets" were to be further elicited, described as concretely as possible, and thoroughly systematized, they could be very promising for automation, perhaps a goldmine for AI.

Fact-checkers' websites are often used by the AI development community as a source for data-mining and constructing "gold standards"[7] for machine learning, and for technical competitions between systems built to separate truth from deceptive/misleading information. However, little attention is paid to how "messy" these data are, what assumptions they are produced with, or what they actually contain (for example, just news stories or additions of fact-checkers' rendition of events, corrections, assessments, or explanations).

Even more alarming is the use of fact-checkers' data to develop lie-detection technology that is built on the law enforcement assumption that an interviewee contributes directly to composing the testimony with either a complete awareness of the truth, or by making it up and imagining it, and thus falsifying (see Sect. 5.2 above). Fact-checker narratives are not remotely similar to law enforcement testimonies that use direct speech describing directly witnessed or imagined events. For an example,

[7] 'A gold standard' for an NLP system development refers to a dataset, or a collection of texts in this case, that are accepted as a reliable standard against which system decisions can be verified. For AI systems like disinformation detectors, such gold standards should contain positive and negative instances of disinformation, as confirmed by human judges.

let us look at Snopes.com debunking a claim that Ted Cruz, a politician serving as the junior United States Senator for Texas, has allegedly eaten a fly.[8] There is a certain structure to the site's presentation of the fact-checker's post: its *title*, *lead* sentence (quite funny in this case), the fact-checker's *name*, *date* posted, and an *image* with its *credit*. The substance of the *claim* indicates the substance of the doubt and fact-checking investigation, immediately followed by the verdict: The rating is "false" in this case. Further, the post offers a restatement of the *origins* of the report including the trigger tweet. The fact-checkers carefully choose their wording and hedge what "supposedly" happened, in best journalistic traditions of signaling an allegation prior to its verification. It is then immediately followed by a definitive verdict, with an almost clipped tone and the simplest language imaginable that gets straight to the findings of verification, presumably for those in a hurry to click out of the sight: "This is not a genuine video of the Texas politician eating a fly. This video was digitally edited to make it look as if there was a fly on Cruz's face (and then in his mouth). In the original video, you'll see no fly. This clip comes from an appearance on the Fox News show "Hannity" on June 27, 2019, when the senator had to take a sip of water to clear his throat. While we don't know why Cruz's voice was hoarse at this moment, it wasn't because he ate a fly" (as per https://www.snopes.com/fact-check/did-ted-cruz-eat-a-fly/ accessed on May 28, 2021).

Only once you scroll past these mandatory metadata elements do you see further details about Senator Cruz taking a sip of water to clear his throat, and an explanation of a pattern with other instances of common flies stealing the spotlights of politicians in real and doctored videos. This is but one example which demonstrates the presence of a predefined metadata structure that fact-checkers may be required to follow, a repetitiveness in restatements within the fact-check post, the addition of "own wording" to the original claim (the trigger tweet), and simplicity in language that has the target audience at heart (respectful of their time and option to get the scoop or read further).

Much of AI-based lie-detection technology (as we will see in Chap. 7) relies on the stylistic content and the pragmatic use of the language, and mere statistical predictions of the likelihood of the patterns in the speech of truth tellers versus deceivers' speech. Fact-checkers' narratives, as described above in the Snopes post, do not appear to be suitable for the direct application of these systems without undergoing prior data cleaning of a dataset collected by simply crawling this or similar fact-checkers' websites.

AI can certainly be trained to distinguish in a binary fashion between two kinds of texts, including those scooped from fact-checkers' results and deemed to be false and those deemed accurate. Be mindful, however, that in such cases, there is no solid theoretical background as to whether the two groupings of fact-checkers narratives that AI could distinguish can be interpreted as simply truthful or deceitful (as in traditional lie detection).

---

[8] See https://www.snopes.com/fact-check/did-ted-cruz-eat-a-fly/ (accessed on May 28, 2021).

5.3.7 Lessons Learned from Journalistic Practices

In sum, veteran journalists urge beginners in their trade and news readers alike to develop "skeptical ways of knowing" in order to deal with uncertainties and ambiguities in the digital news. Journalists should continue asking the right questions, honing on who, what, when, where, how, and why in their news reports. They should double-check how the evidence was presented, whether it was tested or vetted, and what possible alternative explanations or understandings there may be. Meanwhile, we are all are advised to question the kind of content we encounter, particularly for its completeness and sources, as part of news literacy.

A quick sidestep into the history of "yellow journalism," a term coined in the circulation war between newspaper magnets Hearst and Pulitzer in the late 1890s, reveals its tactics—sensationalism, scandal-mongering, exaggeration, and fabrication. Debunking unethical journalism is proving to be difficult. Even if newsrooms issue corrections, they are largely ineffective due to speed of social media's news propagation and the fact that readers are more likely to retain frequently repeated information, even if it is false.

Straightforward automation of fact-checking procedures and investigative reporting verification may prove to be too difficult due to insufficient specificity and transparency of journalists' procedures. Some aspects of fact-checking—such as verifying the concrete details of whether someone in fact said what the news reports claim, name spellings, dates, locations, and other concrete specifics—are more amenable to automation. The grander art of verification when seeking the day-to-day journalistic truths may be harder to accomplish with AI. This elusive journalistic art appears to hold a great promise for AI. Much like any other type of wholistic approaches that require general intelligence, it is currently inaccessible to specialized AI tools.

5.4 Scientific Method of Inquiry

> *"Scientists must keep reminding society of the importance of the social mission of science—to provide the best information possible as the basis for public policy. And they should publicly affirm the intellectual virtues that they so effectively model: critical thinking, sustained inquiry and revision of beliefs on the basis of evidence."*
> (Higgins, 2016)

Of all the kinds of systematic inquiry known to humanity, science is arguably the most successful one and it stands out for its rigorous methodology. "Most scientists would not say that science leads to an understanding of the truth. Science is a determination of what is most likely to be correct at the current time with the evidence at our disposal… The scientific method, it could be said, is a way of learning or a process of using comparative critical thinking" (McLelland, 2021, p. 1). This section briefly describes the underlying principles of the "scientific method" and the basic steps of the process, familiar to any scientist. Many of us in the North American (and many other) educational systems have learned its basics in our first (natural or empirical social) science classes in elementary schooling. The operational nuances are often tackled more scrupulously in methodology classes as prerequisites to graduate research, reinforcing the skills for the kinds of comparative critical thinking necessary to advance scientific knowledge. So, in a nutshell, what is it that scientists do, and how do we establish facts or find out the most likely truths about the world using the scientific method?

5.4.1 Scientific Process

Any empirical scientific enterprise requires carrying out a definitive series of steps or procedures, prescribed to a great degree by disciplinary tradition, but they are invariable in their foundation. Procedurally, a researcher starts with an observation of some phenomenon, regularity, or a pattern that they can make a prediction about. This insight allows for a conjecture, also called a hypothesis, which leads to logical consequences, were the conjecture(s) to be true. A hypothesis in science is a carefully constructed explanation of the phenomenon or a set of phenomena, and it must be, in principle, falsifiable. After a hypothesis, or a set of them, are formulated, they are then tested by researchers using discipline-appropriate and content-specific measures. Such tests often take the shape of controlled experiments but are not strictly limited to those methodological practices. Researchers then collect and analyze the resulting data so that they can establish whether the empirical observations conflict or agree with the outcomes predicted by the hypothesis, or the set thereof. Thus, "in a typical application of the scientific method, a researcher develops a hypothesis, tests it through various means, and then modifies the hypothesis on the basis of the outcome of the tests and experiments. The modified hypothesis is then retested, further modified, and tested again, until it becomes consistent with observed phenomena and testing outcomes" (Editors of Encyclopedia Britannica, 2021). A generic scientific inquiry typically is an iterative cycle of observations, testing, and refinement of hypotheses. The stress on refuting the hypothesis, stemming from Karl Popper's view, is essential to the scientific endeavor: No number of experiments can ever verify a theory or prove a law, but a single counterinstance, an exception, can conclusively refute the corresponding universal law (Thornton, 2018). (See also Chap. 4, Sect. 4.3.4 on pragmatist theories of truth).

Exact traditions of how empirical data are collected and analyzed and what constitutes appropriate procedures and possible interactions in the process are often predetermined by disciplinary affiliation or further enforced customs of research institutions, or even by an individual researcher's preferences and idiosyncrasies. Most of the time, however, the process itself is not a straightforward linear progression, but it often involves false starts, iterations, and sidesteps to be able to reach meaningful valid conclusions. "Scientific knowledge is necessarily contingent knowledge rather than absolute, and therefore must be evaluated and assessed, and is subject to modification in light of new evidence" (McLelland, 2021, p. 1).

Explanations of what happens in the scientific process are often reconstructed to smooth the progression of the steps, often by avoiding some routine iterations and by omitting irrelevant facts, to make the descriptive scientific narrative more succinct and to the point, while retelling a fair logical reconstruction of what was done and accomplished in the research. Such are the expected norms of scientific communication.

5.4.2 *Scientific Principles*

Much of scientific inquiry depends on specific traditions in the field, but a set of general scientific principles carry throughout. Abstracted from institutional practices, variations in the field's customs, instruments, and subjects of study, the sciences place a great emphasis on verifiability through testing and reliability of results. We will discuss three aspects of verifiability in turn: the falsifiability, reproducibility, and replicability of scientific hypotheses. Each aspect can be considered a cornerstone to the empirical methodology of acquiring scientific knowledge.

5.4.3 *Falsifiability*

The idea of falsifiability refers to the capacity of a statement to be refuted, that is, to be contradicted by evidence. Hypotheses are essentially educated guesses by an expert based on years of training, research, and immersion in specialized literature. Hypotheses are formulated in such a way that the scientist can attempt to make them fail and prove them wrong rather than right. Falsifiability entails the ability of hypotheses to be tested. In other words, if hypothetical statements (or propositions offered) are not testable in some way, they would not be considered scientific. For example, supernatural explanations or phenomena (such as the speed of angels or density of ghosts), or broader, "anything that cannot be observed or measured or shown to be false is not amenable to scientific investigation. Explanations that cannot be based on empirical evidence are not a part of science (National Academy of Sciences, 1998)" (cited in McLelland, 2021, p. 1).

Such a metaphysical stance is inextricably tied to realism as a philosophical position (see also Chap. 4 for overview of philosophical positions and truth theories). Empirical sciences are grounded in the tendency toward methodological naturalism whose adherents (such as realists) hold an assumption that the world is real, consistent, objectively knowable or discoverable, and that people can accurately perceive this reality and describe it logically.

5.4.4 Reproducibility (Transparency) and Replicability (Generalizability)

Reproducibility and replicability of experimental results are what make scientific findings transparent and reliable. Noticing the difference in the use of these two terms across disciplines, a cross-disciplinary Consensus Report (National Academies of Sciences, Engineering, and Medicine et al., 2019) surmises that confidence in scientific knowledge comes down to answering the following four questions about any reported scientific study:

1. "Are the data and analysis laid out with sufficient transparency and clarity that the results *can be checked*?
2. If checked, do the data and analysis offered in support of the result *in fact* support that result?
3. If the data and analysis are shown to support the original result, can the result reported be found again in the *specific study context* investigated?
4. Finally, can the result reported or the inference drawn be found again in a *broader set of study contexts*?" (Emphasis by the authors, numbering added here) (National Academies of Sciences, Engineering, and Medicine, 2019).

Reproducibility, used synonymously with *computational reproducibility* in computational sciences, refers to "obtaining consistent results using the same input data; computational steps, methods, and code; and conditions of analysis" (National Academies of Sciences, Engineering, and Medicine et al., 2019). In other words, if a study is reproducible, it contains enough information to confirm the study results: A study is transparent enough to be repeated by other researcher(s) or at another time (by the same researchers(s)). Thus, reproducibility is strongly associated with *transparency*, answering Questions 1 and 2 above.

Replicability, on the other hand, is about "obtaining consistent results across studies aimed at answering the same scientific question, each of which has obtained its own data" (National Academies of Sciences, Engineering, and Medicine et al., 2019). In other words, replication attempts are consequent studies aimed at the same or similar scientific research question(s). When a study is being replicated, it may be performed under different conditions, in different contexts, or with different study populations. Replicability is associated with another almost synonymous term, generalizability, answering Questions 3 and 4 above. Generalizability, sometimes used as a measure of replicability, for example by the U.S. National Science Foundation (NSF), refers to "the extent that results of a study apply in other contexts or populations that differ from the original one" (National Academies of Sciences, Engineering, and Medicine, 2019).

While reproducibility of results is expected as the norm of good scientific practice, non-replicability of results can occur for a number of reasons. Benign and even helpful unexpected non-replicable results can further our understanding and may extend to reveal previously unforeseen sources of variability. "Science is, however, a human endeavor and is subject to personal prejudices, misapprehensions, and bias" (McLelland, 2021, p. 1). So, in some cases, non-replicability can occur due to errors, a lack of methodological diligence, built-in misperceptions, or even intentional academic fraud. The NSF advises its scientific community in its cross-disciplinary Consensus Report (National Academies of Sciences, Engineering, and Medicine, 2019) to assure the reliability of scientific knowledge specifically with "reviews of cumulative evidence on a subject, to assess both the overall effect size and generalizability" as measures for gaining confidence in replicability.

5.4.5 Systematicity

Philosophers and historians of science continue to debate the nature of science as its practices have morphed over the centuries and moved away from its more absolute roots in antiquity (of Plato and Aristotle's philosophies) to accommodate other perspectives on the world that account for the social, cultural, and political factors that inevitably shape scientific practice and its products. Some scholars have further suggested that what distinguishes contemporary science from other kinds of knowledge is its *degree of systematicity*. Citing Paul Hoyningen-Huene's (2013) main argument, Andersen and Hepburn (2015) write: "Systematicity can have several different dimensions: among them are more systematic descriptions, explanations, predictions, defense of knowledge claims, epistemic connectedness, ideal of completeness, knowledge generation, representation of knowledge and critical discourse. Hence, what characterizes science is the greater care in excluding possible alternative explanations, the more detailed elaboration with respect to data on which predictions are based, the greater care in detecting and eliminating sources of error, the more articulate connections to other pieces of knowledge, etc. On this position, what characterizes science is not that the methods employed are unique to science, but that the methods are more carefully employed."

5.4.6 Lessons Learned from the Scientific Method

Science takes great care in excluding possible alternative explanations, detecting and eliminating sources of error, elaborating on the data on which predictions are based with great detail, and for articulating connections to other pieces of knowledge. Some argue that scientific inquiry does not necessarily differ in kind from other kinds of inquiry, but rather in the degree of care it takes. It also requires broad and detailed background knowledge and familiarity with specialized vocabulary.

The lesson of scientific thinking is rational discernment by observing, asking questions, and seeking answers. It also teaches us to be flexible and open-minded. We need to be willing to refine our original predictions if they do not work out.

5.5 Conclusions

The predominant interviewing techniques and analytical procedures, overviewed in this Chapter, speak to their practitioners' abilities to assess the likelihood of truth and deception in written and spoken statements under the North American justice system. These lie detection and credibility assessment professional practices have thus far served as the most concrete and detailed methodologies for detecting deception automatically. Several state-of-the-art systems are presented in Chap. 7 with the aim of combatting online disinformation at scale.

Journalism has been under pressure to change its practices with recent newsroom production demands resulting from the industry going digital. The underpinnings of ethical reporting are laid out in many a journalism textbook and practical guide. Journalists are urged to pursue "skeptical ways of knowing" by asking "wh-" questions to examine the nature of the content they encounter and the reliability of their sources. The desire for sensationalism and the urgency to report on newsworthy stories may override the traditional newsroom fact-checking safeguards, processes that are now regularly eroding and can often be outsourced. There is a greater onus on the news readers to be news literate, ask similar "wh-" questions with a healthy dose of skepticism, traditionally reserved for investigative journalists. A relatively new genre of externally outsourced fact-checking is leading the way in establishing credibility criteria for the verification of sources and assertions, a potential goldmine ready for AI exploration. The nascent field of automated factchecking is exemplified in Chap. 7 with R&D efforts originating from computational labs worldwide.

While journalism presents us with the discipline of verification, science teaches us the discipline of mind. Thinking scientifically, we can discern facts rationally by observing, asking systematic questions, and seeking valid answers with greater care. We can be open to refining our original predictions in the face of facts. Science, as an immensely successful human enterprise, offers us "standards of good, strong, supportive evidence and well-conducted, honest, thorough and imaginative inquiry are not exclusive to the sciences, but the standards by which we judge all inquirers. In this sense, science does not differ in kind from other kinds of inquiry, but it may differ in the degree to which it requires broad and detailed background knowledge and a familiarity with a technical vocabulary that only specialists may possess" (Andersen & Hepburn, 2015). Following the principles of scientific methodology (rigor, objectivity, systematicity, transparency, replicability, and reproducibility) is necessary but not sufficiently detailed enough for potential AI application until it is applied to a concrete research problem. Applying the generic scientific method, AI R&D needs to further elaborate on its specific subtasks for combatting disinformation.

Each group of experts discussed in Chap. 5—investigators, journalists, and scientists—rely on their systematic methodologies to make professional calls. However, none of these inquiries—in the court of law, in journalism, or in scientific research—is infallible as a system. Many things can go wrong with an ever-present human component in these methods. Cases of false convictions, journalistic fraud, and academic dishonesty are failings of the systems or particular individuals, but their ramifications may go beyond the personal impact to those involved. They may derail proper justice, scientific discoveries, public awareness, and democracy. Their sum total rather undermines trust in these institutions of justice, damaging the credibility of dissemination channels and the validity of scientific discoveries for the public good.

On the beneficial side, what unites the three ways of thinking—by detectives, journalists, and scientists at their best—is the inquisitive critical mindset. Each best practice is armed with a system for a purposeful investigation to answer well-formulated questions, based on the experts' knowledge and the observable evidence. I argue in this book that such a systematic frame of mind, is a cornerstone to news and media literacy in approaching any novel information online and can help media users spot and mitigate the impact of encountering disinformation online.

For automation attempts with AI-based techniques—whether they are based on checklists from police interrogations, fact-checking credibility indicators, or the rigors of systematic scientific procedures—human perusal and validation of any AI-based conclusions is strongly advised. Again and again, I underscore that the ultimate decision should be in the mind of the user of the AI tools: Technology advises and assists us but never replaces the human judgment in determining what is truthful and what is disinformative.

References

Adams, S. H. (1996). Statement analysis: What do suspects' words really reveal? *FBI Law Enforcement Bulletin, 65*(10), 12.

American Polygraph Association. (2021). *Polygraph frequently asked questions*. Retrieved from https://apoa.memberclicks.net/polygraph-frequently-asked-questions.

American Press Institute. (2021). *What is journalism? Definition and meaning of the craft*. American Press Institute. Retrieved from https://www.americanpressinstitute.org/journalism-essentials/what-is-journalism/

American Psychological Association. (2004, August 5). *The truth about lie detectors (aka Polygraph Tests)*. Retrieved from https://www.apa.org. https://www.apa.org/research/action/polygraph

Andersen, H., & Hepburn, B. (2015). Scientific method. *Stanford Encyclopedia of Philosophy Archive* (Winter 2020 edition). Retrieved from https://plato.stanford.edu/archives/win2020/entries/scientific-method/

Axciton Systems, Inc. (2021). *What is a polygraph?* Retrieved from https://www.axciton.com/whatispolygraph.htm

Birks, J. (2019). Fact-checking, false balance, and 'fake news': The discourse and practice of verification in political communication. In *Journalism, power and investigation* (pp. 245–264). Routledge. https://doi.org/10.4324/9781315181943-17

Burns, L. (2017, January 31). *Additions to the five journalistic "W"s*. The New Yorker. Retrieved from https://www.newyorker.com/humor/daily-shouts/additions-to-the-five-journalistic-ws

Dean, W. (2021). *API's journalism essentials guide: The elements of journalism*. American Press Institute. Retrieved from https://www.americanpressinstitute.org/journalism-essentials/what-is-journalism/elements-journalism/

Editors of Encyclopedia Britannica. (2021). Scientific method: Definition, steps, and application. In *Encyclopedia Britannica*. Retrieved from https://www.britannica.com/science/scientific-method

Editors of Merriam-Webster's Dictionary. (2021). Definition of technology. In *The Merriam-Webster Dictionary*. Retrieved from https://www.merriam-webster.com/dictionary/technology

Editors of New World Encyclopedia. (2021). *Yellow journalism*. Retrieved from https://www.newworldencyclopedia.org/entry/Yellow_journalism

Editors of Wikipedia. (2021). Yellow journalism. In *Wikipedia*. Retrieved from https://en.wikipedia.org/w/index.php?title=Yellow_journalism&oldid=1000332663

Fitzpatrick, E., Bachenko, J. C., & Fornaciari, T. (2015). *Automatic detection of verbal deception*. Morgan & Claypool. https://doi.org/10.2200/S00656ED1V01Y201507HLT029

Frankfurt, H. G. (2005). *On bullshit*. Princeton University Press. Retrieved from https://press.princeton.edu/books/hardcover/9780691122946/on-bullshit

Government of Canada Tri-Agency: CIHR, NSERC, and SSHRC. (2015, June). Public communications policy of the Federal Research Funding Organizations. *Innovation, Science and Economic Development Canada*. Retrieved from https://science.gc.ca/eic/site/063.nsf/eng/h_711200B1.html.

Griesel, D., Ternes, M., Schraml, D., Cooper, B. S., & Yuille, J. C. (2013). The ABC's of CBCA: Verbal credibility assessment in practice. In B. S. Cooper, D. Griesel, & M. Ternes (Eds.), *Applied issues in investigative interviewing, eyewitness memory, and credibility assessment* (pp. 293–323). Springer. https://doi.org/10.1007/978-1-4614-5547-9_12

Guarino, B. (2019, June 24). Misinformation is everywhere. These scientists can teach you to fight BS. *Washington Post*. Retrieved from https://www.washingtonpost.com/science/2019/06/24/misinformation-is-everywhere-these-scientists-can-teach-you-fight-bs/.

Higgins, K. (2016). Post-truth: A guide for the perplexed. *Nature, 540*(7631). https://doi.org/10.1038/540009a

Hoyningen-Huene, P. (2013). Systematicity: The nature of science. In *Systematicity*. Oxford University Press. Retrieved from https://oxford.universitypressscholarship.com/view/10.1093/acprof:oso/9780199985050.001.0001/acprof-9780199985050.

Huff, D. (1993). *How to lie with statistics*. Illustrated edition. WW Norton.

Huynh, S., & Balcetis, E. (2014). Hearst, William Randolph. In T. Levine (Ed.), *Encyclopedia of deception* (Vol. 1, pp. 436–437). SAGE Publications. https://doi.org/10.4135/9781483306902

Jahng, M. R., Eckert, S., & Metzger-Riftkin, J. (2021). Defending the profession: U.S. journalists' role understanding in the era of fake news. *Journalism Practice*, 1–19. https://doi.org/10.1080/17512786.2021.1919177

Johnson, M. K., Foley, M. A., Suengas, A. G., & Raye, C. L. (1988). Phenomenal characteristics of memories for perceived and imagined autobiographical events. *Journal of Experimental Psychology. General, 117*(4), 371–376. https://doi.org/10.1037/0096-3445.117.4.371

Johnson, M. K., & Raye, C. L. (1981). Reality monitoring. *Psychological Review, 88*(1), 67–85. https://doi.org/10.1037/0033-295X.88.1.67

Köhnken, G. (2004). Statement validity analysis and the 'detection of the truth'. In P. A. Granhag & L. A. Strömwall (Eds.), *The detection of deception in forensic contexts* (pp. 41–63). Cambridge University Press. https://doi.org/10.1017/CBO9780511490071.003

Kovach, B., & Rosenstiel, T. (2001). *The elements of journalism: What newspeople should know and the public should expect* (1st ed.). Crown Publishers.

Kovach, B., & Rosenstiel, T. (2007). *The elements of journalism: What newspeople should know and the public should expect* (1st rev. ed., Completely updated and rev.). Three Rivers Press.

Kovach, B., & Rosenstiel, T. (2010). *Blur: How to know what's true in the age of information overload:* Vol. 1st U.S. Bloomsbury.

Kovach, B., & Rosenstiel, T. (2014). *The elements of journalism: What newspeople should know and the public should expect*. (Rev. and updated third edition.). Three Rivers Press.

Liu, B. (2012). *Sentiment analysis and opinion mining*. Morgan & Claypool Publishers. Retrieved from http://www.cs.uic.edu/~liub/FBS/SentimentAnalysis-and-OpinionMining.html

McLelland, C. V. (2021). The nature of science and the scientific method. *The Geological Society of America*. Retrieved from https://www.geosociety.org/documents/gsa/geoteachers/NatureScience.pdf

National Academies of Sciences, Engineering, and Medicine. (2019). Executive summary. In *Reproducibility and replicability in science: Consensus study report.* The National Academies Press. doi: https://doi.org/10.17226/25303.

National Academies of Sciences, Engineering, and Medicine, Policy and Global Affairs, Engineering Committee on Science, Board on Research Data and Information, Division on Engineering and Physical Sciences, Committee on Applied and Theoretical Statistics, Board on Mathematical Sciences and Analytics, Division on Earth and Life Studies, Nuclear and Radiation Studies Board, Division of Behavioral and Social Sciences and Education, Committee on National Statistics, Cognitive Board on Behavioral, & Committee on Reproducibility and Replicability in Science. (2019). Understanding reproducibility and replicability. In *Reproducibility and replicability in science*. National Academies Press (U.S.) Retrieved from https://www.ncbi.nlm.nih.gov/books/NBK547546/

National Academy of Sciences. (1998). *Teaching about evolution and the nature of science*. Doi: https://doi.org/10.17226/5787.

National Association for Civilian Oversight of Law Enforcement. (2021). *Innocence project*. National Association for Civilian Oversight of Law Enforcement. Retrieved from https://www.nacole.org/innocence_project.

Niehaus, S. (2008). Merkmalsorientierte Inhaltsanalyse [criteria-based content analysis]. In R. Volbert & M. Steller (Eds.), *Handbuch der Rechtspsychologie* (pp. 311–321). Hogrefe.

Pang, B., & Lee, L. (2008). Opinion mining and sentiment analysis. *Foundations and Trends in Information Retrieval, 2*(1–2), 1–135. https://doi.org/10.1561/1500000001

PBS Online Documentary Film. (1999). *Crucible of empire: The Spanish-American war*. Retrieved from https://www.pbs.org/crucible/frames/_journalism.html.

Reiss, J., & Sprenger, J. (2020). Scientific objectivity. In E. N. Zalta (Ed.), *The Stanford encyclopedia of philosophy (winter 2020)*. Metaphysics Research Lab, Stanford University. Retrieved from https://plato.stanford.edu/archives/win2020/entries/scientific-objectivity/

Sagan, C. (1996). *The demon-haunted world: Science as a candle in the dark*. Random House.

Sanders, K. (2003). *Ethics & journalism*. Sage.

Schmuhl, R. (1984). *The responsibilities of journalism*. University of Notre Dame Press.

Shapiro, I., Brin, C., Bédard-Brûlé, I., & Mychajlowycz, K. (2013). Verification as a strategic ritual. *Journalism Practice, 7*(6), 657–673. https://doi.org/10.1080/17512786.2013.765638

Silverman, C. (2015). Lies, damn lies, and viral content: How news websites spread (and debunk) online rumors. *Unverified Claims and Misinformation. Tow Center for Digital Journalism*. Retrieved from http://towcenter.org/wp-content/uploads/2015/02/LiesDamnLies_Silverman_TowCenter.pdf.

Social Sciences and Humanities Research Council of Canada (SSHRC). (2012, May 11). *Guidelines for effective knowledge mobilization*. Retrieved from https://www.sshrc-crsh.gc.ca/funding-financement/policies-politiques/knowledge_mobilisation-mobilisation_des_connaissances-eng.aspx

Sporer, S. L. (1997). The less travelled road to truth: Verbal cues in deception detection in accounts of fabricated and self-experienced events. *Applied Cognitive Psychology, 11*(5), 373–397. https://doi.org/10.1002/(SICI)1099-0720(199710)11:5<373::AID-ACP461>3.0.CO;2-0

Steller, M., & Köhnken, G. (1989). Criteria-based statement analysis: Credibility assessment of children's testimonies in sexual abuse cases. In D. Raskin (Ed.), *Psychological methods for investigation and evidence*. Springer.

Stencel, M., & Luther, J. (2020, October 13). Fact-checking count tops 300 for the first time. *Duke Reporters' Lab. Fact-Checking News Archives*. Retrieved from https://reporterslab.org/category/fact-checking/

Synnott, J., Dietzel, D., & Ioannou, M. (2015). A review of the polygraph: History, methodology and current status. *Crime Psychology Review, 1*(1), 59–83. https://doi.org/10.1080/23744006.2015.1060080

Editors of Encyclopedia Britannica. (2021). Scientific method: Definition, steps, and application. In *Encyclopedia Britannica*. Retrieved from https://www.britannica.com/science/scientific-method

The International Fact-Checking Network (IFCN). (2016). International fact-checking network fact-checkers' code of principles. *Poynter*. Retrieved from https://www.poynter.org/ifcn-fact-checkers-code-of-principles/

Thornton, S. (2018). Karl Popper. In E. N. Zalta (Ed.), *The Stanford encyclopedia of philosophy (fall 2018)*. Retrieved from https://plato.stanford.edu/archives/fall2018/entries/popper/#ProbDema

Tuchman, G. (1972). Objectivity as strategic ritual: An examination of Newsmen's notions of objectivity. *The American Journal of Sociology, 77*(4), 660–679. https://doi.org/10.1086/225193

Undeutsch, U. (1954). Die Entwicklung der gerichtspsychologischena Gutachtertätigkeit. [the historical development of the use of expert psychological testimony]. In A. Wellek (Ed.), *Bericht uber den 19, Kongress der Deutschen Gesellschaft für Psychologie*. Verlag für Psychologie.

Undeutsch, U. (Ed.). (1967). *Beurteilung der Glaubhaftigkeit von Aussagen [veracity assessment of statements]* (Vol. 11). Hogrefe.

Volbert, R., & Steller, M. (2014). Is this testimony truthful, fabricated, or based on false memory?: Credibility assessment 25 years after Steller and Köhnken (1989). *European Psychologist, 19*(3), 207–220. https://doi.org/10.1027/1016-9040/a000200

Vrij, A. (2005). Criteria-based content analysis: A qualitative review of the first 37 studies. *Psychology, Public Policy, and Law, 11*(1), 3–41. https://doi.org/10.1037/1076-8971.11.1.3

Wikipedia Contributors. (2021). Poynter institute. In *Wikipedia*. Retrieved from https://en.wikipedia.org/w/index.php?title=Poynter_Institute&oldid=1011406393.

Zhang, A. X., Robbins, M., Bice, E., Hawke, S., Karger, D., Mina, A. X., Ranganathan, A., Metz, S. E., Appling, S., Sehat, C. M., Gilmore, N., Adams, N. B., Vincent, E., & Lee, J. (2018). A structured response to misinformation: Defining and annotating credibility indicators in news articles. *Companion of The Web Conference 2018 on The Web Conference 2018—WWW '18*, 603–612. doi: https://doi.org/10.1145/3184558.3188731.

Chapter 6
Manipulation in Marketing, Advertising, Propaganda, and Public Relations

Propaganda "regiment[s] the public mind every bit as much as an army regiments the bodies of its soldiers."
(Bernays, 1928, p. 25).

Virality has become the holy grail of digital marketing
(Akpinar & Berger, 2017, p. 318)

[*]*Cui bono?*

Abstract Many practices in marketing, advertising, and public relations, presented in **Chapter 6**, have the intent to persuade and manipulate the public opinion from the onset of their endeavors. I lay out marketing communications strategies and dissect the anatomy of the ad revenue model. I review key ideas in advertising standards and self-regulation policies that do not allow misleading ads. Advertising techniques in marketing campaigns and political propaganda abound in truth-bending, but regulatory bodies such as the U.S. Federal Trade Commission distinguish between puffery and materially harmful misleading ads. Digital media users may be constantly bombarded with ads—puffed-up, borderline deceptive, emotionally suggestive, or plainly misleading— and they may grow weary, distrustful, and resistant to ads. Advertisers get more creative with variants of covert advertising exemplified in Chap. 6 with native ads, sponsored links, or branded content. I discuss the elusive concept of virality, and explore what makes us vulnerable to viral conspiracy theories. Masters of persuasion may exploit human biases and logical fallacies including the bandwagon appeal, glittering generalities, and bait and switch, to name a few. Propagandists, advertisers, and public relations experts know when and how to appeal to emotions or individuality, use wit and humor, or finetune who delivers the message as a relatable credible source. More AI countermeasures and stricter regulation are needed to curb the existing ad revenue model and unscrupulous financing that instigates the spread of mis- and disinformation. AI-based technologies such as spambots, paybots, autolikers, and other inauthentic accounts are farmed out to create hype and social engagement by propagating falsehoods, clickbait, provocations, misleading, or otherwise inaccurate messages. Policymakers and legislators are broadly encouraged to focus on regulating algorithmic transparency, platform

[*]This phrase in Classical Latin translates into English as "Who benefits?" or "To whom is it a benefit?"

V. L. Rubin, *Misinformation and Disinformation*,
https://doi.org/10.1007/978-3-030-95656-1_6

accountability, digital advertising, and data privacy, while avoiding crude measures of controlling and criminalizing digital content or stifling free speech. The ultimate goal is to reestablish trust in the basic institutions of a democratic society by bolstering facts and combatting the systematic efforts at devaluing truth. Professional manipulators, propagandists, and their unscrupulous technologies— at the service of few "deep pockets— require more public oversight of their unscrupulous disinformation campaigns that misuse commerical and public discousre to manipulate the general public. Establishing a global regulatory framework and venues for enforcement may be key to addressing the problem across nations, cultures, and language boundaries. Nation-wide educational efforts in digital literacy should meanwhile instruct digital media users in recognizing manipulative techniques to resist the powers of propaganda and advertising.

Keywords Persuasion · Manipulation · Influence · Truth bending · Advertising · Public relations · PR · Marketing communications · Digital advertising · Digital marketing · Internet marketing · Search engine marketing · Paid-search marketing · E-mail marketing · Social media marketing · Online PR · Digital PR · DAGMAR · AIDA · Puffery · Misleading ads · False advertising · Advertising standards · Regulatory antitrust bodies · Covert advertising · Native advertising · Native ads · Advertorials · Sponsored content · Sponsored links · Branded content · Propaganda · Persuasive tactics · Logical fallacies · Conspiracy theories · Propaganda wars · Mass propaganda · Commercial campaigns · Political campaigns · Advertising campaigns · Virality · Salesmanship in print · Advertising paradigms · Marketing automation · Programmatic advertising · Online behavioral advertising · Data-driven marketing · Targeted marketing · Chat bots · Paybots · Social bots · Autolikers · Inauthentic accounts · Fake accounts · Rumor detectors · Rumor busters · Ad-blockers · Regulation · Content take-downs · Advertising transparency · AI transparency · Removal · Blocking · Filtering · Surveillance · Data protection measures · Legal framework redefinition · Bills · Revisions · Advocacy · Watchdogs · Private sector mobilization · Media accreditation · Journalistic controls · Parliamentary inquiries · Congressional hearings · Cybersecurity · Monitoring · Reporting

6.1 Introduction

Marketing and advertising communications, political propaganda, and public relations aim to influence public opinion by presenting their views, products, and beliefs in the most favorable light to benefit their goals. Their messages often mimic the ways in which scientific, journalistic, or law enforcement inquiries present their findings. However, ads and marketing claims should certainly not be confused with reported science, verified news reports, or factual verdicts. Scientists typically aim to inform the public of their justified true beliefs which they formed based on strong evidence and specialized knowledge. Marketers and advertisers aim, rather, to persuade the public about what to think and how to act, no matter how justified their claims are, and at times, in contradiction with the advertising industry codes and standards that prohibit false or misleading advertising practices.

Advertising communications as a marketing strategy and promotional tool aims to persuade the public and, at times, entice them to act. Many formats of digital marketing and advertising are in existence and have become increasingly popular due to the ease with which they reach large audiences and the mechanisms that allow them to target individual users. Longstanding advertising paradigms of persuasion, seduction, spin, and showbiz are now combined with the capabilities of new technologies and the infrastructure of social media platforms. Marketers compete to overcome consumer[1] resistance and indifference.

I lay out several marketing communications strategies and the anatomy of the ad revenue model. The mechanisms for capturing digital media users' attention underpin the existence of the so-called attention economy: users' "eye-balls" are commodified through the time they spent on various apps and platforms, allowing advertisers to promote their content under various payment models. In an effort to keep their users' attention, advertising tactics evolve to make promoted content irresistible. Manipulative tactics drip messages in our ears to inform us of enticing products, covertly promote brand names, or instill beliefs and perceptions of reality to advance the agenda that is paid-for. Sensational content and conspiracy theories shift the user experience of digital media toward perpetual attempts at manipulation and getting users addicted to online social engagement.

The classic manipulative and persuasive techniques of mass propaganda go back a hundred years to the pioneers of propaganda and public relations. In the 1920s, propaganda was seen as enduring efforts in creating circumstances, shaping events, and instilling pictures in the minds of millions to influence the public's relations to an enterprise, idea, or group. In the 2020s, to achieve a similar effect in commerce, marketing campaigns continue to exploit mass insecurities, fears, and unfulfilled desires. Political campaigns direct their attention to inflating personas and branding policies. The persuasive tactics may be essentially interchangeable in political and commercial campaigns, often with a deceptive, misleading, or otherwise manipulative slant. Techniques in these areas may require creativity, a mastery of language, and knowledge of psychological triggers and human biases to achieve widespread impact, but their inventory is surprisingly limited.

Persuasive means of achieving organic virality compete with clever automated means of faking user engagement. AI-based technologies such as spambots, paybots, autolikers, and other inauthentic accounts are farmed out to create hype and propagate content. What sells best is often promoted by including falsehoods,

[1] *Consumers* is a common term referring to people as targets of marketing and advertising in the digital media, or potential customers with purchasing powers. Terminologically, my personal preference lies with terms *digital media users*, *news readers*, or simply *us*, *people* who engage with digital technologies. For the sake of simplicity, I will continue using of the term *consumers* when reviewing the specialized literature, and especially for direct quotes, but will replace it with less consumeristic equivalents, whenever possible.

clickbait, provocations, misleading statements, or otherwise inaccurate messages. Chap. 6 discusses some potential AI countermeasures and stricter legislations and regulations that are needed in this AI-infused tug-of-war to stop and deter marketing and advertising communications from instigating the spread of mis- and disinformation as part of income-generating digital business strategies (e.g., ad revenue model).

6.2 Marketing and Advertising

6.2.1 Marketing Communications as a Promotional Tool

Experienced marketing practitioners list advertising as part of marketing communications, alongside other promotional tools such as sales promotions, public relations, sponsorship and celebrity endorsement, direct marketing, digital marketing, and personal selling (e.g., in Shaw, 2018), as well as events and experiences, and (electronic) word of mouth (WOM) marketing (Bianchi, 2020). According to the model known by its acronym (DRIP) marketing communications serve four distinct purposes: to differentiate, reinforce, inform, or persuade (see elaborations in Table 6.1. as per Shaw (2018)).

The idea of marketing for any organization is to communicate to its customers or stakeholders its "consistent image" with preplanning and integration across channels of communications: "While companies may target a number of different audiences and stakeholder groups using a range of communications channels and media, they need to represent a consistent set of values and core messages to establish their identity and positioning so that all audiences know who they are and what they stand for. This is known as *integrated marketing communications* (IMC). If organisations present very different images to their customer groups this undermines credibility and leaves positioning unclear. Organisations must develop a strong positioning, which supports their overall marketing strategy, and resonates with their target audiences" (Shaw, 2018, p. 142). Promotional tools to accomplish these marketing goals can be employed across a range of media including print, television, radio, and digital channels, as well as personal communications.

Table 6.1 Four purposes of marketing communications. [Adapted from Fill (2013, p. 15), as cited in Shaw, 2018]

Task.	Explanation
Differentiate	To make product or service stand out from the competition
Reinforce	To consolidate previous customer experiences and messaging
Inform	To draw attention to product or service and its features
Persuade	To move potential customers toward purchase decision or further enquiry

6.2.2 Digital Advertising[2] Formats

In practice, *digital marketing*[3] formats are increasingly popular due to their immediacy and their large audience reach. These formats encompass several techniques, each fiercely competitive and potentially lucrative. To decode the marketing and advertising jargon, let us briefly go over digital advertising activities, often referred to in the literature in accordance with the technologies or media they use, following the order of terminology in Fig. 6.1 compiled by Shaw (2018) from previous research.

Internet advertising is, just as it sounds, advertising using various internet affordances such as directing web traffic from search engines based on keywords. The idea is to encourage users to click on links leading to promoted websites. In online marketing communications having a website constitutes "an online presence," considered an essential element. A website is the modern "business card" of an organization or brand which fulfills crucial functions for every company, such as conveying information, building an image, and facilitating sales (Bianchi, 2020). Once any website is up online, *search engine marketing (SEM)* is used to generate traffic to it. SEM refers to "promoting an organisation through search engines to meet its objectives by delivering relevant content in the search listings for searchers and encouraging them to click through to a destination site. The two key techniques of SEM are *search engine optimisation (SEO)* to improve results from the natural listings, and *paid-search marketing* to deliver results from the sponsored listings within the search engines" (Chaffey & Ellis-Chadwick, 2018, p. 484). SEO involves achieving the highest position or ranking in the natural or organic listings on the main body of the *search engine results pages (SERPS)* in response to combinations of keywords entered by the search engine user. *Paid search (pay-per-click) marketing (PPC),* similarly to conventional advertising, displays a link to a company page in a relevant text ad when the user of a search engine types in a specific phrase (usually labeled as "sponsored links" to distinguish them from "natural listings" (Chaffey & Ellis-Chadwick, 2018). "In a nutshell, in paid search a company pays for clicks on the link to its website that appears at the top of search results. The average *click-through rate (CTR)* for Google AdWords is 4.1%, meaning that only four out of a hundred users click on a sponsored link in Google (Chaffey & Ellis-Chadwick, 2018). This rate can seem low, but the average CTR in paid search is actually one of the highest among the elements of online marketing communications" (Bianchi, 2020, Chap. 2, p. 13). Paid search delivers immediate results, has a high reach, and allows for precise audience targeting: for example, by demographic, geolocation, or device used,

[2] *Internet advertising* is also referred to as *online advertising* or *digital advertising*, used interchangeably here.

[3] *Digital marketing* and *e-marketing* are used interchangeably in marketing communications, emphasizing the electronic nature of the tools and activities.

Activity type	Explanation
Internet advertising PPC, display ads	Advertising using the internet to increase website traffic by encouraging recipients to click through to a web page. This includes display advertising on home pages, social media advertising on Facebook and Instagram, rich media advertisements, pay-per click and post-roll advertising on YouTube.
Search marketing	Techniques used to raise the profile of a webpage in the results of search engines. Known as Search Engine Optimization (SEO), this includes paid entries in search engines, contextual advertising and ensuring that web pages contain frequently used search words.
Email marketing	The use of email and other electronic means to send marketing communications directly to targeted groups. This is used to encourage repeat purchase and build relationships with customers. This can be a very effective way of targeting specific customer groups, but can only be used where customers have opted-in to receive electronic communications.
Mobile marketing	Similar to email marketing but using mobile devices such as smart phones to engage with an audience often with interactive content.
Viral marketing	Online marketing intended to encourage consumers to pass a message on to others, hence the expression 'gone viral'. Often used with entertaining or emotional content such as YouTube clips, promoted tweets, Buzz feed.
Online retailing	Selling to customers online from a website for delivery or collection from a local store.
Advergaming	Online games and videos used to promote a product or brand. Also used by not-for-profit organisations to promote causes and ideas.
Social media marketing	Using social networking sites for marketing purposes either by generating content that users will share or by building engagement with target groups through interaction with the brand.
Content marketing and online PR	Publishing content online which consumers will find informative, useful or entertaining in order to position the brand as a trusted or respected source of knowledge in its field. A means of engaging consumers with the brand without being seen to try to sell. There may be ethical considerations here.
Marketing automation	Using software to automatically manage customer interactions from lead generation to email marketing, campaign management and analytics.

Fig. 6.1 Digital marketing activities and their explanations (Adapted from Shaw (2018))

but users may be skeptical of the relevance of sponsored ads and thus natural search results[4] powered by SEO may be seen as more trustworthy (Bianchi, 2020).

E-mail marketing reaches users via e-mail with newsletters and other periodic e-mail blasts, *social media marketing* relies on engaging users via their social media platforms, *mobile marketing* pushes special offers via SMS and other smart phone technologies, *advergaming* engages users with games, and other self-evident terms and their explanations are included in Fig. 6.1.

[4] Search results are considered *organic*, or *natural*, if they are primarily based on or relevant to the user's search query.

The last two advertising activity types in Fig. 6.1—online public relations (PR) and marketing automation—stand out from the rest in the lineup, since they could be employed across media types and over time. *Online or digital PR* is a set of conscious, planned, and continuous efforts to establish, shape, and maintain a positive consumer opinion, image of the organization, and/or brand image, specifically by maximizing positive mentions of an organization, brand, product, or service on third-party websites which are visited by specific target groups (such as consumers, media/journalists, employees, investors, bloggers, online influencers, or other stakeholders (Bianchi, 2020, Chap. 1, p. 15)).

Marketing automation refers broadly to enabling various marketing and sales processes in computerized ways to support the work of marketers and advertisers, such as by delivering "relevant communications as personalised emails and website messages" (Chaffey & Ellis-Chadwick, 2018, p. 664). *Programmatic advertising* is another term that refers to "a software-based method of buying, displaying and optimizing advertising space subject to the available audience data in order to better target certain potential users" (Bekh, 2020, p. 533).

The various kinds of user behavioral targeting and profiling based on users' personal data are equivalent to technological surveillance measures, and their use raises questions of whether tech and social media companies are overreaching their authority. Specifically, "*online behavioral advertising (OBA)* is a form of marketing that uses behavioral targeting technology based on gathering and aggregating information to infer user preferences from online behavior—pages visited and searches made. The data involved may extend over a single browsing session, or it may extend over a considerable time, involving cookies and click tracking. *Location targeting* involves targeting audiences based on their location or their geographic area of interest. In the context of the Google ad network, the terms used are *Location of Presence (LOP)* or *Area of Interest (AOI). Semantic data targeting* relies on embedding structured data (such a price, availability, physical specifications, etc.)" in HTML[5] or XML[6] web pages in the RDFa[7] format, invisible to users, but of great use to automated systems for the optimization of their performance (Weller, 2012, pp. 6–8). "*Data-driven targeting* lets advertisers efficiently find their target audience, so when a publisher is connected to the relevant ad networks and their selling

[5] *HTML* refers to the HyperText Markup Language, the standard markup language for documents designed to be displayed in a web browser.

[6] *XML* refers to the eXtensible Markup Language, which is another markup language that defines a set of rules for encoding documents in a format that is both human-readable and machine-readable.

[7] RDFa is short for Resource Description Framework-in-attributes, a structure indented as a description of online data (i.e., metadata) for the Semantic Web, or Web 3.0. At its inception metadata was seen as a form of lightweight e-commerce vocabulary that would tag content with attributes, enriching HTML or XML documents. It is done practically "behind the scenes" or unbeknown to the casual user. Schema.org offers a unified standard of such set of vocabulary tags (also known as metadata or microdata) in RDF formats, that can further be added to online data describe its content (primarily to machines). The library and database management community is seeing a resurgence in the use of such metadata.

process is automated, advertising turns out to be one of the most profitable revenue models for digital media." (Bekh, 2020, p. 532).

6.2.3 Ad Revenue Model

The term "revenue model" describes how an organization generates its income as part of its business strategy. Six popular revenue models have been developed for digital media companies, including advertising, affiliate marketing and sales, subscription, freemium (basic free services with subscription premium services), one-time purchase, and pay per use (see details in Bekh (2020)). *Advertising-based*, or simply, *ad revenue model* implies that a company creates a digital media product which attracts interaction and engagement by users and then sells access to this user audience to advertisers. "Digital media users want useful content for free or at a low price, while publishers accelerate the growth and adoption of their digital product" (Bekh, 2020, p. 532). This way of monetizing content apparently dates back to the early "media wars" between newspaper magnates, William Randolph Hearst and Joseph Pulitzer. (See Sect. 5.3 in Chap. 5 on Hearst's and Pulitzer's role in the creation of "yellow journalism"). The ad-revenue model has been widely adopted, for example, by such e-publishing giants as "*BuzzFeed, Vox Media, HuffPost, Mashable, Insider Inc., Vice Media, the Washington Post, the Wall Street Journal, Fox Corporation, etc*." (Bekh, 2020, p. 532).

Internet advertising is huge in terms of spending among advertising markets. At the global level, the Internet is the medium with the highest share in advertising spending and the highest growth rate as compared to other advertising markets (newspapers, magazines, TV, radio, out-of-home, and cinema). "In 2019, the TV share was 29.2% while the Internet accounted for 49.5% but it is expected to account for 54% in 2022 (Zenith, 2019). In 2018, Internet advertising revenues in the U.S. totaled $107.5 billion and increased by 21.8% over 2017 (IAB & PwC, 2019). The gross digital advertising expenditure in Europe amounted to €55.1 billion (around $65 billion) in 2018, up 13.9% from €48.4 billion in 2017 and the market has more than doubled since 2012 (IAB Europe, 2019). Social media have driven the growth of the Internet share in advertising spending (IAB Europe, 2019; IAB & PwC, 2019; Statista, 2015) and have dramatically changed the role of consumers" (as cited in Bianchi (2020, pp. 18–19)).

6.2.4 Advertising Paradigms: Salesmanship in Print, Seduction, and Others

Advertising is an attention-getting communicative act. According to some advertising textbooks, it has been used to deliver messages across history from the times of stone slabs inscribed with hieroglyphics by the Ancient Egyptians to the digital ads at bus stops in modern transit systems (Shaw, 2018). "Advertising in many forms was both persuasive and pervasive from the first days of a mass market printed media in the 1600s" and as early as the 1700s, Victorian entrepreneurs in the U.K. pottery business were acutely aware of the prestigious effect of having their products placed in royal portraits, books and London stage plays. In the U.S., the advertising industry had so grown in importance by the 1920s that many educators and social scientists were trying to develop a scientifically based advertising discipline (Hackley, 2010).

Two distinctly opposing paradigms have been in use by advertisers since that time. The idea of advertising as "salesmanship in print" appeals to logic. This paradigm of reason is from a rational view of advertising, like that held by influential ad-men including John E. Kennedy, Albert Lasker, Hopkins, Daniel Starch, and Russell Colley. Advertising, seen as a process of decision-making in this view, is based on rational persuasion, comprehension, and conviction. Prominent models of the paradigm (such as DAGMAR and AIDA) reduce advertising to logical procedures. AIDA, the acronym for Awareness-Interest-Desire-Action, was a theory introduced by Daniel Starch for print advertising. AIDA explains how marketing communications work, implying that they can be powerful determinants of consumer attitudes and buying behaviors, and "therefore these communications can be construed as persuasive and manipulative" (Bannock, 2002, p. 9). Consistent with AIDA, DAGMAR is an acronym for setting objectives in advertising and it stands for "Defining Advertising Goals for Measured Advertising Results." Coined by Russell Colley in 1961, the DAGMAR approach involves four specific measurable communication tasks: *awareness* (making consumers aware of the brand), *comprehension* (helping them understand what the brand is and does), *conviction* (developing a disposition toward the brand), and *action* (enabling consumers to make the purchase) (Bannock, 2002, p. 92).

In the 1920s–1930s, these sequential models of persuasion in advertising assumed that consumers respond to marketers' solicitations in much the same way they would respond to a salesman's pitches in person. Consumers are indifferent until their attention is captured, their interest is further excited, then a desire for the product is stimulated, and finally they are persuaded to act on those intentions by purchasing the product. "At each of these stages the marketer would encounter resistance or indifference. The gradual accumulation of persuasive messages would eventually tip the consumer into each subsequent stage; hence there was a 'hierarchy of effect' as the weight of persuasion built up at each stage to tip the consumer into the next stage, like a champagne glass fountain" (Hackley, 2011, p. 92). Theories of communication and cognition were evoked to give this intuition firmer intellectual grounds in works like Edward Strong's (1925) *book, The Psychology of Selling and Advertising*, and Harry D. Kitson's (1923) article, *Understanding the Consumer's Mind.* They implied that persuasive communication obeyed the same psychological principles and information transfer process, no matter the medium (Hackley, 2010). The AIDA model, capturing the essence of the method, was adapted by marketing and advertising textbooks and it "remains highly influential in the field for its clarity, economy and universalism" (Hackley, 2010, 2011).

In the 1960s, famous practitioners popularized several evolutional ideas in the advertising industry: Bill Bernbach's ideas on "advertising creativity," David Ogilv's notion of "brand personality" and Leo Burnett's use of "dramatic realism," all of which were counterpoints to the "hard sell" school of advertising associated with Claude Hopkins (Hackley, 2010). Paul Feldwick, using his 30+ years of experience, thoroughly reflects on the logical and emotional approaches in his book, *The Anatomy of Humbug* (Feldwick, 2015). He calls this contrasting paradigm of advertising a process of "seduction" of the audience. The goal is to make ads interesting and enjoyable by using the power of visual symbolism in images and connotative language, and the emotional associations created in the mind of the audience.

According to Wilson (2015), Feldwick concludes that beyond the science and the art of advertising, four other elements are necessary: salience, social connection, spin, and showbiz. *Salience* is about the advertising message being popular and noticed ubiquitously; *social connection* creates a friendly, approachable, and interactive image of the advertised brand; *spin* provides normative reinforcement of perceptions, and *showbiz* is important since exceptional showmanship is distinctive, it grabs attention with flashiness or even stuntsmanship (Wilson, 2015). The four elements would seem indespensable for manufacturing virality (see Sect. 6.5 below), a contemporary spin for the digital content that can side-step some more traditional persuasion tactics.

6.3 Persuasion in Advertising

6.3.1 Classic Modes of Persuasion

The three well-known *means of persuasion* based on Aristotle's art of rhetoric are *logos, ethos,* and *pathos.* These three correspond, respectively, to the use of logic, emotions, and the speaker's good character in addressing the audience. The point of rhetorical language use is to make the argument more compelling with various rhetorical devices and figurative speech, and overall to be persuasive to the audience. Since Aristotle, *a rhetorician* has long been defined "as someone who is always able to see what is persuasive (*Topics* VI.12, 149b25)" though they will not necessarily be able to persuade everyone under all circumstances, and rhetoric, correspondingly is "the ability to see what is possibly persuasive in every given case (*Rhet.* I.2, 1355b26f)" (Rapp, 2010). The art of rhetoric can be used for either good or bad purposes. In an effort to promote products and brands to consumers, advertisers, in effect, become skillful rhetoricians, without a moral compass. Their true north is perpetually commercial profit (see Chap. 4, Sect. 4.2 on what constitutes morality in applied philosophy and ethics, and on the distinction between amorality and immorality).

6.3.2 Advertising Strategies

An insightful analysis of the strategies used in persuasive advertising is offered by Spence and Van Heekeren (2005), specifically in terms of their effectiveness and the possible complications of three strategic possibilities. *Strategy 1* is to provide complete and true information about the advertised products or services. This is a descriptive strategy that purports the truth and works perfectly well when the products indeed live up to the advertising claims. For example, advertising the most luxurious car brand in the world is persuasive if it happens to be true, but advertising fast food as the most nutritious options on the market will not convince anyone, since the claim is bogus. Providing only accurate information is not always possible and making it complete is not always practical since it may be boring, plus the airtime is costly, especially in TV ads (Spence & Van Heekeren, 2005). Notice the inescapable parallel of such informative and descriptive persuasion to the use of *logos* in the art of rhetoric.

According to Spence and Van Heekeren (2005), *Strategy 2* is to present advertised products or services in the best and most attractive possible light to maximize appeal. How can this be achieved? Without any scruples, apparently. Exaggerate or invent the merits, and underplay or leave out most, if not all, shortcomings to make the product look more attractive. However, ethically and legally it may amount to deception (see Chap. 2, Sect. 2.3 on various faces of deception). When deception is found, it creates pragmatic problems: consumers may clue in and the brand may lose its credibility which in turn leads to the loss of sales and further repercussions from industry regulators or legal sanctions. The use of rhetorical *ethos* and visual suggestion can lead to the loss of *pathos.*

Strategy 3 is to focus on the consumer. It is a sneaky way to avoid disclosing nothing but the truth, which gets advertisers around the ordeal of deceiving consumers. Strategy 3 creates in the consumer *positive feelings* and thoughts about the products or services advertised, or alternatively, associates *negative feelings* with the absence or lack of the advertised products. Both "feel-good" or "feel-bad" variants of this strategy may avoid the use of explicit statements and resort to suggestion, appealing to values and lifestyles that consumers instantly recognize and consider worthy (e.g., love, freedom, independence, family, friends, comfort, pleasure, security, happiness, humor, imagination, and fantasy). Over time, the associations are formed in the minds of consumers. Catchy simple slogans *"Happiness is Hyundai,"* or images of people jumping for joy *"O what a feeling ... Toyota!"* associate car manufacturers with these positive values or others such as the notion of freedom, while perfumes may start to associate with the notions of romance or independence (Spence & Van Heekeren, 2005, p. 50).

Spence and Van Heekeren's (2005) analysis concludes that "truth is only marginally relevant to advertising" (p. 50). Namely, truthful information is purported under Strategy 1, but is hardly relevant under Strategy 2, while Strategy 3 provides almost no information about the product or service itself, but rather appeals to consumers' sentiments, values, and lifestyle aspirations. While advertisers accustom consumers to certain frames of communication—fictional, metaphorical, or fantastical storylines—consumers are willing to *suspend disbelief* and are unlikely to be surprised by an ad conjuring an image of a chocolate cookie genie granting someone a wish. With no intent on the part of the sender of the information, and no willingness to believe on the part of its receiver, there cannot be possibly an act of deception, or at least this is the message that the advertising industry projects. Without the intention to provide descriptive truth in the message—both completely and accurately—truth becomes irrelevant. While consumers suspend their disbelief, advertisers evoke sentiments and associations through the use of imagery, humor, fiction, and fantasy. Deception can only occur, on the face of it, if proper care is not taken to avoid misleading ads, those that intend to deceive. While the logic is clearly explicated in advertising textbooks and in antitrust lawsuits, it is hard to believe most consumers are fully aware of the suggestive manipulative games that the industry is playing. Neither do most lay people make a clear distinction between the nuances of the three persuasive strategies.

6.3.3 Codes, Standards, and Regulation in Advertising Industry

Advertisers operate in fiercely competitive environments, whether they are engaging in digital advertising techniques or in more traditional media such as radio, TV, or outdoor public spaces. Considering the pressures that they feel, it is tempting for advertisers to resort to less than honorable ways of capturing their audiences' attention and set moral scruples aside. To restrict advertisers' ability to mislead consumers, various regulatory bodies have developed codes and standards for advertising practices to govern advertising activities with consumers' well-being at heart. The principles of honesty and truth are rooted in philosophy and applied ethics (see Chap. 4). For example, "The British Codes of Advertising and Sales Promotion is representative of the parameters established to create a more truthful and socially acceptable advertising environment:

- Advertising Code.
  - Principles (2.1) All advertisements should be legal, decent, honest, and truthful.
  - Truthfulness (7.1) No advertisement should mislead by inaccuracy, ambiguity, exaggeration, omission, or otherwise" (Spence & Van Heekeren, 2005, p. 45).

Advertising claims' accuracy can be adjudicated in various civil courts for the protection of consumers from misleading, exaggerated, untruthful, or unsubstantiated claims in marketing and advertising communications. In the U.S., the Federal Trade Commission (FTC) is an independent agency for the protection of consumers and the regulation of market competition at the federal level. The FTC's mission is accomplished "by preventing anticompetitive, deceptive, and unfair business practices through law enforcement, advocacy, and education without unduly burdening legitimate business activity" (Federal Trade Commission, 2013). It also provides information to help spot, stop, avoid, and prevent fraudulent, deceptive, and unfair business practices. For example, the FTC issues, maintains, and updates various guides, such as the Green Guides, to assist marketers in making non-deceptive claims, and outlines general principles and provides guidance to the industry for administrative interpretations of the law. Such industry guides "do not have the force and effect of law and are not independently enforceable but the Commission, however, can take action under the FTC Act if a marketer makes an environmental claim inconsistent with the Guides. In any such enforcement action, the Commission must prove that the challenged act or practice is unfair or deceptive" (Federal Trade Commission, n.d.).

To adequately substantiate their marketing claims, marketers are advised by the FTC Guides to obtain "competent and reliable scientific evidence" such as "tests, analyses, research, studies or other evidence based on the expertise of professionals in the relevant area, conducted and evaluated in an objective manner by persons qualified to do so, using procedures generally accepted in the profession to yield accurate and reliable results" (Federal Trade Commission, n.d.). (See also Chap. 5, Sect. 5.4 for the scientific method and key scientific principles underpinning empirical inquiries.)

Procedurally, if issues of *false advertising* and other forms of *fraud* are raised by consumer and business reports, pre-merger notification filings, congressional inquiries, or reports in the media, the FTC may investigate these issues pertaining to a single company or an entire industry. "If the results of the investigation reveal unlawful conduct, the FTC may seek voluntary compliance by the offending business through a consent order, file an administrative complaint, or initiate federal litigation" (Editors of Wikipedia, 2021b).

"Because purportedly independent comparative ratings can be highly material to consumers, advertisers need objective evidence to support what they say" (Fair, 2021). According to the FTC's Deception Policy Statement, "an ad is deceptive if it contains a statement or omits information that is likely to mislead consumers acting reasonably under the circumstances; and is "material," that is, important to a consumer's decision to buy or use the product (Spence & Van Heekeren, 2005, p. 45). Spence and Van Heekeren (2005) suggest that the FTC is clear in drawing the line between truth and untruths when the latter lead to harmful consequences: "An advertisement that is not literally true is not the key issue, but an advertisement that is designed to deceive or mislead a consumer is a different matter. This is the situation in which the use of puffery in advertising comes under scrutiny" (p. 45).

6.3.4 Puffery

Puffing up products or brand names is a common practice in advertising, and it is not considered illegal (Spence & Van Heekeren, 2005). In advertising and marketing, the law requires that objective claims be truthful and substantiated (Starek III,

1996). While in everyday language, *puffery* means "undue or exaggerated praise" (Editors of Dictionary.com, n.d.), in law, it is "a promotional statement or claim that expresses subjective rather than objective views, which no 'reasonable person' would take literally" (*Newcal Industries, Inc. v. IKON Office Solutions, 513 F.3d 1038, 1053* (ninth Cir. *Newcal Industries v. Ikon Office Solution*, 2008), cited by (Editors of Wikipedia, 2021a)). "Metaphors, similes, and exaggerations are considered to be accepted by consumers, who are now more than familiar with the genre, as less than harmful" (Spence & Van Heekeren, 2005, p. 45). A statement that is quantifiable and makes a claim as to the "specific or absolute characteristics of a product" may be an actionable statement of fact while "a general, subjective claim about a product is nonactionable puffery" (Newcal Industries, Inc. v. IKON Office Solutions, 513F.3d 1038, 1053 (ninth Cir. *Newcal Industries v. Ikon Office Solution*, 2008)).

Puffery is "the key reason why consumers have the expectation that advertising will stretch the truth rather than express the truth. In essence, the notion of puffery refers to exaggerated claims, comments, commendations, or hyperbole, and in its most common usage, puffery is based on subjective views and opinions. It is generally considered to be part of the artfulness and playfulness of advertising and as such is not considered to be taken very seriously by reasonable consumers. It is for this reason that most regulatory bodies disregard it as a code-violating activity" (Spence & Van Heekeren, 2005, p. 45).

The Federal Trade Commission (1983) Statement on Deception stated: "The Commission generally will not pursue cases involving obviously exaggerated or puffing representations, i.e., those that the ordinary consumers do not take seriously." Since puffing is only an expression of opinion and not a representation of fact, "a seller has some latitude in puffing his goods, but he is not authorized to misrepresent them or to assign to them benefits they do not possess. Statements made for the purpose of deceiving prospective purchasers cannot properly be characterized as mere puffing. *Wilmington Chemical, 69F.T.C. 828, 865 (1966)*" (Federal Trade Commission, 1983).

In contrast with mere subjective puffery, the FTC may pursue claims with an objective component—such as "more consumers prefer our hairspray to any other" or "our hairspray lasts longer than the most popular brands" —to make sure that such claims have adequate substantiation, competent and reliable scientific evidence is generally required as a reasonable basis for health and safety claims (Starek III, 1996).

6.3.5 *False Claims and Other Acts of Misleading Advertising*

"Misinformation and unsubstantiated claims pose a more critical set of problems that have made the practice not only unethical but also illegal" (Spence & Van Heekeren, 2005, p. 45). The FTC and other regulatory bodies alike distinguish puffery from *false* or *unsubstantiated claims, scams,* or other *deceptive sales pitches*. Specific types of claims can lead consumers to make erroneous judgments, for example, *incomplete comparisons* (meaningless claims of "higher" quality of a product without a clear referent, as in "Acme brand is faster acting") and other *implied superiority claims* (without explicitly and directly stating it, as in "no toothpaste fights cavities better") (Darke & Ritchie, 2007, p. 114). Other logical fallacies (see Chap. 4, Sect. 4.7) could just as easily be put to use to evoke erroneous conclusions in the minds of consumers. The line between acceptable puffed-up and unacceptable forms of suggestions may be firmly drawn in the minds of the regulatory bodies but may appear less clear for an average consumer, especially if they are already pre-disposed to be susceptible to mis- and disinformation (see also Chap. 2, Sect. 2.4 for psychological predispositions, human tendencies, and biases).

During the COVID-19 pandemic, the Commission issued warning letters to multiple companies, both in the United States and other countries, in relation to the efficacy of treatments or prevention of the coronavirus disease. These companies were warned to cease making unsubstantiated claims about their products, with the FTC threatening a federal court injunction and an order requiring money to be refunded to consumers. The warned companies advertised and/or sold their products online ("everything from a bundle of supplements called an "ANTI-VIRUS KIT" to "Sonic Silicone Face Brushes" and intravenous (IV) "therapies" with high doses of "Vitamin C"), or alternatively, purported treatments were offered in clinics or consumers' homes. "'It's shameful to take advantage of people by claiming that a product prevents, treats, or cures COVID-19,' said Andrew Smith, Director of the FTC's Bureau of Consumer Protection. 'We're seeing these false claims for all sorts of products, but anyone who makes them simply has no proof and is likely just after your money'" (Federal Trade Commission, 2020).

Since the 1980s, Section 5 of the FTC Act declared *unfair or deceptive acts or practices* unlawful with Section 12 specifically prohibiting *false ads* likely to induce the purchase of food, drugs, devices, or cosmetics and Section 15 defining a false ad for purposes of Section 12 as "one which is misleading in a material respect" (Federal Trade Commission, 1983, p. 1). In its elaboration on the FTC Act, the Commission explained that the following three elements must undergird all deception cases: *representation*, *reasonableness,* and *materiality*. First, there must be a representation, omission, or practice that is likely to mislead the consumer, as exemplified by "false oral or written representations, misleading price claims, sales of hazardous or systematically defective products or services without adequate disclosures, failure to disclose information regarding pyramid sales, use of bait-and-switch techniques, failure to perform promised services, and failure to meet warranty obligations." Second, in examining cases the FTC considers "the practice from the perspective of a consumer [group] acting reasonably in the circumstances." Finally, the third key element examined by the FTC is "whether the act or practice is likely to affect the consumer's conduct or decision with regard to a product or service. If so, the practice is material, and consumer injury is likely, because consumers are likely to have chosen differently but for the deception. In many instances, materiality, and hence injury, can be presumed from the nature of the practice. In other instances, evidence of materiality may be necessary" (Federal Trade Commission, 1983, pp. 1–2). Thus, the FTC investigates the three elements and will find a representation, omission, or practice deceptive if it is likely to mislead the consumer acting reasonably under the circumstances, to the consumer's detriment.

6.3.6 Regulatory Actions and Antitrust Investigations

As a concrete example of the FTC's application of its three elements in determining misleading or deceptive advertising, let us look at an ad by a company that offered older consumers some health products and peddled their unproven health benefits. In a 2020 press release, the FTC distinguished this ad from "mere puffery" explaining the material consequences to the puffed-up promises as established in the details of the case (Fair, 2020). The Boca Raton-based Renaissance Health Publishing company invoked a name of an entity, called the American Journal of Pain and Inflammation, in their direct mail brochure and claimed that *Isoprex can relieve* pains in "as little as one day."[8] Glowing endorsement by "Geoff," "Christy," "John A.," and others attested to the cure's effectiveness, appearing in ads next to active "mature" consumers paying tennis, golf, or biking. "Other ads featured a laboratory professional next to a graph that purported to show the results of clinical testing of the supposed active ingredient in *Isoprex*. According to a handwritten notation, 'Wow! Severe painful knee swelling disappears. 100% effective!'" (Fair, 2020).

The FTC challenged the company's claims about relieving pain, reducing inflammation, and rebuilding joints as false or unsubstantiated, as well as their puffed-up promises of *Isoprex* pills providing relief comparable or superior to over-the-counter drugs with a 100% effectiveness at relieving inflammation and swelling. The proposed FTC's order requires that the defendants, Renaissance Health Publishing and its managing member, pay a financial settlement and send a letter to customers describing the FTC's lawsuit. The FTC's proposed settlement also requires that the company conduct human clinical testing to support its wide range of health claims and clearly disclose the unexpected material connections to endorsers such as paid friends and relatives employed by the company ("Geoff," "Christy," "John A.") (Fair, 2020).

Regulatory actions were also taken by the European Union's Commission (the EU's antitrust enforcer) as well as the U.K.'s Competition and Markets Authority, as part of a 2021 wave of antitrust enforcement in and around Europe. Formal antitrust law investigations are being pursued and charges formally filed against advertising practices by Facebook Inc.'s classified-ads service Marketplace for their alleged repurposing of Facebook data gathered from advertisers who bought ads in order to give illegal advantages to the Marketplace service (Schechner, 2021).

[8] *See* this misleading ad example with unsubstantiated claims evoking the non-existing journal's name at the FTC's Business Blog post: https://www.ftc.gov/news-events/blogs/business-blog/2020/04/no-pain-relief-no-gain-ftc-challenges-claims-aimed-older. (Accessed on 28 May 28, 2021).

Advertising is largely a self-regulated industry that is also overseen by various global and national agencies such as Advertising Standards Authorities (in the U.K., New Zealand, or Ireland) or the Better Business Bureau (BBB). These agencies set criteria for acceptable advertising, provide mechanisms for collecting consumer complaints, adjudicating, and resolving disputes, and educating and protecting consumers and businesses against fraud. Although consumers often fall prey to subtle inferences implied in misleading advertising claims or claim-fact discrepancies (when products do not live up to ads' promises), an analysis of complaints received by such Ad Standards and BBBs "indicate that many people who are initially misled by advertising eventually recognize that they were duped" (Darke & Ritchie, 2007, p. 114). The threat of being found out in their wrongdoing, however, drives advertising to new heights of inventiveness. Meanwhile, people who are constantly bombarded with ads—puffed-up, borderline deceptive, emotionally suggestive, or plainly misleading—grow weary of the giant psychological experiment that continues to play out, both on- and offline.

6.3.7 *Negative Attitude toward Ads, Distrust, and Ad Avoidance*

Advertising deception may lead consumers to become defensive and broadly distrustful of further advertising claims, leaving them with a negative bias in their attitudes toward subsequent advertisements by evoking their now-negative stereotypes about ads, and thus reducing the persuasive impact of future ads (Darke & Ritchie, 2007). As early as 1986, Richard Pollay, a curator of advertising archives at the University of British Columbia, wrote of the unintended societal consequences of advertising, finding its practices intrusive and its effects "inescapable and profound." He reviewed the works of humanities and social sciences scholars and concluded that advertising reinforces "materialism, cynicism, irrationality, selfishness, anxiety, social competitiveness, sexual preoccupation, powerlessness and loss of self-respect" (Pollay, 1986, p. 18). Pollay's indictment could not be more to the point in the 2020s. Moreover, ads are disruptive to the minds of individuals and culturally, to the fabric of society. "Virtually all citizens seem to recognize this tendency of ad language to distort, advertising seems to turn us into a community of cynics, and we doubt advertisers, the media, and authority in all its forms" (Pollay, 1986, p. 29). Distrust engenders perceptions of conspiracy so that people overestimate the coordination of different advertisers' influence attempts; and confirmation bias often reinforces this distrust by undermining later-received truthful information, which leads to judgments that verify the initial distrust (Darke & Ritchie, 2007, p. 115). People distrust advertising since manipulative persuasion tactics can also "activate persuasion knowledge[9] and generate reactance" (Akpinar & Berger, 2017, p. 319).

[9] *Persuasion knowledge* can be roughly equated to our remembered life experiences in encountering and recognizing a situation as persuasive such as someone intention to sell you something. See

Nonetheless, advertising as a form of communication is so pervasive, so incessant in capturing our attention, that it unwittingly occupies a chunk of our lives. We become skeptical of the ways organizations attempt to grab our attention, manipulate us into believing their message, and persuade us to act on cue. Recent research on advertising notes a decline in trust and a rise in negative attitudes toward advertising: "people do not like and avoid advertisements" (Bianchi, 2020, p. 8).

In digital media, advertisers are increasingly creative in capturing Internet users' attention, but users still catch on to the advertising patterns and formats as these new tricks become predictable. Some users may get frustrated, others may become too accustomed to ads and find ways to ignore and actively avoid looking at ads that appear in predictable places or formats. These kinds of behaviors are known as *ad avoidance* or *banner blindness* (Wojdynski, 2016). In addition, users may turn to ad-blocking technology to proactively suppress ads in their favorite web browsers and apps. "Advertisers are left with a dilemma. Do they make their advertisements more obnoxious and noticeable: popups, blinking, noisy ads? Or do they make their ads more deceptive, unnoticed and therefore un-avoided? Native advertisements represent a step toward the latter choice, creating a genre of covert advertising alongside product placement and advergames (a portmanteau of 'advertising' and 'games'). While creators tout them as being content-like and entertaining for consumers, the difficult truth is that these ads work more effectively when readers do not realize that they are looking at an advertisement (Schauster et al., 2016; M. Wu et al., 2016)" (cited in Cornwell & Rubin, 2019, p. 5206). Ad avoidance behaviors in turn make it more challenging for marketers to capture our imagination, so they redouble their efforts and spending.

6.4 Covert Advertising in Many Guises

6.4.1 Irresistibility of Native Advertising

Native advertising has become ubiquitous in social media, increasingly present in online media, and popular with advertisers. Its popularity equals, if not supersedes, the spending on other advertising forms. According to a 2017 digital publishers survey by Reuter's Digital News Project, there was a clear move away from traditional advertising toward sponsored content (42%), as well as direct reader payment (45%) and membership (14%) (Newman, 2017). Wojdynski and Golan in (2016) also saw native ads as the "primary driving engine of the Internet marketing economy," and had predicted native ad spending at $21 billion for 2018. According to the Statista Research Department (2021a), native ad spending in fact exceeded the 2016 prediction, amounting to $35 billion, and they predict the figure will have reached $53 billion by the end of 2020. Another independent think tank, the Native Advertising

further clarifications on the original *Persuasion Knowledge Model* by Friestad and Wright's (1994) a few pages below.

Institute, also marked the year 2018 as showing "a dramatic increase in the financial importance of native advertising," expecting the revenue generation from native advertising to increase by another 46% by the end of 2021 (Laursen & Hewes, 2018).

"The ad industry has been serving up a sneaky solution: make ads look less like ads—and more like the articles, videos and posts around them. An ad that matches the typeface, design and layout of the real articles feels less like a tacky intrusion. ...[E]ven the Web sites of journalistic bastions such as the *New York Times* and the *Wall Street Journal* are incorporating it [native advertising]. Social-media companies have signed on as well. On Facebook and Twitter, every tenth item or so is an ad; only the subtle subtitle "Sponsored," appearing in light gray type, tells you which posts are ads" (Pogue, 2015). A content analysis by Warnick in (2016) concurs: advertising articles from *the New York Times*, *Wall Street Journal*, *Washington Post*, and *Slate* mirror the content of traditional journalism. In Canada, as in the U.S., newspapers have opened their own Native Advertising production houses—for instance, *Postmedia's Polar* and *The Globe and Mail's Globe Edge.* Newspaper watchdogs have raised ethical concerns about the deceptive quality of native ads, calling them "fraudulent" and "a form of prostitution" by Canadian magazine editors (Russell, 2005) and as a means of advertisers putting pressure on editorial news coverage to avoid controversy and sell with sensationalism (Carlson, 2015; King, 2016) (as cited in Cornwell & Rubin, 2019).

Why are various forms of native ads hugely popular with advertisers? One simple explanation is that people have difficulties immediately identifying them as advertisements due to their ambiguous nature or deceptive labels such as "From around the web." In fact, consumers are known to be frequently tricked despite the labels: even when labeled, they can be apparently recognized as advertising by only 7% of those exposed to them in one study and by only 18.3% in another study by Wojdynski and Evans (2016). More generally it is clear that covert advertising in its various forms has gained momentum: favored by advertisers, frequently used by tech giants, and often overlooked or misunderstood by consumers. The latter point deserves further elaboration.

The question at the heart of the problem is why people are easily fooled by native ads. The answer may lie in our uneven ability to recognize and interpret covert ads. One possible explanation is in people's preconceptions. We have certain ideas about what frames or schemata are used by persuaders, referred to as the Schema Theory, or Framing Theory in communications research (Entman, 1993). According to the Persuasion Knowledge Model (PKM) (Friestad & Wright, 1994) our *persuasion knowledge* is updated through experience across our lives: when we encounter a situation that we recognize as persuasive, we activate our persuasion knowledge in concert with our *topic knowledge* at hand and our knowledge of the convincer, or *agent knowledge*. Cornwell and Rubin (2019) elaborate: "For example, when Helen reads the newspaper, she will activate her persuasion knowledge in the classifieds section. This persuasion knowledge, along with her agent knowledge of newspaper ads and topic knowledge of jobs, will let Helen use *persuasion coping behaviors* to decide which job ads are legitimate or scams. However, if Helen does not realize that some job ads can be illegitimate, she will not activate her persuasion knowledge, and she will not be able to access her persuasion coping behaviors. Instead, she would likely assume that what she reads is true. *Truth bias*—experiencing situations as truthful by default—is a common cognitive heuristic (Levine et al., 1999). The inactivation of persuasion knowledge has been hypothesized to be the working model of covert advertisements (Amazeen & Muddiman, 2017; Evans & Park, 2015). When readers do not realize that they have encountered an advertisement, they do not "defend" against the persuasion using their coping behaviors, and so are more easily convinced (Amazeen & Muddiman, 2017)."

6.4.2 Native Ad Varieties: Advertorials, Sponsored Content, Branded Links, and Others

Native advertising in digital media is basically ads made to look like something else (like organic search results or authentic journalistic editorials). In the early 2010s, native advertising was hailed as the next big trend in the strategic communication field as a marketing technique with its persuasive intent obscured, and the ads' content sources left ambiguous (M. Wu et al., 2016). The word *native* refers to this coherence in the form and content with the other media that appears on the platform, be it a video, article, or editorial, often functioning like an advertorial (Editors of Wikipedia, 2021c). "*An advertorial* is an advertisement that masquerades as editorial comment or opinion, usually on a matter of public interest" (Spence & Van Heekeren, 2005, p. 51). *Sponsored content* or *branded content* are other labels applied to this type of covert advertising.

Bekh (2020) observes that in contrast to the flashy banner format (more "traditional" in digital spaces), the native format returns the status of useful information to advertising. When digital users' needs for all sorts of life occasions are expressed through the search engine key words they input and then neatly aggregated in giant profiles, it is easy to provide them with such a just-in-time solution, manipulatively responding to users' unfulfilled needs or intentions. A solution served to users at the perfect time is hard to pass by. Without recognizing *sponsored content* for the ads that they are, users share these content or links with friends and followers via social media networks. They find it useful since the timing is right and the content matches their needs. This perceived usefulness increases the virality of the ad content. Advertisers tracking user engagement have seen clear results. "In September 2015, the *Financial Times* began to publish native ads on their website under the heading of *Paid Post*. 74% of newspaper readers found the information they receive in sponsored content useful. After the launch of the Paid Post column, the active engagement time has increased by 123% and the click-through rate by 30%" (Bekh, 2020, p. 534).

Sponsored links are located in the places where a user is used to reading content. "The brand message is embedded in the website's content, but when users click on it, they get to the advertiser's site. It lets brands and media get away from the problem of banner blindness. Native ad statistics show that the CTR[10] for native ads is as much as eight times higher than for traditional display ads. … In the case of sponsored content, a publisher shares with an advertiser the authority of its publication. For example, *Vice* and *BuzzFeed* efficiently integrate sponsored content in their business strategy. This format provides great opportunities for targeting and does not limit the advertiser in formats. Today, as a way to monetize their websites, native advertising embedded in the feed is used by most major media companies, such as *The Guardian, The Independent, Forbes, 9GAG, The Sun,* etc." Behk (2020) (p. 534).

Thus, a three-prong classification (by Wojdynski, 2016)—sponsored content, hyperlinks, and social media posts—covers the formats of the sponsorship and promotion. The (2013) classification by the Interactive Advertising Bureau (IAB)[11] is more elaborate and it is valuable for recognizing visual variations[12] in native ads across prominent digital media. Here is the six-prong IAB classification with each category's representative samples, some of which are lesser known than others: (1) in-feed units (*Forbes, Yahoo, Facebook,* and *Twitter*); (2) paid search units (*Yahoo, Google, Bing,* and *Ask*); (3) recommendation widgets (*Outbrain, Taboola, Disqus,* and *Gravity*); (4) promoted listings (*Etsy, Amazon, Foursquare,* and *Google*); (5) in-ad (IAB Standard) with native element units (*Appssavy, Martini Media, EA, Onespot,* and *Federated Media*); (6) Custom/"Can't be contained" (*Hearst, Flipboard, Tumblr, Spotify,* and *Pandora*).

As evident from the discord around labeling and the disagreement in classification categories, there is no global standardization in place to label the various types of covert ads. Plus, as in any other creative and lucrative enterprise, the varieties are likely to morph in ever more elusive guises with time. The point of showing such a variety or having these multiple disguises is that it keeps digital media users on their toes: this masquerade reduces potential immediate recognition, increases the chance of a click, visit, or reshare, and avoids the banner blindness without infuriating potential customers.

Acknowledging the many guises covert advertising may take, one of our studies aimed to comparatively analyze the labeling, content, and context in which native ads were presented in two Canadian English-language national papers (*The Globe and Mail*; *The National Post*) as well as the highest circulating paper from each of Canada's two largest cities (*The Toronto Star*; *The Vancouver Sun*) (Cornwell & Rubin, 2019). Ten native ads and ten editorial pieces were collected

---

[10] CTR, or clickthrough rate, is a ratio showing how often people who see an ad click on it. CTR can be used to gauge the performance of certain keywords, ads, or free listings. CTR is calculated with the number of clicks that an ad receives divided by the number of times it is shown: clicks ÷ impressions = CTR. For example, with 5 clicks out of 100 impressions, the CTR is 5% (Google Ads Help, 2021).

[11] IAB, or the Interactive Advertising Bureau, is another prominent advertising media coalition that develops industry standards for Europe and USA, headquartered in New York.

[12] *See* the visuals at https://www.iab.com/wp-content/uploads/2015/06/IAB-Native-Advertising-Playbook2.pdf (Accessed on May 28, 2021) (Interactive Advertising Bureau, 2013).

from each of the four newspapers, for a total of 80 articles. The editorial articles collected from each paper were on the same topic, for example, a *Toronto Star*'s native advertisement for the Building Industry and Land Development Association (BILD) "Development: What's in it for you" was paired with the editorial article "Toronto considering doubling development charges for new projects." The study found that one of the mechanisms for the success of native ads is their ability to be practically indistinguishable from editorials in the same paper, specifically in respect to their content. The grammatical moods and emotional affect used in the sample editorials and native ads were very similar, as were the types of images included in both native advertisements and editorials. Consistent with other studies (Amazeen & Muddiman, 2017; Campbell & Evans, 2018; Wojdynski, 2016; M. Wu et al., 2016), these stylistic similarities help to explain the high levels of reader confusion found in other studies (Cornwell & Rubin, 2019). In addition, the labeling of native ads was suboptimal for readers. The labels were present, but inconsistent across news outlets (see Fig. 6.2), small in size, and in low contrast with the background, possibly increasing their chance of being overlooked (Cornwell & Rubin, 2019).

The surrounding context in which native ads were served (images, links, banners, etc.) differed in one salient respect from the context surrounding editorials: native ads were almost exclusively associated with a single advertiser. Campbell and Evans already pointed out in (2018) that digital news readers use this cue to recognize advertising content. However, "recommendation widgets" (in IAB's terms) may be intentionally placed to prod people to click on what they may misconstrue as further readings. See visual similarities of the elements in Fig. 6.3. This type of mimicry is likely to deceive users who are accustomed to finding links to other journalistic pieces in article sidebars.

6.4.3 Overdue Measures: Regulation, Clearer Labeling, and Awareness of the Masquerade

The advertising industry rarely sees anything particularly wrong, to the best of my knowledge, with the covert advertising tactics employed to persuade, nudge, or manipulate digital media users. Marketing communications strategies pamphlets propound them in their best practice advice, boosted by the rosy projections of native ad expenditures as cited above. Ironically, automated solutions to blocking native ads—ad-blockers—only seem to raise the bar in "the media wars." For instance, the closer the ads are to the editorials in form and content (ethical concerns of erosion of journalistic integrity aside), the harder such advertorials are to detect and block automatically. The more that ad-blockers are successful, the harder advertisers have to work on innovative and creative tricks to put up the smoke and mirrors. Additionally, since ad-blockers sometimes fail at filtering out covert ads, the nature of covert advertising may be even less obvious to the average digital media user. Furthermore, when native ads are mixed in with legitimate news reporting in an aggregated

Newspaper	Text	Location	Readability		Example (cropped to include only label and headline)
			Color	Font Size	
The Globe and Mail	Sponsor Content	above the headline	grey text on pastel yellow or blue horizontal bar	similar to body font size	SPONSOR CONTENT What's in store for the future of mobile payments?
The Toronto Star	Partner Content	above the headline	blue text on white horizontal bar	similar to body font size	Partner ContentBuilding Complete Communities Our neighbourhoods are intensifying
The Vancouver Sun	Presented by [advertiser]	above the headline	black text on white background	similar to body font size	Colliers PRESENTED BY COLLIERS INTERNATIONAL The death of retail storefronts greatly exaggerated
The National Post	This content is sponsored by [advertiser]	above the headline	black text on white background	smaller than body font size	She was diagnosed with Stage 4 cancer on Mother's Day. Now, 15 years later, this mom is still fighting.

Fig. 6.2 Native advertising labeling practices are shown across four sample Canadian online newspapers (Cornwell & Rubin, 2019)

Fig. 6.3 Recommended content around *the National Post* (Canada) native ads (on the left) mimics the links to other news content around editorials (on the right) (Cornwell & Rubin, 2019)

newsfeed (such as when they are shared on social media), the labels themselves (such as sponsored content, branded content, etc.) may be less prominent, if not completely removed. The context in which the ad appears may also be disassociated from the reliable cue of the surrounding ads, and users may be even further confused and disinformed. Research into the effectiveness of native ads and the human flaws that native ads exploit has implications for media literacy. The more awareness media users have about the covert tactics, the schemes or frames used by advertisers, the more likely people will have their defenses up against unwanted manipulative ad content.

Regulation of clearer standardized labels that reveal the sponsorship of promotions more explicitly may make users see advertisers' motivations (financial or otherwise) more clearly. While the current growth of various lucrative forms of covert advertising is underpinned by users' relative unawareness of the almost "magic tricks," a complete unveiling of the practice and persistent reminders to consumers may reveal the magicians' true intentions. Stricter regulations of such "social engineering" hacks (by the FTC and other antitrust regulatory bodies) may also eventually render covertly persuasive advertising pointless.

6.5 Virality

6.5.1 Elusive Properties of Virality

Virality is defined by the Urban Dictionary (2021) as "the degree to which a piece of internet content has been or might be shared in a short amount of time." By analogy with epidemiological quality, it is a property of how quickly and widely any digital content (be it video, image, tweet, post, article, news, or a combination of the above) can circulate among Internet users.

What are the characteristics of virality? Well-staged explicit or covert messages can go viral, a rare feat for professional advertisers: one which requires much thought, preparation, and perhaps a stroke of luck. Can anyone definitively articulate the illusive properties of virality, or how to orchestrate a guaranteed word-of-mouth marketing campaign? It seems like many have tried and keep trying. Some are able to dispense advice, but few are able to achieve the desired buzz. As Terry O'Reilly in his (2014) episode of *Under the Influence* puts it: "These days, marketers can't necessarily spend their way to consumer attention, they have to earn it. And a viral video is one of the most powerful ways to do it." In his *CBC* radio episode on viral content, O'Reilly brings up multiple examples of virality, wondering whether creative ideas can go too far in their pursuit of success. For example, *Nivea* staff ambushed and pranked a series of people in the waiting area of a German airport. The prank involved an elaborate scheme in which the selected victims were made to believe that they were mistakenly suspected of a crime and were wanted by German authorities. When victims realize the gravity of the situation—underscored by the "Wanted" signs on airport TV screens — see the approaching "authorities," and imagine the imminent arrest, the *Nivea* staff reveal their anti-stress deodorant product and hidden cameras, to much relief of the prank victims. The pitch is that the deodorant will protect them in future stressful situations. O'Reilly reports this *Nivea* deodorant marketing campaign video going viral on the day of its release, attracting over 7 million views to the date of the last airing of the episode by the CBC (on 20 August 2018) and surpassing 25 million views by the time of the writing of this book (on 13 July 2021).[13]

O'Reilly laments about the ever-increasing threshold for advertisers' measure of virality. "To make the Advertising Age Viral Video Chart, for example, it takes at least 1.5 million views. Four years ago, it only took 220,000. That's significant, because it means advertisers have to fight much harder to get noticed today, which can lead to risky decisions. According to Advertising Age magazine research, audiences in 2012 chose to watch video ads 4.6 billion times or about 13.2 million times per day. That is incredibly alluring for the advertising industry, because those people actively sought out those videos. Very few people actively seek out television commercials. Plus, a lot of paid advertising dies the moment you stop paying for it. Viral videos live on" (O'Reilly, 2014).

6.5.2 Compelling Word-of-Mouth (WOM)

As one of the goals of viral marketing as a communications strategy, a literature review (Rodić & Koivisto, 2012) names positive word-of-mouth communications among desired audiences in both off- and online circles. The emphasis should be on *compelling the receiver to share the message with others*. To increase the likelihood of a message going viral, Rodić and Koivisto (2012) suggest the tactics of turning the message content into a *social object*—the message being relevant to the target audience, entertaining, interactive (a conversation starter), and positive (inspiring, transparent, responsible, etc.) *Viral mechanics* should also be in place—an uncluttered message with multiple options of sharing it using the right networks. Other practitioners and marketing gurus dispense advice on how to achieve this elusive success, much of it boiling down to the use of emotional appeal, some abstract talk of unpredictability, and the value of the product intended for sale. It is hard to stay optimistic about this advice, especially since after a quick Google search, most of it comes in a clickbait listicle format (e.g., "9 best practices in viral marketing").

6.5.3 Emotional Vs. Informative Appeals in Advertising

Emotional appeals in advertising are designed to evoke strong emotions (like anger, fear, disgust, happiness, sadness, and surprise) by using mood, music, drama, or images, while informational appeals present select descriptions of the products' features or rationale of the brands' benefits. Compared to straightforward

[13]You can see a more up-to-date tally of views of this viral video entitled "Funniest Airport Prank" at https://youtu.be/es2krnRZ2Ko

descriptive information, emotional appeals are more likely to be shared. For example, Heath et al. (2001) found that people were more willing to pass along urban legends that elicited strong disgust. Psychological arousal is known to increase social transmission (Berger 2011). People may share surprising or interesting content because it makes them look good (Akpinar & Berger, 2017). The basic human motive to self-enhance leads consumers to generate positive word-of-mouth (WOM) (i.e., share information about their own positive consumption experiences) but transmit negative WOM about others' negative experiences (De Angelis et al., 2012). By contrast with emotional pitches, the persuasion attempts of informative appeals are direct by nature (e.g., explicit presentation of product features), so inferences about persuasive attempts should be more positive, which should boost brand evaluations and purchase intent, because consumers may evaluate informative appeals' persuasion attempts as fairer and less manipulative (Akpinar & Berger, 2017).

6.5.4 Viral Marketing as a Measure of Success in Digital Advertising

What strikes me as odd is that in research that purports to be investigating viral properties there is little distinction made between digital marketing and viral marketing. For example, one paper in *Entrepreneurial Executive* recommends having "good products," advertised "with humor, style, visual impact, free giveaways, or other interesting content," representing them in an interactive and shareable way, and to make sure the company "seems genuine and friendly" (Rollins et al., 2014, p. 10). I fail to see how these practices are any different from "just marketing," and at this point, finding no evidence-based practical advice for best practices, I must resort to the idea that virality requires ingenuity, originality, creativity, and a strike of luck to achieve the mega threshold of success since there seem to be no clearly identifiable systematically prescribed methods for achieving it.

6.6 Mass Propaganda

6.6.1 Consistent and Enduring Efforts to Influence Minds

The classic manipulative and persuasive techniques of mass propaganda have been laid out since the work of Edward Bernays, the father of public relations, also known for orchestrating elaborate advertising campaigns. In 1928, he described *propaganda* as a consistent, enduring, continuous effort to create circumstances and pictures in the mind of the millions, to shape events to influence the relations of the public to an enterprise, idea, or group. In Bernays' mind, no important undertaking, even at that time, could be carried out without propaganda: whether the enterprise was to build a cathedral, endow a university, market a moving picture, float a large bond issue, or elect a president. Bernays wrote he was aware that the word "propaganda" "carries to many minds an unpleasant connotation. Yet whether, in any instance, propaganda is good or bad depends upon the merit of the cause urged, and the correctness of the information published" (Bernays, 1928, p. 20). He justifies the use of propaganda as "a perfectly legitimate form of human activity. Any society, whether it be social, religious or political, which is possessed of certain beliefs, and sets out to make them known, either by spoken or written words, is practicing propaganda" (Bernays, 1928, p. 22).

Bernays continues: "Truth is mighty and must prevail, and if any body of men believe that they have discovered a valuable truth, it is not merely their privilege but their duty to disseminate that truth. If they realize, as they quickly must, that this spreading of the truth can be done upon a large scale and effectively only by organized effort, they will make use of the press and the platform as the best means to give it wide circulation. Propaganda becomes vicious and reprehensive only when its authors consciously and deliberately disseminate what they know to be lies, or when they aim at effects which they know to be prejudicial to the common good" (Bernays, 1928, p. 22).

Propaganda involves the intentional sharing of facts, opinions, and ideas designed to change behavior or motivate action; a great emphasis is placed on intentionality and motive on the part of the sender of the message, the impact on the receiver's actions and behaviors, as well as the receiver's level of free will in accepting or rejecting the message (Hobbs & McGee, 2014, p. 57). The issue that should be raised time and again is: whose goals do the actions of propagandists serve? *Cui bono?* Whose ideas or options are being promoted?

Marketing and advertising campaigns promote their products or brands by exploiting mass insecurities, fears, wants, and desires by suggesting a direct link to their imagined better lives. Political campaigns direct the public's attention by branding policies, inflating personas, or justifying or preparing the public for government actions. As we saw earlier (in Chap. 2, Sect. 2.2), if a goal is unachievable with honest and truthful means, people may risk achieving their goals through deceit (see also Chap. 2, Sect. 2.3 for many deception varieties.) Thus, it is not surprising that peddling disinformation has become part of social and political promotional campaigns.

6.6.2 Propagandist Techniques

What techniques of persuasion and influence can propagandists resort to, more specifically? The options are unlimited in theory, but in practice many tactics, strategies, and tricks are more commonly in use by promoters and manipulators of public minds who realize the powers of compelling language and know how to exploit human weaknesses.

In the 1930s, the Institute for Propaganda Analysis (IPA), an independent organization that included journalists, college faculty, and secondary school teachers, functioned as a proto-media literacy group in the U.S. Hobbs and McGee (2014) review some educational materials created by the IPA in the 1930s, explicitly presenting their contents as the knowledge needed to avoid being victimized by presumably powerful and manipulative persuaders. The IPA materials go over seven popular propaganda techniques in Miller and Edwards (1936). See also Hobbs and McGee's (2014) succinct summary of the seven with brief explanation and historical example for each. The seven popular mass propaganda techniques used in IPA counter-propaganda educational efforts were: (1) name calling (a trick attempting to make us accept conclusions without full consideration of the facts); (2) band wagon (e.g., "everybody is doing this" logic); (3) glittering generalities (attempts to sway emotions through the use of shining ideal and virtues such as freedom, justice, truth, democracy, and education in a large general way); (4) flag waving (a trick of the propagandist holding up a symbol or flag that we recognize and respect); (5) "plain folks" (a trick of the propagandist demonstrating that they are just like the rest of us); (6) testimonial (best represented by the straw vote, a trick involving getting social and business leaders endorsing the party or candidate); and (7) stacking the cards against the facts (intentionally or unintentionally) (Hobbs & McGee, 2014).

Many of these seven devices sound surprisingly fresh and are amply used in contemporary political rhetoric, both on- and offline. The U.S. Presidential Election debates in 2016 and 2020 witnessed an escalation of such rhetoric, moving away from reasoned discussions and short-circuiting our ability to stay emotionless (see, for example, an opinion piece on Trumpisms and his cheap name-calling strategy to grab headlines (Buzenberg, 2016)).

The roots of these propagandist techniques are in the classic studies of rhetoric for persuasive argumentation, taught extensively in contemporary rhetoric classes for speech writing and public speaking. They were described by the IPA as "devices as of 'folk origin,' but tremendously powerful weapons for the swaying of popular opinions and actions," with the awareness of these devices keeping us from "having our thought processes blocked by a trick," being fooled or manipulated (Miller & Edwards, 1936, p. 24). Another term for those devices is logical fallacies and the list of seven is just a start (see also Chap. 4, Sect. 4.7 for suggested links to philosophy and rhetoric resources that compiled extensive lists of logical fallacies, some upward of one or two hundred specific types).

Both propaganda and advertising can be deceptive, or at least misleading and manipulative, if they intentionally uses these persuasive tactics. For instance, the "bandwagon" appeal uses the argument that a person should believe or do something because everyone else does. It exploits the desire of many people to join the crowd, to not miss out, or to be on the winning side. "Glittering generalities" may evoke ideas of youth, patriotism, freedom, peace, or child welfare. Advertisers have to be creative in their craft and superb in their language mastery to achieve widespread impact, but the inventory of such techniques, while exceeding the 1930s group of seven, is still surprisingly limited to a commonly-used two or three dozen.

Yet, they work when the audience is not overly critical or is prone to logical fallacies and indiscriminately accepts any promotional claims.

In their efforts to help American voters recognize professional propagandists' tricks, Miller and Edwards (1936) enumerate the IPA's five "ABC's of propaganda," or "five things to remember about propaganda:"

(a) *"Propaganda is opinion or action intended to influence the thoughts and actions of others.* Many of us unconsciously become the tools of clever propaganda tricksters and help them to do their work. Remember, anybody who expresses an opinion is a propagandist.
(b) *All propaganda centers on some conflict—some issue.* Make for yourself a list of issues-local, national, international to see the direction of your own opinions or propaganda relating to them.
(c) *Our own opinion or propaganda* with respect to every issue is determined by our environment, our training, our education, by the extent to which we, ourselves, have been influenced by the propaganda of others.
(d) *Opinion or propaganda is good or bad*—if we as Americans believe in the Declaration of Independence—insofar as it promotes or hinders life, liberty, and the pursuit of happiness of the great majority of our fellow citizens.
(e) *The best way to deal with propaganda* whether it is our own or that of others is to subject it to searching criticism and appraisal. This is the way of the intelligent man. Ask who holds the opinion—*who* utters the propaganda—and *why*" (Miller & Edwards, 1936, p. 71).

The bullet list calls for a reflective critical mind with historical and cultural awareness. These ideas seem to be timeless and still valid today: ready to be posted, on any wall or website, for the next election cycle in the U.S., Canada, or anywhere else in the world. Such educational efforts could anticipate some of the core concepts and instructional practices of media literacy in the twenty-first century (Hobbs & McGee, 2014). (See also Chap. 2, Sect. 2.6 on interventions to disinformation in the form of literacy education and inoculation of the public.)

6.6.3 Public Relations

Marketing campaigns focus on advertised products and brands, political propaganda may brand policies or inflate personas, while public relations campaigns focus on organizations. Going by the definition of *public relations (PR)* in the Merriam-Webster Dictionary Online (2021), similarly to the enterprises of advertising or political propaganda, PR is "the business of inducing the public to have understanding for and goodwill toward a person, firm, or institution." It is a game of deliberate control of the public perception of a certain enterprise. Digital PR is "conscious, planned and continuous efforts to establish and maintain mutual understanding between the organization and its environment, with a positive image of the organization in the environment implemented via the Internet" (Bianchi, 2020). While good (digital) PR are always welcome, bad publicity in digital media may quickly inflict damages to an enterprise's image, credibility, financial standing. When the image of an enterprise does not correspond to the descriptive truth, PR may carry undertones of distortions or cover-up. PR can be, in such cases, a politically correct term for smoke and mirrors. "While many scholars distinguish between public relations, advertising and propaganda, Bernays treated public relations and propaganda as equivalent, noting that propaganda may be either beneficial or harmful to the public, depending on the context and point of view of the interpreter" (Hobbs & McGee, 2014, p. 58). The parallels between the rhetorician's use of persuasive tactics—for the good or bad—are also undeniably transparent. Furthermore, Bernays, who is often credited with inventing the term PR, believed that that only thing that distinguishes propaganda (i.e., PR) from education is the point of view: "The advocacy of what we believe in is education. The advocacy of what we don't believe is propaganda" (Hobbs & McGee, 2014, p. 58). The same can be said of the politically charged and weaponized use of the term "fake news" in the twenty-first century. (See alternative terms and their nuances discussed in Chap. 1, Sect. 1.4 and Sect. 1.6).

6.6.4 Conspiracy Theory Peddling

Propaganda wars can be waged through the mass distribution of conspiracy theories. The term *conspiracy theory* almost always carries a negative connotation and is often used pejoratively by outsiders, while those who are misguided by conspiracy theories may defend them with passion and ardor. Conspiracy theories start with planting a seed message and theorists work on promoting the message much as mass propagandists do, via digital media channels, especially by appealing to non-critical minds which may not exert much scrutiny. What is characteristic of the spread of conspiracy theories is that their believers become propagandists themselves, and often feel compelled to share their beliefs, even when the message was not studied critically. The emotional persuasive appeal (*pathos*) may outweigh the rational reasoning (*logos*). When someone dismisses sound scientific advice in favor of a popular conspiracy, just because "it feels right," the material harm may be comparable to that of misleading digital advertising. The harm can be impactful if personal metadata is used to profile private intentions and target the profiled people with direct advertising. Conspiracy theorists may use advertising techniques in attempts to manipulate people's attitudes and behaviors. Some ideas may take on viral qualities. The mechanism exploits human biases, many of which have already been discussed (see Chap. 2, Sect. 2.4).

Belief in conspiracy theories has emerged in the past two decades as an area of socio-psychological research which characterizes this phenomenon with its four distinct properties: Conspiracy theories are "[1] *consequential* as they have a real impact on people's health, relationships, and safety; [2] they are *universal* in that belief in them is widespread across times, cultures, and social settings; [3] they are *emotional* given that negative emotions and not rational deliberations cause conspiracy beliefs; and [4] they are *social* as conspiracy beliefs are closely associated with psychological motivations underlying intergroup conflict" (van Prooijen & Douglas, 2018, p 897). Several specific mechanisms explain how one might be lured into believing conspiracy theories. *Cascading logic* is a belief that as new evidence comes up, it can be dismissed easily by claiming that more and more people are involved in the cover-up of an alleged misdeed. People may self-insulate themselves, that is, be resistant to questioning, and be antagonistic to those who do question them. The absence of evidence that would expose the conspiracy may be interpreted as a part of the plot and as a conspiratorial deception, since the belief is that there must be an explanation, rather than the lack of evidence pointing to the lack of a basis for the conspiracy existing. Questioning the "establishment," be it governments, scientists, or other authorities—is used as a typical logic to challenge the absence of definitive and immediate answers to the complicated issues, questions, and phenomena at hand. "Meanwhile, conspiracy theorists offer their own alternative theories with the flimsiest of evidence, challenging the authorities to prove them wrong" (Goertzel, 2010). The dangers may lie in the habit of shallow meme-like thinking which predisposes people to believe in other conspiracy theories once they are tricked into one of them. While advertising pre-conditions us negatively—we may end up avoiding and distrusting future ads, belief in conspiracies coupled with distrust in authorities may pre-condition us positively toward magicians' trickery, attention manipulation, and propagandist techniques. These claims are purely speculative, based on some analogies and a few links that attempt to explain why conspiracy theories are magnets for some of us but not others. It is a fascinating area of research that deserves a more nuanced investigation in the future.

6.7 Implications for Artificial Intelligence and Counterintelligence

6.7.1 Virality, Popularity, and Rumor Propagation in AI-Research

It is possible to conduct evidence-based studies of the properties of *virality* retroactively by looking at the propagation of a message through online networks and analyzing the message's features in order to make future predictions. No guarantees can be given that the discovered "tricks" will work again, though they can be ported to other contents, contexts, media, or products—specifically and deliberately—to attempt to re-create the message's virality.

Though a range of theoretical models has been proposed for modeling *online popularity*, online users typically do not have access to "readily available software that allows regular users to easily examine the popularity" of messages over time and to forecast their future popularity (Kong et al., 2018, p. 1). Advertisers and other promoters in the Internet attention economy are strongly financially incentivized to achieve the best results from their marketing, advertising, and promotional campaign dollars. There is no doubt that online influencers and manipulators, individuals and organizations alike, strive for the competitive advantage that such predictive tools can give them. Some tech enterprises aim to develop such predictive systems "in house," outsource, or otherwise acquire them. For example, a commercially developed strand of AI-based research taps into attention mechanisms, predicting in essence the future popularity of a video and the attention it would harness, based on the video's headline and video frame features (Bielski & Trzcinski, 2018).

Closely related to the evidence-based area of prediction of popularity is *rumor propagation* research in AI. This research aims to detect and bust viral rumors (Liu et al., 2015; Qazvinian et al., 2011; K. Wu et al., 2015; Yang et al., 2012). Researchers have long argued that rumors spread most in the "3 Cs—times of conflict, crisis, and catastrophe" such as wars or natural disasters (Turner & Koenig, 1986). The generalized anxiety or apprehension about negative outcomes from the 3 Cs has been traditionally used as an explanation for why rumors flourish in times of panic (Berger, 2011, p. 891). More recently, it has been demonstrated that the physiological arousal of either negatively or positively valanced emotions can explain the transmission of news or information in a wide range of settings. "Situations that heighten arousal should boost social transmission, regardless of whether they are positive (e.g., inaugurations) or negative (e.g., panics) in nature" (Berger, 2011, p. 892). Such mechanisms are at the core of virality, popularity, and the propagation of messages (both truthful and misleading) and are undoubtedly of much interest to industries in the business of manipulating public perceptions. (See also on how automated rumor debunkers work in Chap. 7, Sect. 7.5.)

6.7.2 Ad-Blockers Wars

From the point of view of advertisers, PR experts, or promotional campaigners, ad-blockers are an obstacle that may prevent their reach to (a portion of) the target audience and that brings ad revenue losses. An *ad-blocker* is "a software designed to remove or alter online advertising in a web browser or an application, preventing page elements containing ads from being displayed. [...] The average global adblocking rate in 2018 is estimated at 27%. Greece and Poland are masters of adblocking: their rates are 42% and 36% respectively. Among other leaders are the following countries: France (34%), Turkey (33%), Germany (33%), Hungary (32%), Croatia (32%), Sweden (32%), Portugal (31%), Spain (31%) and Austria (31%). Ireland and the USA have the exact same penetration rate of 27%." (cited in Bekh, 2020; from Statista Research Department, 2021b).

Thus, one type of AI-enabled tug-of-war consists of digital media users trying to ignore, avoid, or block ads, and platforms responding with attempts to deactivate or eliminate users' ad-blockers. Some apps also incentivize users to opt out of ad-blockers or influence them to whitelist certain content providers. They often argue that their ad revenue is essential to the enterprise, and that ad-blockers harm their content quality in the absence of alternative funds for staffing. Such arguments are more believable for smaller local digital newsrooms that have had their ad revenue essentially eliminated since giant tech platforms took over the distribution of news and lured advertisers away from legacy newsrooms.

6.7.3 Google's AdWords & AdSense

What kinds of technologies are used for advertising? Around 2012, software developer Bart Weller declared that Google is predominantly an advertising company and not the search or technology company that most users see it as. He wrote a book-length explanation for how small business owners with limited ad budgets could set and tune up a Google's AdWords campaign to compete on the same Google search results page with "the big guns" (i.e., companies with large advertising budgets). He explains that "the real-time auction-based system used by AdWords for determining ad position uses both keyword and landing page quality scores, on the one hand, and keyword bids, on the other. And quality scores count for a lot in this system" (Weller, 2012, p. xviii). "AdWords—and the Google Search Network—is the system responsible for presenting the short clickable ads you often see on the right or at the top of Google's organic search results page. AdSense—and the Google Display Network—on the other hand, is Google's paid ad placement technology responsible for presenting the AdWords ads you see on individual web pages. Google then pays the publishers of these web pages based either on user clicks on the ad (PPC) or impressions (PPM), depending on the type of ad. In general, AdWords is the interface used by advertisers and AdSense is the system used by publishers of web sites" (p. 16. Chap. 1) (Weller, 2012). This is but one of the ad models that have been around for almost 10 years. Digital advertising has no doubt evolved and innovated as it is driven by intense market competition. Perhaps more creatively stitched-together advertising "products" may be offered as of the day you are reading this book, but it is probably safe to assume that the principle of bidding wars by some quality factors remains.

Weller describes in general strokes what happened in most online paid placement marketing in 2012. A user enters search keywords in a search box, expressing their information need or search intent, and a search engine results page appears with some ads on the side, in addition to the *organic search results*. Which ads users see is determined not only by their search keywords, but also by the ad campaign's settings as selected by advertisers (such as the bid cost-per-click (CPC); the keyword phrases advertisers selected for their ad groups and campaigns; negative keywords; topic targeting; geographic targeting; time, day, location settings, or other constraints). In addition, Google's AdWords technology uses some undisclosed algorithms applied to the two previous items to determine which ads will appear and in which order. Regulatory standards and state and federal statutory frameworks specific to the areas of online marketing are also considered in the determination of what to show the user.

6.7.4 Facebook's Multiplatform Targeted Advertising

Google is obviously in competition with other giant tech companies for their ad revenue share and ad services. In 2021, Facebook reached out to businesses with an appeal to bring their business online with advertising options across their five platforms (Facebook, Instagram, Messenger, WhatsApp, and Audience Network). Facebook encourages any user, developer, or creator in media and publishing, as well as in small or large businesses, agencies, or startups to advertise on their five Platforms. For example, Facebook promotes its own ad services suggesting ad-related business development ideas by instructing businesses how they can presumably leverage their ads. There are more than two dozen reasons given to businesses, framed as marketing goals "within reach," a promotional campaign in and of itself, and a promise of providing "the right tools and resources."[14] Building an online presence (implied with Facebook) promises to help, for example, with raising awareness for the business, finding more customers, and increasing sales. Varieties of ad formats are presented in appealing ways and are accompanied by training materials and free workshops on how to, for instance, accomplish A/B testing of your ads (by varying the demographics of the targeted audience, creative elements, or types of placements including automatic ones[15]).

Facebook's advertising practices have been under criticism. Mark Zuckerberg's personal mentor from 2006 to 2009 and Facebook early investor, Roger McNamee, in his book (2019) *Zucked: Waking Up to the Facebook Catastrophe* delivers an emphatic verdict on the misuse of targeted advertising. He directly states that Facebook's manipulative practices prey on human biases and emotions, and moreover Facebook's purposeful engineering of persuasive technologies are built for surveillance, attention getting, addiction, and behavioral modification. This engineering was done without regard for the harms associated with undermining democratic processes and the impact on the individuals in democratic societies (McNamee, 2019). McNamee blames Mark's youthful idealism, his inability to imagine improper uses of his social media's networks, and his unwillingness to take responsibility for Facebook's nefarious uses. McNamee points out that Zuckerberg is especially overconfident in the righteousness of the position that anything engineered in innovative and profitable ways is justifiable, and for the most part, beneficial to mankind. Such a position, typical of the Silicon Valley mentality, unfortunately leads to technologically-enabled manipulation of the masses by the select few with deep pockets.

---

[14] See details on advised strategies at https://www.facebook.com/business/goals (Accessed on May 28, 2021).

[15] See https://www.facebook.com/business/help/1962159924052051 (Accessed on May 28, 2021).

6.7.5 Bot Wars: Automation of Advertising, Customer Services, and Social Engagement

The area of *marketing bots* is still poorly explored and scantly described in AI research literature but in principle can extend the capabilities of marketing automation and programmatic advertising (see Sect. 6.2.2). *Ad bots* are a type of software that automatically mimics some human tasks for the purpose of promoting products, content, or organizations. *Chat bots* can mimic human conversations, while other ad bots may simulate posts, likes, or other engagement, or may provide customer services. Promoters of such technologies seem to be banking on the switch from "the one-to-many, 'yell-and-sell' style [advertising] campaigns to one-on-one, 'chat-and-connect' conversations" that can even include the saved histories of prior interactions with such chatbots (AdWeek Editorial Staff, 2017). The advertising industry may still be working up to the ultimate impersonal con it has to offer, but the AI community has been exploring the use of conversational AI systems for several decades to enhance the ease of human-computer interactions. Chat bots, or intelligent conversational assistants, in their basic commercially available forms are still far from passing the Turing Test.[16] In other words, they do not sound or otherwise compose language intelligently enough to pass seamlessly for a human being. The risk of such impersonations being discovered may prevent advertisers from widely adopting such bots secretively. Or, users may start to actively avoid such interactions if they become the norm while still lacking in quality and frustrating their users. Complex and nuanced tasks still require human comprehension. (See a brief overview of conversational agents' capabilities and their use history in Rubin et al. (2010).)

Another form of *social bots* spreading on social media are *sybil accounts* which rely on computer algorithms to imitate humans by automatically producing content and interacting with other users. Such social bots pollute authentic content and spread misinformation by manipulating discussions, altering user popularity ratings, and "even perform[ing] terrorist propaganda and recruitment actions" (Davis et al., 2016). For example, Subrahmanian et al. (2016) identified three types of Twitter bots that engage in deceptive activities: (1) *spambots* spread spam on various topics; (2) *paybots* illicitly make money by copying tweet content from respected sources like @CNN and pasting URLs that re-direct traffic to sites which pay the bot creators, and (3) *influence bots* attempt to influence Twitter conversations on a specific topic to favor a political agenda (Rubin, 2017). Subrahmanian et al. (2016) also felt that influence bots "pose a clear danger to freedom of expression," citing examples of their effect on the spread of radicalism, political disinformation, and propaganda campaigns. Bots can also be easily used (by interested parties and their

[16] The *Turing Test* is an AI inquiry to determine whether or not a computer is capable of "thinking" like a human being, or at least can pass for one. It is named after Alan Turing, a mathematician and computer scientist, considered by some the father of modern computer science, and the founder of the Test.

promotional or PR campaigns) to propagate conspiracy theory beliefs with customization from personal data and search histories to tap into pre-estisting biases, emotional triggers, and other psychological motivations in underlying inter-group conflicts.

Some social media companies ban fake accounts outright and put effort into removing them regularly by employing armies of content moderators and engaging in AI detection algorithms. (See also Chap. 2 on efforts in content moderation and other interventions in Sect. 2.6.) Zhang (2021) describes her experience working on one of Facebook's fake engagement teams to remove inauthentic accounts. In this op-ed piece, she positions herself as a whistleblower frustrated with Facebook's unwillingness to disable the accounts of real users who share their accounts with a bot farm: "Most of these accounts were commandeered through *autolikers*: online programs which promise users automatic likes and other engagement for their posts. Signing up for the autoliker, however, requires the user to hand over account access. Then, these accounts join a bot farm, where their likes and comments are delivered to other autoliker users, or sold en masse, even while the original user maintains ownership of the account. Although motivated by money rather than politics—and far less sophisticated than government-run human troll farms—the sheer quantity of these autoliker programs can be dangerous" (Zhang, 2021). It is clear that autolikers rig the system to advantage a select few, but the industry has not yet self-corrected to ban those practices.

Spambots, autolikers, fake or inauthentic accounts, or other types of automated agents personifying human beings may appear to be an extremely lucrative area of R&D from the point of view of advertisers, propagandists, influencers, and influencer-wannabes. AI-based tools are still fairly obscure to the average Internet user but as a commodity they are becoming increasingly sought out by advertisers and their clients who strategize to gain competitive marketing advantages. AI-supported countermeasures are urgently required, and more importantly, so is the R&D and the public will to clean up the rigged systems. AI-aware policies and more stringent regulation should be enacted—and not just in "the Western bubble" but worldwide—to avoid the escalation of this AI-infused tug-of-war that privileges entities with access to these technologies. The next section explores a few options and associated concerns with regulatory oversight.

6.8 Broader Regulations and Legislation as Countermeasures

6.8.1 Avoiding Governmental Overreach in Freedom of Speech and Citizens' Liberties

Let us consider what is involved in taking on more stringent regulations in terms of the potential stumbling blocks and known precedents of measures taken across the world. The need to balance the freedoms of expression are often put on the counterbalance to regulating online communications, a common concern among many democracies especially addressed to restrictive regimes. In the U.S., for example, information in the public discourse online, regardless of its accuracy has previously received substantial First Amendment protection,[17] while professional speech, commercial speech, and court testimony are considered non-public discourse, and are thus subject to rules and regulation. Distributor liability for technological architecture or digital content intermediaries, while generally afforded greater protection than other distributors as a result of §230 of the Communications Decency Act,[18] is not absolute and is not constitutionally required (Baron & Crootof, 2017). Governmental regulation starts with testing the boundaries of the depth of protection of false speech with the First Amendment (specifically, in the U.S. context), regulation of fraud in non-public discourse (such as commercial or professional speeches), and tech companies' liabilities, as well as options for the enforcement of the above measures at national levels.

Discussing viability and desirability of possible regulatory measures to curb mis- and disinformation, a (2017) Information Society Project at Yale Law School considered both "the negative state" and "the positive state" governmental actions. "The negative state involves the government engaging in coercive actions, such as fining, taxing, and imprisoning. The positive state involves creating institutions and incentives, like land grant colleges or tax subsidies" (Baron & Crootof, 2017). These are the proverbial "stick and carrot" methods, but the legal experts and academics in the (2017) workshop were reluctant to propose negative state regulations. They noted key tradeoff between tighter controls is the freedom of speech in public discourse and the harms coming from the loss of trust in democratic institutions and the loss of the value of facts (Baron & Crootof, 2017).

A (2019) group of foreign law analysts for the U.S. Congress surveyed the state of governmental responses to disinformation across the European Union (EU) and fifteen selected countries from around the world. They also sounded out a few warnings, in particular, given that the interpretation of what constitutes mis- and disinformation is fluid (see also Chap. 1, Sect. 1.4 on "fake news" and other terminology) and some measures may be construed as governmental censorship of online communications (Ahmad, 2019). They cautioned against the "broad application of emergency powers to block content based on grounds of national or public security," and against applying "draconian penalties for alleged offenders without the ability to present an effective defense," "strict enforcement of defamation laws in the absence of journalistic defenses" (Ahmad, 2019). They were concerned that such measures may be seen as "potential threats to the principle of free speech and the administration of the rule of law posed by overreaching regulations concerning disinformation." This position echoes the earlier legal experts' reluctance to

[17] "The [U.S.] Supreme Court has repeatedly held that false speech enjoys full First Amendment protection. See, e.g., United States v. Alvarez, 567U.S. ___ (2012)." (Baron & Crootof, 2017, p. 7).

[18] "47U.S.C. § 230 (1996). Sect. 230 provides: "No provider or user of an interactive computer service shall be treated as the publisher or speaker of any information provided by another information content provider." (Baron & Crootof, 2017, p. 8).

institute "the negative state" measures (Baron & Crootof, 2017). At the same time, the public attitude exhibits increasing concern about the mis- and disinformation problem and is shifting toward favoring acceptance of regulatory actions. For example, "roughly half of U.S. adults (48%) now say the government should take steps to restrict false information, even if it means losing some freedom to access and publish content, according to the survey of 11,178 adults conducted July 26–Aug. 8, 2021. That is up from 39% in 2018" (Mitchell & Walker, 2021).

European experts seem to concur and are encouraging policymakers and legislators to avoid crude measures of controlling and criminalizing content. Instead, they suggest to focus on issues surrounding algorithmic transparency, platform accountability, digital advertising, and data privacy: "As algorithms and artificial intelligence have been protective of their innovations and reluctant to share open access data for research, technologies are black-boxed to an extent that sustainable public scrutiny, oversight and regulation demands the cooperation of platforms." (Nothhaft et al., 2018, p. 12). This position aligns, in principle, with the U.S. experts (Baron & Crootof, 2017) on the overarching goals that "reestablishing trust in the basic institutions of a democratic society [...] to combat the systematic efforts being made to devalue truth" and figuring out how to bolster facts (p. 11).

The NATO Strategic Communications Center of Excellence analyzed the regulatory measures across 43 world governments proposed or implemented between 2016 and 2018 (Nothhaft et al., 2018). Their 2018 insight is that "in the current highly-politicized environment driving legal and regulatory interventions, many proposed countermeasures remain fragmentary, heavy-handed, and ill-equipped to deal with the malicious use of social media. Government regulations thus far have focused mainly on regulating speech online—through the redefinition of what constitutes harmful content, and measures that require platforms to take a more authoritative role in taking down information with limited government oversight. However, harmful content is only the symptom of a much broader problem underlying the current information ecosystem, and measures that attempt to redefine harmful content or place the burden on social media platforms fail to address deeper systemic challenges, and could result in a number of unintended consequences stifling freedom of speech online and restricting citizen liberties" (Nothhaft et al., 2018, p. 12).

Journalists and fact-checking organizations keep track of other governmental actions against mis- and disinformation, see for example, the story put out by *Poynter* with frequent updates into mid-2019 (Funke & Flamini, 2019). More recently, there have been further calls for legislative reforms and tougher regulation of social media companies by self-regulation by the platforms (e.g., (Ghosh, 2021)), while others are explicitly are concerned about the human right perspective of content regulation triggering "censorship, internet shutdowns and the prosecution of dissenting voices all ostensibly in the name of fighting disinformation" (in the context of EU actions within the EU and in third countries) (Colomina et al., 2021, p. 56).

Overall, in term of their practical implementations, regulations and legislations can be grouped into four categories by their targets: social media platforms, offenders, governments, and civil society. Below are a few examples from around the world (informed primarily by the insightful analysis by Nothhaft et al. (2018)). This four-pronged survey is informative in terms of available options, in principle, and their outcomes, in practice.

6.8.2 Regulatory and Legislative Measures Targeting Social Media Platforms

(a) **Content Take-Downs.** "Government-monitored removal, blocking, and filtering of illegal content online is a well-established practice" but it lacks transparency which in turn "could lead to chilling effects in the digital public sphere" such as "collateral censorship" (such as seen in Russia, Vietnam and Zimbabwe) (Nothhaft et al., 2018, p. 6).
(b) **Advertising Transparency.** Political advertising (in print and broadcast media) is subject to laws, tight regulations and standards, in several countries striving to ensure the fairness of democratic processes, with controls on campaign spending, messages, scope, and timing. "With billions of dollars spent on advertising, engagement campaigns, and the curation of voter profiles, lawmakers have yet to extend the same scrutiny to digital contexts." (Nothhaft et al., 2018, p. 7). For online advertising, some laws aim to improve transparency around the purchasers of advertising space and target audiences, as well as block foreign spending on domestic political campaigns, for example with the Honest Ads Act in the U.S. (Nothhaft et al., 2018, p. 7). The U.S., France, and Ireland require social network companies to collect and disclose information to users about who paid for an advert or piece of sponsored content, and to share information about the audience that advertisers target. "While some platforms have begun to self-regulate, their self-prescribed remedies often fall short of providing efficient countermeasures and enforcement mechanisms" (Nothhaft et al., 2018, p. 12).
(c) **Data Protection.** Only a few countries implemented data protection measures to combat data-driven targeting in advertising that relies on data collected from users. EU citizens' data are protected by the General Data Protection Regulation but it has gaps in coverage and enforcement that limit its effectiveness. Some countries may overreach their governmental control over citizen data, for example, Vietnam's data localization law keeps data stored within its state borders and accessible to the state (Nothhaft et al., 2018, p. 7).

6.8.3 Regulatory and Legislative Measures Targeting Offenders

(a) **Criminalization of Disinformation and Automation.** Several countries opt to legally deter or prosecute offenders with monetary fines and increased prison sentences (such as in Egypt, Indonesia, and Kuwait), in spite of experts' cautionary words on crude negative state measures. Other bills (e.g., in Ireland and

California) do not merely prosecute the originators of online disinformation but also malicious disseminators and amplifiers through automation. "Rooting their countermeasures in various legal arguments surrounding national security, the disturbance of national order, hate speech, and the provision of false and misleading information, Australia, Indonesia, Ireland, Italy, Malaysia, and the Philippines are among the countries that rely on criminal penalties and fines for producing or sharing disinformation, or for creating and launching a bot campaign targeting a particular political issue. Instances of the misuse of these frameworks to crackdown on political dissidents, minorities, and human rights defenders have already taken place in Iran, Malaysia, Russia, Saudi Arabia, and Tanzania" (Nothhaft et al., 2018, p. 8).

(b) **Expanding the Definition of Illegal Content.** The world's democracies thus far pioneered the redefinition of legal frameworks to propose novel definitions of "illegal content online" and revision of bills to sharpen their enforcement. For example, Australia has "strict punishment for anyone found guilty of communicating information against the "national interest," particularly with regard to false or distorted content. Germany's Network Enforcement Act explicitly extends the application of the German Criminal Code in cases where freedom of speech and constitutional values are in conflict. And France defers the legal interpretation of fake news and online content to its judiciary, whereby judges rule on prominent untruthful content on a case-by-case basis." Worrisomely, countries like Saudi Arabia and Egypt introduced even broader definitions of illegal content online (Nothhaft et al., 2018, p. 8).

6.8.4 *Regulatory and Legislative Measures Targeting Government Capacity*

(a) **Media Literacy and Watchdogs.** Long-term educational and advocacy efforts have been implemented nationally, around the world. For example, Croatia instituted new literacy initiatives, and "France is expanding the obligations of media watchdogs to improve public information literacy and exercise scrutiny over non-governmental institutions" (Nothhaft et al., 2018, p. 9).

The Government of Canada's multi-pronged approach, the Digital Citizen Initiative, funds citizen education in critically assessing online information, to "build resilience against disinformation" and build partnerships with the academic community and other Canadian civil society stakeholders "to support a healthy information ecosystem" (Canadian Heritage, the Government of Canada, 2021).

In the European Commission's Action Plan against Disinformation "four key pillars" were established: "(1) a coordinated response by the EU, mobilizing all government departments; (2) improving detection, analysis and exposure capabilities; (3) mobilizing the private sector; and (4) building societal resilience and raising awareness through conferences, debates, specialized training and media literacy programmes to enable citizens to spot disinformation" (Ignatidou, 2019).

(b) **Media Accreditation and Journalistic Controls.** Tighter controls over national media have also been put in place in several countries. "Strategies, such as the United States' enforcement of the Foreign Agents Registration Act, seek to bolster quality journalism while improving transparency regarding information sources." Media accreditation strategies were observed in restrictive regimes to exercise control over journalistic production of all content (e.g., in Iran and Tanzania) (Nothhaft et al., 2018, p. 9).

6.8.5 Regulatory and Legislative Measures Targeting Citizens, Civil Society, and Media Organizations

(a) **Parliamentary and Congressional Hearings** are legal tools and the starting point for action through the creation of policy briefs and recommendation documents (e.g., the Cambridge Analytica scandal-related inquiries in several countries: the UK's Digital, Culture, Media, and Sport Committee's inquiry, the Canadian House of Commons' investigation into data breaches and election integrity, the U.S. Congressional Hearings on election interference). "At a regional level, the European Union also established a High-Level Expert Group that brought together government representatives, academics, and issue-area experts to put forward recommendations" (Nothhaft et al., 2018, p. 9). Since private data are used extensively for targeted ad campaigns, its use is implicated in the dissemination of targeted mis- and disinformation as well.

(b) **Security and Defense.** Cybersecurity and information security units are established within several countries' militaries for the security and defense of informational infrastructure, strategic cyber warfare operations, and the prevention of foreign interference (e.g., Australia's Election Integrity Task Force, units in Brazil, the Czech Republic, Sweden, and Vietnam). "Countermeasures include systematic observation of the online space, identifying offenders, analyzing strategies of offense, reporting on problematic information as it rises to prominence, and debunking falsehoods." The scope of security, defense, and surveillance operations and their intervention remaining opaque and little is known to the public (Nothhaft et al., 2018, p. 10). Involving military counterintelligence aimed at foreign interferences is illustrative of the state of apprehension in governments about the potential harms of propagandist manipulations across national boundaries.

(c) **Monitoring and Reporting** includes providing users with portals to report misinformation, for example, G7's Rapid Response Mechanism and Italy's monitoring portal for investigation in the run up to their next election. Taxing citizens for using social media has been used to generate revenue and limits the amount of rumors on social media: in Uganda, for instance, "to access certain online platforms, citizens are expected to pay approximately 200 Uganda shillings (0.05 EUR) per day to use the platforms" (Nothhaft et al., 2018, p. 11).

Overall, regulatory and legislative countermeasures to mis- and disinformation implemented within the national boundaries can be restricting of citizens' liberties in their respective societies. Experts advise against tight controls but the problem is persisting not just within individual national boundaries, but also internationally, and it has been ranked highly among the global challenges of the twenty-first century (Gray, 2017; Heaven, 2017). Establishing a global regulatory framework and venues for enforcement may be key to addressing the problem across nations, cultures, and language boundaries.

6.8.6 Cutting the Finances of "the Disinformation for Hire" Industry

Most of the regulatory measures discussed thus far target citizens' behaviors and put the burden of verification and self-regulation primarily on the tech companies. Less emphasis has been placed on controlling the "the deep pockets" that finance unscrupulous disinformation campaigns. While foreign government interferences have been investigated and discussed at length, the same cannot be said of interferences by corporations, industries, or political entities that can instigate the mass dissemination of falsehoods by hiring a workforce to spread the word on their behalf.

Max Fisher of *the New York Times* reported that French and German social media influencers received a strange proposal in May 2021: "A London-based public relations agency wanted to pay them to promote messages on behalf of a client. A polished three-page document detailed what to say and on which platforms to say it. But it asked the influencers to push not beauty products or vacation packages, as is typical, but falsehoods tarring Pfizer-BioNTech's Covid-19 vaccine. Stranger still, the agency, Fazze, claimed a London address where there is no evidence any such company exists. Some recipients posted screenshots of the offer. Exposed, Fazze scrubbed its social media accounts. That same week, Brazilian and Indian influencers posted videos echoing Fazze's script to hundreds of thousands of viewers." (Fisher, 2021). "The false message claimed that "Pfizer's Covid-19 vaccine is deadly and that regulators and the mainstream media are covering it up," according to another New York Times exposé (Alderman, 2021).

Unscrupulous PR firms, advertisers, or individual online content generation workers struggling to make ends meet may accept a lucrative offer for PR campaigns, regardless of their content. "Private firms, straddling traditional marketing and the shadow world of geopolitical influence operations, are selling services once conducted principally by intelligence agencies. They sow discord, meddle in elections, seed false narratives and push viral conspiracies, mostly on social media. And they offer clients something precious: deniability" (Fisher, 2021). "The French health minister, Olivier Véran, denounced the operation … calling it "pathetic and dangerous" [but] did not elaborate on whether the government was investigating the matter" (Alderman, 2021). "Most trace to back-alley firms whose legitimate services resemble those of a bottom-rate marketer or email spammer" (Fisher, 2021). The question remains open about who is behind the attempt to derail the mass immunization in France. The New York Times reporters implicated the usual suspects, the Russian Government and Chinese media, but it stands to reason that corporations stand to benefit from the defamation of their competitors as well.

Employing armies of inauthentic bots, fake reviewers, dishonest advertisers, clickbait writers, and ideologically promiscuous influencers who knowingly and intentionally deceive should not be considered legitimate business practices. Neither should these activities bring profit. No doubt, determining what is false or inaccurate is not always straightforward, but there is precedence in the advertising industry against similar outlawed acts (Sect. 6.3)

Cutting off funding to disinformation campaigns as important. Those who create false content "professionally," for profit on behalf of others, are implicated intermediaries. They are communicators and amplifiers on social media with the "influencer" status, but the real influence comes from commercial or political agendas.

To address the underlying problems, nations need to regulate more stringently all aspects of "the disinformation for hire" industry: those that put in an order for and who finance unscrupulous ads and propaganda under guises of public discourse, those that implement such campaigns, and those that allow for falsehoods to trickled down within their platforms. For example, "articles on investment news sites are frequently based on corporate news releases, which suggests that these firms may have been attempting to artificially inflate stock prices. In 2017, the Securities and Exchange Commission (SEC) charged 27 separate entities and individuals with running stock promotion schemes that involved hiring firms and freelancers to write and publish bullish articles about the publicly traded companies" (Stansberry, 2021).

The lucrative "disinformation for hire" business is not limited to manipulating investment behaviors. Many areas of private lives can be re-shaped by incorrectly formed impressions and attitudes which subsequently manipulate people's behaviors. Over 644 brands, in the banking, retail and luxury good, airline, and cosmetic industries to name a few, have been linked[19] to advertising on "questionable sites ranging from hardcore hyper-partisan fake news sites to clickbait venues hosting bogus content with no particular agenda, except making a quick buck" (Filloux, 2017). Why it is legally allowed to seek such services for hire is beyond my understanding and regulating industries against using them should be an obvious step to tighter controls on online communication.

[19] *Storyzy* is a French start-up that scans the web for disinformation sources and reported this information (Accessed at https://storyzy.com/ on 1 September 2021).

6.9 Conclusions

It is reasonable to conclude that there are no salient differences between propagating social, political, scientific, or religious beliefs, and propagating commodities. The various types of campaigns—promotional advertising, propagandist, or PR—are likely to use similar persuasive tactics. They aim to capture attention and generate interest and engagement. But in essence, manipulators of public perceptions want to compel digital users to act in ways that are consistent with their campaign goals: to purchase, to vote, or to support their cause. It is wise to consider whose interests they are speaking for and who is to benefit (*Cui bono?*). Tighter control of commercial discourse and misuse of public discourse should be the next logical step in policy-making and legislation. The counter-disinformation measures should be considered at national levels, as they have already been in many states, but also extended globally, seeking for cooperation across nations to institute a single international framework to curb the spread of false information online.

Propagandists, advertisers, and public relations experts are well versed in the psychology of persuasion and know to appeal to emotions, appeal to individuality, and how to use wit and humor. They may finetune who delivers the message: a celebrity, a common folk, or a perceived medical professional. Unmasking such verbal, visual, and emotional techniques in ads, social or political campaigns, and PR messages is invaluable to liberate yourself from their influence.

To resurrect advice from 1930s anti-propaganda campaigns: "An educated man, if he is wise, just when a crowd is filled with enthusiasm and emotion, will leave it and will go off by himself to form his own judgement" (Miller & Edwards, 1936, p. 69). This emphasis on critical thinking and perspicacity is what they called being *an intelligent citizen* then, and it could not be a more timely goal almost a hundred years later. Another proclamation is equally simple: "Education is a powerful antidote to propaganda" (Hobbs & McGee, 2014). Preparedness and a practiced eye may help the vulnerable and the susceptible to liberate themselves from the powers of manipulative persuasion. A conscious awareness inoculates a potential target of trickery; warning the mark of the con. Any magic trick loosens its grip on the audience, once they know exactly where to look, and what the illusion is, no matter how skillful the magician.

References

AdWeek Editorial Staff. (2017, December 11). *There can be no denying it: Bots are now marketers*. Retrieved from https://www.adweek.com/performance-marketing/beerud-sheth-gupshup-guest-post-bots-are-now-marketers/.

Ahmad, T. (2019, September). *Government responses to disinformation on social media platforms* [Web page]. Retrieved from https://www.loc.gov/law/help/social-media-disinformation/canada.php

Akpinar, E., & Berger, J. (2017). Valuable virality. *Journal of Marketing Research, 54*(2), 318–330. https://doi.org/10.1509/jmr.13.0350

Alderman, L. (2021, May 26). Influencers say they were urged to criticize Pfizer vaccine. *The New York Times*. Retrieved from https://www.nytimes.com/2021/05/26/business/pfizer-vaccine-disinformation-influeners.html.

Amazeen, M. A., & Muddiman, A. R. (2017). Saving media or trading on trust? *Digital Journalism, 1–20*. https://doi.org/10.1080/21670811.2017.1293488

Bannock, G. (2002). *The new Penguin business dictionary*. Penguin Books. Retrieved from https://catalog.hathitrust.org/Record/102028705

Baron, S., & Crootof, R. (2017). *Fighting fake news: Workshop report*. Yale Law School Publications. Retrieved from https://law.yale.edu/isp/publications

Bekh, A. (2020). Advertising-based revenue model in digital media market. *Ekonomski Vjesnik, 33*(2), 547–559.

Berger, J. (2011). Arousal increases social transmission of information. *Psychological Science, 22*(7), 891–893. https://doi.org/10.1177/0956797611413294

Bernays, E. L. (1928). *Propaganda*. H. Liveright.

Bianchi, A. (2020). *Driving consumer engagement in social media: Influencing electronic word of mouth*. Routledge. Retrieved from https://doi.org.proxy1.lib.uwo.ca/10.4324/9781003125518

Bielski, A., & Trzcinski, T. (2018). Pay attention to Virality: Understanding popularity of social media videos with the attention mechanism. *ArXiv:1804.09949 [Cs]*. Retrieved from http://arxiv.org/abs/1804.09949

Buzenberg, B. (2016, June 1). Why we Shouldn't call trump an "ignorant bully" (even when we really want to). *YES! Magazine*. Retrieved from https://www.yesmagazine.org/opinion/2016/06/01/why-we-shouldnt-call-trump-an-ignorant-bully-even-when-we-really-want-to

Campbell, C., & Evans, N. (2018). The role of a companion banner and sponsorship transparency in recognizing and evaluating article-style native advertising. *Journal of Interactive Marketing*. https://doi.org/10.1016/j.intmar.2018.02.002

Canadian Heritage, the Government of Canada. (2021, 20). *Online disinformation*. Retrieved from https://www.canada.ca/en/canadian-heritage/services/online-disinformation.html

Carlson, M. (2015). When news sites go native: Redefining the advertising-editorial divide in response to native advertising. *Journalism, 16*(7), 849–865. https://doi.org/10.1177/1464884914545441

Chaffey, D., & Ellis-Chadwick, F. (2018). *Digital marketing: Strategy, implementation and practice* (6th ed.). Pearson Education. Retrieved from http://www.dawsonera.com/depp/reader/protected/external/AbstractView/S9780273746225

Colomina, C., Margalef, H. S., & Youngs, R. (2021). *The impact of disinformation on democratic processes and human rights in the world*. 64. Retrieved from https://www.europarl.europa.eu/RegData/etudes/STUD/2021/653635/EXPO_STU(2021)653635_EN.pdf

Cornwell, S., & Rubin, V. L. (2019). What am I reading?: Article-style native advertisements in Canadian newspapers. *Truth and lies: Deception and cognition on the internet* (10 pp). doi: https://doi.org/10.24251/HICSS.2019.625.

Darke, P. R., & Ritchie, R. J. (2007). The defensive consumer: Advertising deception, defensive processing, and distrust. *Journal of Marketing Research, 44*(1), 114–127. https://doi.org/10.1509/jmkr.44.1.114

Davis, C. A., Onur Varol, O., Ferrara, E., Flammini, A., & Menczer, F. (2016, February 2). *BotOrNot: A system to evaluate social bots*. WWW'16 companion, Montréal, Québec, Canada. Retrieved from http://arxiv.org/pdf/1602.00975.pdf

De Angelis, M., Bonezzi, A., Peluso, A. M., Rucker, D. D., & Costabile, M. (2012). On braggarts and gossips: A self-enhancement account of word-of-mouth generation and transmission. *Journal of Marketing Research, 49*(4), 551–563. https://doi.org/10.1509/jmr.11.0136

Editors of Dictionary.com. (n.d.). Definition of puffery. Www.Dictionary.Com. Retrieved June 4, 2021, from https://www.dictionary.com/browse/puffery

Editors of Urban Dictionary. (2021). Virality: Definition. In *Urban Dictionary*. Retrieved from https://www.urbandictionary.com/define.php?term=Virality

Editors of Wikipedia. (2021a). Puffery. In *Wikipedia*. Retrieved from https://en.wikipedia.org/w/index.php?title=Puffery&oldid=1015952434

Editors of Wikipedia. (2021b). Federal Trade Commission. In *Wikipedia*. Retrieved from https://en.wikipedia.org/w/index.php?title=Federal_Trade_Commission&oldid=1025828293
Editors of Wikipedia. (2021c). Native advertising. In *Wikipedia*. Retrieved from https://en.wikipedia.org/w/index.php?title=Native_advertising&oldid=1026047009
Entman, R. M. (1993). Framing: Toward clarification of a fractured paradigm. *Journal of Communication, 43*(4), 51–58. https://doi.org/10.1111/j.1460-2466.1993.tb01304.x
Evans, N. J., & Park, D. (2015). Rethinking the persuasion knowledge model: Schematic antecedents and associative outcomes of persuasion knowledge activation for covert advertising. *Journal of Current Issues & Research in Advertising, 36*(2), 157–176. https://doi.org/10.1080/10641734.2015.1023873
Fair, L. (2020, April 16). *No pain (relief), no gain? FTC challenges claims aimed at older consumers*. Federal Trade Commission. Retrieved from https://www.ftc.gov/news-events/blogs/business-blog/2020/04/no-pain-relief-no-gain-ftc-challenges-claims-aimed-older
Fair, L. (2021, March 15). *Avoid mixed signals when advertising antennas*. Federal Trade Commission. Retrieved from https://www.ftc.gov/news-events/blogs/business-blog/2021/03/avoid-mixed-signals-when-advertising-antennas.
Federal Trade Commission. (1983, October 14). FTC policy statement on deception. Federal Trade Commission. Retrieved from https://www.ftc.gov/public-statements/1983/10/ftc-policy-statement-deception.
Federal Trade Commission. (2013, March 1). About the FTC. Federal Trade Commission. Retrieved from https://www.ftc.gov/about-ftc.
Federal Trade Commission. (2020, April 13). *FTC announces latest round of letters warning companies to cease unsupported claims that their products can treat or prevent coronavirus*. Federal Trade Commission. Retrieved from https://www.ftc.gov/news-events/press-releases/2020/04/letters-warning-companies-cease-unsupported-coronavirus-claims
Federal Trade Commission (n.d.). *The Federal Trade Commission (FTC) GreenGuides: Statement of basis and purpose*. Retrieved June 4, 2021, from https://www.ftc.gov/sites/default/files/attachments/press-releases/ftc-issues-revised-green-guides/greenguidesstatement.pdf
Feldwick, P. (2015). *The anatomy of humbug: How to think differently about advertising*. Matador.
Fill, C. (2013). *Marketing communications: Brands, experiences and participation* (6th ed.). Pearson Canada.
Filloux, F. (2017, August 22). *More than 600 global brands are advertising on fake news websites—And they don't seem to care*. Quartz. Retrieved from https://qz.com/1059158/more-than-600-global-brands-still-feed-the-fake-news-ecosystem-and-they-dont-seem-to-care/
Fisher, M. (2021, July 25). Disinformation for hire, a shadow industry, is quietly booming. *The New York Times*. Retrieved from https://www.nytimes.com/2021/07/25/world/europe/disinformation-social-media.html
Friestad, M., & Wright, P. (1994). The persuasion knowledge model: How people cope with persuasion attempts. *Journal of Consumer Research, 21*(1), 1–31. https://doi.org/10.1086/209380
Funke, D., & Flamini, D. (2019, August 13). A guide to anti-misinformation actions around the world. *Poynter.* Retrieved from https://www.poynter.org/ifcn/anti-misinformation-actions/.
Ghosh, D. (2021, January 14). Are we entering a new era of social media regulation? *Harvard Business Review.* Retrieved from https://hbr.org/2021/01/are-we-entering-a-new-era-of-social-media-regulation
Goertzel, T. (2010). Conspiracy theories in science. *EMBO Reports, 11*(7), 493–499. https://doi.org/10.1038/embor.2010.84
Google Ads Help. (2021). *Clickthrough rate (CTR): Definition*. Retrieved from https://support.google.com/google-ads/answer/2615875.
Gray, R. (2017, March 1). Lies, propaganda and fake news: A challenge for our age. *BBC Futures*. Retrieved from https://www.bbc.com/future/article/20170301-lies-propaganda-and-fake-news-a-grand-challenge-of-our-age
Hackley, C. (2010). Advertising (Vol. 1–3). doi: https://doi.org/10.4135/9781446260807.

Hackley, C. (2011). Theorizing advertising: Managerial, scientific and cultural approaches. In *The SAGE handbook of marketing theory* (pp. 89–108). SAGE Publications Ltd.. https://doi.org/10.4135/9781446222454

Heath, C., Bell, C., & Sternberg, E. (2001). Emotional selection in memes: The case of urban legends. *Journal of Personality and Social Psychology, 81*(6), 1028–1041. https://doi.org/10.1037/0022-3514.81.6.1028

Heaven, D. (2017, February 18). A guide to humanity's greatest challenges. *BBC Future.* Retrieved from https://www.bbc.com/future/article/20170228-a-guide-to-humanitys-greatest-challenges.

Hobbs, R., & McGee, S. (2014). Teaching about propaganda: An examination of the historical roots of media literacy. *Journal of Media Literacy Education, 6*(2), 56–67. https://doi.org/10.23860/JMLE-2016-06-02-5

Ignatidou, S. (2019). *EU–US cooperation on tackling disinformation.* International Security Department. Retrieved from https://www.chathamhouse.org/2019/10/eu-us-cooperation-tackling-disinformation

Interactive Advertising Bureau. (2013). *The native advertising playbook* (p. 19). Interactive Advertising Bureau. Retrieved from https://www.iab.com/wp-content/uploads/2015/06/IAB-Native-Advertising-Playbook2.pdf

IAB Europe (2019). AdEx Benchmark 2018. Retrieved from https://www.iabeurope.eu/research-thought-leadership/resources/iab-europe-report-adex-benchmark-2018/

IAB & PwC (2019). IAB Internet Advertising Revenue Report - 2018 full year results. Retrieved from https://www.iab.com/wpcontent/uploads/2019/05/Full-Year-2018-IAB-Internet-Advertising-Revenue-Report.pdf

Kitson, H. D. (1923) *Understanding the consumer's mind.* The Annals of the American Academy of Political and Social Science Vol. 110, Psychology in Business, pp. 131–138

King, R. P. (2016). Popular sources, advertising, and information literacy: What librarians need to know. *The Reference Librarian, 57*(1), 1–12. https://doi.org/10.1080/02763877.2015.1077772

Kong, Q., Rizoiu, M.-A., Wu, S., & Xie, L. (2018). Will this video go viral? Explaining and predicting the popularity of Youtube videos. *Companion of the the web conference 2018 on the web conference 2018—WWW '18* (pp. 175–178). doi: https://doi.org/10.1145/3184558.3186972.

Laursen, J., & Hewes, J. (2018). *Native advertising trends 2018.* Native Advertising Institute. Retrieved from https://nativeadvertisinginstitute.com/wp-content/uploads/2018/12/FIPP_NAI_rapport2018-1.pdf.

Levine, T. R., Park, H. S., & McCornack, S. A. (1999). Accuracy in detecting truths and lies: Documenting the "veracity effect". *Communication Monographs, 66*(2), 125–144. https://doi.org/10.1080/03637759909376468

Liu, X., Nourbakhsh, A., Li, Q., Fang, R., & Shah, S. (2015). *Real-time rumor debunking on twitter.* 1867–1870. doi: https://doi.org/10.1145/2806416.2806651.

McNamee, R. (2019). *Zucked: Waking up to the Facebook catastrophe.* Penguin Press.

Merriam-Webster Dictionary Online. (2021). *Public relations.* Retrieved from https://www.merriam-webster.com/dictionary/public+relations.

Miller, C. R., & Edwards, V. (1936). The intelligent Teacher's guide through campaign propaganda. *The Clearing House, 11*(2), 69–73. https://doi.org/10.1080/00098655.1936.11474219

Mitchell, A., & Walker, M. (2021, August 20). *More Americans now say government should take steps to restrict false information online than in 2018.* Pew Research Center. Retrieved from https://www.pewresearch.org/fact-tank/2021/08/18/more-americans-now-say-government-should-take-steps-to-restrict-false-information-online-than-in-2018/

Newcal Industries v. Ikon Office Solution. (2008). 513 F. 3d 1038 (United States Court of Appeals, Ninth Circuit January 23, 2008). Retrieved from https://scholar.google.com/scholar_case?case=12136909284171376975

Newman, N. (2017). *Journalism, media, and technology trends and predictions 2017* (p. 37) [Digital News Project]. Reuters Institute for the Study of Journalism. Retrieved from https://reutersinstitute.politics.ox.ac.uk/sites/default/files/2017-04/Journalism%2C%20Media%20and%20Technology%20Trends%20and%20Predictions%202017.pdf

Nothhaft, H., Bradshaw, S., & Neudert, L.-M. (2018). *Government responses to malicious use of social media* (p. 19). NATO strategic Communications Centre of Excellence. Retrieved from https://stratcomcoe.org/publications/government-responses-to-malicious-use-of-social-media/125

O'Reilly, T. (2014, March 9). Viral videos: CBC transcript (SS18E9) [podcast]. In *Under the influence with Terry O'Reilly.* CBC. Retrieved from https://www.cbc.ca/radio/undertheinfluence/viral-videos-1.2801778

Pogue, D. (2015, May 1). Truth in digital advertising. *Scientific American.* doi: https://doi.org/10.1038/scientificamerican0515-32.

Pollay, R. W. (1986). The distorted Mirror: Reflections on the unintended consequences of advertising. *Journal of Marketing, 50*(2), 18–36. https://doi.org/10.1177/002224298605000202

Qazvinian, V., Rosengren, E., Radev, D. R., & Mei, Q. (2011). *Rumor has it: Identifying misinformation in microblogs.* 1589–1599.

Rapp, C. (2010). Aristotle's rhetoric. In E. N. Zalta (Ed.), *The Stanford encyclopedia of philosophy (spring 2010).* Metaphysics Research Lab, Stanford University. Retrieved from https://plato.stanford.edu/archives/spr2010/entries/aristotle-rhetoric/

Rodić, N., & Koivisto, E. (2012). *Best practices in viral marketing.* 37.

Rollins, B., Anitsal, I., & Anitsal, M. M. (2014). Viral marketing: Techniques and implementation. *Entrepreneurial Executive, 19,* 1–17.

Rubin, V. L. (2017). Deception detection and rumor debunking for social media. InSloan, L. & Quan-Haase, A. (Eds.) The SAGE Handbook of Social Media ResearchMethods, London: SAGE: (pp. 342–364). https://uk.sagepub.com/en-gb/eur/the-sage-handbook-of-social-media-research-methods/book245370

Rubin, V. L., Chen, Y., & Thorimbert, L. M. (2010). Artificially intelligent conversational agents in libraries. *Library Hi Tech, 28*(4), 496–522.

Russell, N. (2005). *Morals and the media: Ethics in Canadian journalism* (2nd ed.). UBC Press. Retrieved from https://www.ubcpress.ca/morals-and-the-media-2nd-edition

Schauster, E. E., Ferrucci, P., & Neill, M. S. (2016). Native advertising is the new journalism: How deception affects social responsibility. *American Behavioral Scientist, 60*(12), 1408–1424. https://doi.org/10.1177/0002764216660135

Schechner, S. (2021, June 4). Facebook's marketplace faces antitrust probes in EU, U.K. *Wall Street Journal.* Retrieved from https://www.wsj.com/articles/eu-and-u-k-open-antitrust-probes-into-facebook-11622800304

Shaw, K. (2018). Promotion: Marketing communications. In G. McKay, P. Hopkinson, & L. Hong Ng (Eds.), *Fundamentals of marketing.* Goodfellow Publishers Limited. Retrieved from http://ebookcentral.proquest.com/lib/west/detail.action?docID=5433725

Spence, E., & Van Heekeren, B. (2005). *Advertising ethics.* Prentice Hall.

Stansberry, K. (2021, April). The financial drain of misinformation—PRsay. *PRSay, the public relations Society of America (PRSA).* Retrieved from https://prsay.prsa.org/2021/04/22/the-financial-drain-of-misinformation/

Starek, III, R. B. (1996, October 15). *Myths and half-truths about deceptive advertising.* The National Infomercial Marketing Association. Retrieved from https://www.ftc.gov/public-statements/1996/10/myths-and-half-truths-about-deceptive-advertising.

Statista (2015). Social media advertising expenditure as share of digital advertising spending worldwide from 2013 to 2017. Retrieved from https://www.statista.com/statistics/271408/share-of-social-media-in-online-advertising-spending-worldwide/

Statista Research Department. (2021a, January 14). *Native digital display ad spend in the U.S. 2020.* Statista. Retrieved from https://www.statista.com/statistics/369886/native-ad-spend-usa/

Statista Research Department. (2021b, June 11). *Adblocking: Penetration rate 2018, by country.* Statista. Retrieved from https://www.statista.com/statistics/351862/adblocking-usage/

Strong, E. K. (1925). The psychology of selling and advertising. McGraw-Hill.

Subrahmanian, V., Azaria, A., Durst, S., Kagan, V., Galstyan, A., Lerman, K., Zhu, L., Ferrara, E., Flammini, A., & Menczer, F. (2016). The DARPA twitter bot challenge. Retrieved from http://arxiv.org/pdf/1601.05140.pdf.

Turner, R. H., & Koenig, F. (1986). Rumor in the marketplace: The social psychology of commercial hearsay. *Contemporary Sociology, 15*(5), 776–777. https://doi.org/10.2307/2071086

van Prooijen, J. W., & Douglas, K. M. (2018). Belief in conspiracy theories: Basic principles of an emerging research domain. *European journal of social psychology, 48*(7), 897–908. https://doi.org/10.1002/ejsp.2530

Warnick, A. (2016). A qualitative analysis of the native advertising model with reference to the conventions of journalism [thesis]. In *Journalism and multimedia arts: Vol. master of science in media arts*. Duquesne University.

Weller, B. (2012). *The definitive guide to Google AdWords create versatile and powerful marketing and advertising campaigns* (1st ed.). Apress. https://doi.org/10.1007/978-1-4302-4015-0

Wilson, A. (2015). Book review: The anatomy of humbug: How to think differently about advertising. *International Journal of Market Research, 57*(2), 323–324. https://doi.org/10.2501/IJMR-2015-025

Wojdynski, B. W. (2016). Native advertising: Engagement, deception, and implications for theory. In R. Brown, V. K. Jones, & B. M. Wang (Eds.), *The new advertising: Branding, content and consumer relationships in a data-driven social media era* (pp. 203–236). Praeger/ABC Clio. Retrieved from https://www.researchgate.net/publication/281972887_Native_Advertising_Engagement_Deception_and_Implications_for_Theory

Wojdynski, B. W., & Evans, N. J. (2016). Going native: Effects of disclosure position and language on the recognition and evaluation of online native advertising. *Journal of Advertising, 45*(2), 157–168. https://doi.org/10.1080/00913367.2015.1115380

Wojdynski, B. W., & Golan, G. J. (2016). Native advertising and the future of mass communication. *American Behavioral Scientist, 60*(12), 1403–1407. https://doi.org/10.1177/0002764216660134

Wu, K., Yang, S., & Zhu, K. Q. (2015). False rumors detection on Sina Weibo by propagation structures. *IEEE International Conference on Data Engineering, ICDE.*

Wu, M., Huang, Y., Li, R., Bortree, D. S., Yang, F., Xiao, A., & Wang, R. (2016). A tale of two sources in native advertising: Examining the effects of source credibility and priming on content, organizations, and media evaluations. *American Behavioral Scientist, 60*(12), 1492–1509. https://doi.org/10.1177/0002764216660139

Yang, F., Liu, Y., Yu, X., & Yang, M. (2012). Automatic Detection of Rumor on Sina Weibo. In *Proceedings of the ACM SIGKDD Workshop on Mining Data Semantics,* 13.

Zenith, The ROI Agency (2019). Global intelligence: Data & insights for the new age of communication. *Issue 10* (Q4) Retrieved from https://www.zenithmedia.com/wp-content/uploads/2019/12/Global-Intelligence-10.pdf

Zhang, S. (2021, June 9). Ideas | I saw millions compromise their Facebook accounts to fuel fake engagement. *Rest of WORLD*. Retrieved from https://restofworld.org/2021/sophie-zhang-facebook-autolikers/

Chapter 7
Artificially Intelligent Solutions: Detection, Debunking, and Fact-Checking

Anything a typical human can do with a second of thought, we can probably now or soon automate with AI.

(Ng, 2017)

As the pursuit of fighting fake news becomes more sophisticated, technology leaders will continue to work to find even better ways to sort out fact from fiction as well as refine the AI tools that can help fight disinformation.

(Marr, 2021)

Despite all the hype and excitement about AI, it's still extremely limited today relative to what human intelligence is.

Ng (quoted in (Lynch, 2017))

Abstract **Chapter 7** focuses on artificially intelligent (AI) systems that can help the human eye identify fakes of several kinds and call them out for the benefit of the public good. I explain, in plain language, the principles behind the AI-based methodologies employed by automated deception detectors, clickbait detectors, satirical fake detectors, rumor debunkers, and computational fact-checking tools. I trace the evolution of such state-of-the-art AI over the past 10–15 years. The inner workings of the systems are explained in simple terms that are accessible to readers without a computer science background. How do these technologies operate in principle? What important features do developers consider? What rates of success do these systems report? And finally, what are the next steps in improvements, collaboration, and integration of various approaches?

Keywords Automated deception detection · AI-based lie-detection · Falsehood detectors · Satire detectors · Clickbait detectors · Rumor debunkers · Rumor busters · Automated fact-checkers · AI · Artificial Intelligence · Binary classification · Predictive modeling · ML · Machine learning · Computational fact-checking · Feature Engineering · Research and development · R&D · Computational R&D · Basic research terminology · Data · Data collection · MechanicalTurk data · Human intelligence tasks · MTurk HITs · Input data · Output data · Training dataset · Test dataset · Validation · Prediction · Classification · Manual error analysis · System's performance metrics · Accuracy · Recall · Precision · F-score measure · Ground truth · True positives · True negatives · Feature Engineering · False positives · False negatives · Supervised machine

V. L. Rubin, *Misinformation and Disinformation*,
https://doi.org/10.1007/978-3-030-95656-1_7

learning · Unsupervised machine learning · Garbage in garbage out · GIGO · Rubbish in rubbish out · RIRO · Ethical AI · Semi-structure data · Unstructured databases · Structured databases · Textual data · Metadata · Linguistic insight · Human judgements · Data enrichment · Data augmentation · Textual data analytics · NLP pipeline · Lingsuistic analyses · Text properties · Text features · Weight assignment · Normalization · Data reduction · Information extraction · Named entity recognition · Binary classification · Detecting · Alerting · Quarantining · Blocking · Filtering · Minmizing user exposure · Visualizing · Human computer interaction · HCI · User interfaces · UI · UI layout · UI functionality

7.1 Introduction

Unlike physiological analyzers and polygraphs (Chap. 5, Sect. 5.2), AI-enabled algorithms do not require direct contact with a human being. They analyze natural language as the artifact of our communication. AI systems can examine information written in a textual format, as we see it in digital news, blog posts, tweets, or other computer-mediated messages. Some algorithms, such as rumor debunkers, also look at the networks of propagation for user behaviors and profiles. This chapter focuses on written texts, as it is my specialty, but I acknowledge that doctored images, deepfake videos, and altered speech audios are also sources of concern and are areas with active research and development (R&D) outside of pure textual data analytics.

Modern automated detection of various kinds of text-based "fakes" in digital media content relies on insights from psychology, communication, linguistics, and computer sciences. The premise is that the speech of liars exhibits subtle differences from that of truth-tellers. The language of satirical fakes also has unique properties when compared to news reports. Clickbait headlines differ from more traditional ones. Such differences in each category of fakes can be systematically captured and analyzed as detection cues. The field of natural language processing (NLP), at the intersection of those disciplines, has now proven it possible to automatically detect the presence of those cues in digital environments. NLP algorithms can, at times, exceed the accuracy of our notoriously poor human ability to spot lies, and can certainly beat us in the speed and scale of processing. (See also Chap. 2 on psychology and human inability to detect deception, Sect. 2.5.)

To help my readers with some of the more technical explanations, I start by unpacking several very basic, but key, concepts. Many of the terms that follow are used matter-of-factly in AI, NLP, machine learning (ML), and other data sciences, so much so that they hardly require any references. These concepts have a general background knowledge status in these fields, though the explanations are much simplified here and presented in my own words. In contrast to the statistically and mathematically heavy texts for computer science courses, this chapter intentionally avoids mathematical formulas and programming code snippets. Keeping technical vocabulary to the minimum, I favor explaining step-wise procedures in principles, and wherever possible, offer some examples. If the concepts in the subheadings below are already familiar to you, you may want to skip the following subsection altogether, and proceed directly to Sect. 7.2. Alternatively, for basic explanations of NLP and ML capabilities, you may want to back-track to Chap. 1, Chap. 3 for explanations of IR and HCI, or Chap. 4 for AI and philosophy.

7.1.1 AI Basic Terms and Concepts Demystified

Below are some basic concepts that are key to understanding the R&D and AI-based methodologies described in the remainder of this chapter. We have to start with *data* because data are what drive automation.

Data

In one of his video-recorded ML lectures, Andrew Ng, a computer scientist from Stanford's AI Lab and a former Chief Scientist at Baidu,[1] shared a rule-of-thumb he much prefers to the misguided perception some people hold that "AI can do anything." Ng's rule-of-thumb, by his own admission, is imperfect but nuanced: AI "can probably now or soon" automate "anything that a typical human can do with a second of thought" (Ng, 2017). He elaborates that this rule-of-thumb includes three pre-requisites that make programming AI tasks possible. First, it means that it is guaranteed that the task can be done, in principle, since it is feasible for humans to complete it. Second, it means that we can collect some data as a result of such human activity, especially if it is a simple or repetitive task that we can do routinely. Say, the task is to determine whether an image is of good or bad quality. We can easily record human judgments in connection with particular images. If the task is to express whether we like several images or not, we can collect those data too. If humans do it, we can get data out of those human routines, even those that require highly skilled expertise. For example, medical diagnoses based on skin imagery can guide our ML decisions as well—is it a tumor or not? Third, if a human can do it, then researchers have access to human insight into how we, humans, go about accomplishing the task. The human process for task completion is valuable for R&D in AI procedurally because even though systems cannot express likes or dislikes, they can attempt to mimic the human processes of making such judgments.

[1] Baidu is the company that created the largest Chinese search engine.

Seemingly paradoxically, Andrew Ng is also quoted expressing his pessimism for AI, saying that AI is extremely limited today relative to the general human intelligence (Lynch, 2017). It is widely acknowledged that the *general AI* promise remains unfulfilled, while *narrow AI* allows specialized systems to do a certain specific task fairly well. Software engineers and developers have to partition larger problems into narrower or smaller tasks. These tasks are routinely automated by separating them into doable bits. Later, they may be strung together into more integrated systems to mimic stronger or more human-like general intelligence capabilities. So, in principle, solving individual narrow problems is important so that these solutions can be re-combined toward a higher goal. This logic also applies to finding an overall integrated solution to mis- and disinformation: the big problem of the infodemic (Chap. 1) has to be chunked into smaller sub-problems that should be solved first. In the search for solutions, AI systems learn from human insight, or our experience and expertise in routine tasks. These tasks may be as simple as stating our likes or dislikes, or as complex as the medical diagnosis of benign or malignant tumors. Datasets can reflect both the traces of activities (statements, purchases, images, etc.) and human judgments (about the artifacts of these activities). Such data provide records which are the basis to learn from, that is, they are the input and output data for machine learning.

Data Collection

Datasets are of great value, and many AI R&D efforts start from either *collecting* or otherwise looking for suitable data that are fit to the task's needs. Datasets contain some stimulus (the *input data*, such as an e-mail) and a direct relationship resulting in its direct response (the *output data*, such as a label for an e-mail as spam or not). The first gives the program something to consider, and the other is the correct solution. In the absence of readily available datasets, data can be collected "off the web" with specialized IR applications such as scrapers or crawlers that access online information and capture certain types of data according to the specifications set by the system's designers. Another way to obtain data is to *elicit* them from humans in some systematic fashion. This is often done by requesting humans to contribute data during some routine tasks that they are asked to perform. Study participants enter their reactions, interpretations of situations, likes/dislikes, label or sort images, or perform any number of other *human intelligence tasks*. Often, a classification of labels is offered as an inventory to choose from in making judgments (e.g., like or dislike; red, orange, or green). At other times, participants are given numeric scales (such as the five- or seven-point Likert scales used in surveys) corresponding to verbal statements (e.g., to indicate the strengths of their agreement or disagreement with each statement). Other elicitation designs may record human insights as a free-form commentary such as direct verbal responses to questions (e.g., rationale, explanations, or examples).

Crowdsourcing Human Intelligence Tasks

Several websites are specifically set up for collecting data from such human intelligence tasks for computer science applications, including the popular Mechanical Turk (or MTurk).[2] Tasks on sites like MTurk are comparable to sites like SurveyMonkey[3] and others used in the social sciences, but the tasks on MTurk may be simpler and quicker to accomplish. MTurk study participants are typically paid small amounts for each contribution. In essence, any crowdsourcing platform can be used to obtain human-generated data.

Training and Test Datasets

Each ML dataset is usually divided into several parts. A large portion (usually 70% to 80%) of the total data is used for training—*the training dataset*. The remainder (20% to 30%) is reserved for testing—the test dataset. Some studies also reserve a separate intermediate chunk of data for *validation* between the training and testing stages. *Training* in ML typically means that an ML algorithm is given training data to observe direct relationships between the stimulus and response—the *input* and *output*—and to find regularities and patterns based on similarities or differences in the training data. The algorithm can then reach a certain conclusion, or we say, *make a prediction* on a previously unseen data point, or alternatively, assign a *classification* tag to this new data point. For example, in a sentiment analysis classification task some sentences in the training set can be labeled with positive sentiment (e.g., "I'm glad you've made it to the cottage"), others with negative sentiment (e.g., "Too bad your family couldn't join us"), and yet others as neutral in sentiment (e.g., The cottage is on the Lake Huron shores.") An algorithm looks at how each instance of data is annotated in the training set (not just the three sentences above but many more) so that it learns to predict that a sentence from a test set (e.g., "A rainy day is bad for swimming") should be assigned the negative sentiment annotation.

Supervised and Unsupervised Learning

To see how well a trained algorithm does with its *predictions* or *classifications*, developers run the trained model on the test set, and obtain the output for each previously unseen data point (say, in the form of predicted classification labels such as

[2]Amazon Mechanical Turk (MTurk) is a crowdsourcing marketplace available at https://www.mturk.com/ (Accessed on 16 July 2021). MTurk is known for its power to elicit collective intelligence and human insight for human tasks, especially routine ones. It works best for fast and easy connections of stimuli, such as "tell me whether you see a bridge in the image or not." The quality of MTurk annotations has been questioned at times due to its promotion of hurried decisions and the lack of ability to verify workers' expertise and diligence on complex tasks. In the past, MTurk has also drawn criticism for underpaying its workers in some parts of the world.

[3]https://www.surveymonkey.com/ (Accessed on 16 July 2021).

positive, negative, or neutral sentiment). These predictions are compared to correct labels in the test set. This process is not done on individual data points, testing is rather done in a batch mode to determine reliability and accuracy statistically. An exception may be at the stage of *error analysis*, when developers want to analyze data points *manually*, i.e., to see and analyze each individual data point for what errors may have occurred and for what potential reasons. In the testing stage, the correct matches and incorrect mismatches between the trained system's predictions and the actual labels are then compared, and the system's *performance metrics* are reported.

Such a training process is referred to as *supervised machine learning* due to the fact that machines are given output labels to learn from (i.e., their learning is supervised and constrained by the labeling system). In other words, when the data show what the output should be, and the developers know the output too, they use *supervised learning*. *Decision trees* are an example of supervised ML, where there is a binary choice at each point of the path that the algorithm travels in their *if-else* conditions. The decision paths of these trees are still possible to trace with the human eye nowadays, though the steps are often too many and too cumbersome in practice.

By contrast, when the outputs are uncertain, developers are using *unsupervised learning*. ML models that can discover the inherent structure within unlabeled data without any labels to learn from on their own are then called *unsupervised*. Newer, more powerful, machine learning algorithms such as *neural networks* are unsupervised. Their internal workings are considered *a black box*. It means that humans, somewhat paradoxically, cannot interpret and trace how these models made their decisions, so numerous and intricate their steps are. All we see is the prediction or classification label output by the model.

Both types of models—supervised or unsupervised—can usually benefit from *being trained on* more data, since the ML models can continue to be adjusted and improved incrementally as new data are "seen." Having *a trained data-driven model* usually means that some mathematical formulae is used to capture statically valid patterns and regularities (or simply, correlations) in the observations from a large training dataset.

Performance Measures

We typically care about four *system performance metrics*. *Accuracy* is the simplest, most primitive metric that gives us a rough idea of the performance of the model. Namely, it is percent correct out of the total. For example, with 65,430 labels classified correctly out of 100,000, it is an 65.43% accuracy (regardless of what the predictions were, be they positive or negative sentiment label). Accuracies, like many other metrics, can also be expressed as a fraction on a 0.00–1.00 scale (e.g., 0.6543) instead of the corresponding percentile correct, as you will see in some results tables below.

The more informative three metrics are the *precision* and *recall* measures, and their combination in an *F-score* measure. If the test dataset contains two options (e.g., positive and negative labels), a ML algorithm can be trained to perform *a binary classification task,* i.e., assigning each new sentence either of the two labels ("+" or "–"). *Precision* reflects the fraction of relevant instances among the retrieved instances. It accounts for the fraction of the positive predictions that actually belong to the positive class. It is referred to as the *positive predictive value*. *Recall* is the fraction of relevant instances that were retrieved, or in other words, quantifies the number of positive class predictions made out of all positive examples in the dataset. It is sometimes referred to as *sensitivity*. To balance both precision and recall in a single score, we use *F-measure* which provides just one number, also known as the *harmonic mean* of the precision and recall. The highest possible value of an F-score is 1.0, indicating perfect precision and recall, and the lowest possible value is 0, if either the precision or the recall is zero. F-score is especially helpful when the dataset is unbalanced. An *unbalanced dataset* is unevenly distributed between two potential output labels. Say, a dataset contains a majority of regular e-mail messages, and only very few spam e-mail messages. It is easy to remember such cases by the analogy of finding a needle in a haystack. In that case, we want to report F-score because it will take into account both *false positive* and *false negative* instances, the two ways of incorrectly mislabeling data, as opposed to *true positives* and *true negatives*. Accuracy only takes into account true positives and true negatives, which can over-represent the quality of the algorithm performance, especially in such unbalanced datasets where simply labelling all the messages as "regular" could result in high accuracy.

Ground Truth and Ethical AI

Labeled training datasets can also be referred to as *ground truth,* with true correct mappings from input to output data. In the context of mis- and disinformation it can be difficult to determine the ground truth because humans may not necessarily agree on the correct labeling, or may lack the necessary expertise to do so correctly. It is a common practice to seek out undeniable verdicts for ground truth designations. Systems developers seek authoritative sources and rarely question how the ground truth is established, since this is only the starting point for their training, experimentation, and consequent methodological improvements. Social scientists, then, should be the ones to raise red flags when the relations between the input and output are locked in labeled training datasets. In computing, it is paramount to avoid the "garbage in garbage out" situation, often abbreviated as GIGO, or "rubbish in rubbish out" (RIRO). Both refer to the idea that incorrect or poor quality input always produces faulty output (Editors of Oxford Reference Online, 2021), or more broadly, that a flawed premise cannot possibly lead to an sound argument.

The presumed authority behind ground truth decisions should be probed for biases, misconceptions, or mislabeling. Who made the judgements? Whose human insight was emulated? A movement toward *Ethical AI* is on the rise, an AI subfield that is largely concerned with the ways datasets are created and harnessed. Are there stereotypes and biases in the human judgments used for training AI models? Do these models consequently favor certain population segments over others? Ethical

AI also looks at the ways in which data are obtained. Data may be collected in unethical ways: by either openly violating users' privacy or more covertly when users have only a marginal awareness of the use of their personal data in AI. (See also Chap. 4, Sect. 4.2 for connections between AI, philosophy, and applied ethics (morality).)

Text as Unstructured or Semi-Structured Data for NLP Analyses

NLP comes into play when input data are textual in nature, and such data are considered *unstructured* as opposed to *structured* databases (with one-to-one mappings). Textual data reflect the human capacity to use language creatively; groupings of words can have multiple meanings and be combined in multiple ways. Some research considers texts semi-structured because all languages have at least some predictability that can be uncovered, like the typical subject-verb-object sentence structure in English syntax.

NLP gains insight into texts based on such language patterns and regularities through the lexical, semantic, syntactic, and pragmatic analyses of texts (see more on the layers of NLP analyses in the seminal work by Liddy (2001)). Linguistic insights are captured (as a kind of human judgement) and added to the original dataset as additional labels. Such information can be considered a type of metadata appended to appropriate segments of texts, characters, sentences, or entire documents. For example, an end-of-sentence label can be appended to full stops unless they are preceded by an honorific, such as Dr., Mr., or Mrs. Originally, NLP operated according to the rules that linguists and software engineers wrote and formalized based on their original observations of the data. Later, such rules and procedures were captured and generalized in well-known methods and functions (for example, part-of-speech taggers or sentence-segmenters) and those methods are fairly accurate. Additional rules and regularities are derived based on frequently occurring patterns using more complex (un)supervised ML techniques. Thus, NLP and ML are applied to *enrich* or *augment* textual data with more data (or metadata), and then those metadata are used iteratively, as input for machine learning, to find new insights.

NLP Pipeline and Features

NLP has a standard set of linguistic analyses and other procedures to clean, analyze, enrich, and modify data, called a *pipeline*. A pipeline refers to the particular order of methods and functions in which algorithms *discover* certain linguistic *properties of texts*, often called *features*, and then associate text segments with those feature labels. For example, the part-of-speech is important to know for each individual word because then you can conduct operations at the part-of-speech level, ignoring actual words. We know that, for instance, all adjectives and nouns "behave" in a predictable way for a particular language, such that they combine in a prescribed order: first the adjective and then the noun in English. We can then make inferences based on the parts of speech. If you count the number of adjectives and adverbs, you can *engineer* more complex features, that, for example, measure how embellished or descriptive one's speech is compared to others. To get a sense of the amount of embellishment, we can calculate the ratio of nouns plus verbs to adjectives plus adverbs. Thus, the more qualifiers a speaker uses, the higher the ratio reflected in our embellishment feature, the more flourished her speech is when compared to others.

Weighting and Normalization

Features may differ in their importance and NLP developers can, to some extent, influence algorithmic decisions by assigning greater weight to some features over others. *To assign weights* means to assign numeric multipliers to features, to establish the importance of certain cues, or to finetune their positioning in the preference order to be considered by algorithms. Weights are often assigned on the normalized decimal scale of 0.00–1.00, with 0.1 reducing the value of the feature score, and giving it "less weight," while a weight of 0.9 gives more weight to a feature. For instance, in our embellished speech feature example, we can give more weight to adjectives (arbitrarily 0.6) and less to adverbs (0.3) and multiply all counts of those parts of speech by the corresponding weights. These weights would mean that someone who used more adjectives would receive a higher embellishment feature score than someone who used more adverbs. If feature scores are *normalized*, it usually means that they are divided by some sum total, like the total number of words in a sentence. Normalization should make conceptual sense for the task, though. For example, if we care about how flourished each e-mail in a dataset is, we can normalize the feature score according to the number of total words in each e-mail.

Information Extraction

Datasets, in other words, can be enriched using NLP or augmented with metadata. Another option is to *reduce data* from an unstructured text format to only the particular relevant segments and to use only certain pieces of un- or semi-structured data to populate a structured database. This selection and reduction process is referred to as *Information Extraction*. For example, an algorithm may be used to identify and collect geographic names from submitted online resumes, or to collect all people's and organizations' names. This kind of task is called *Named Entity Extraction or Recognition*. Such an extraction algorithm compares the words it finds that start with a capital letter (in English), to names in a database of geolocation, personal, and organizational names, and then assigns one of those three categories (*geo, person, org*) to the word. The better and more nuanced the labeled typology within the database you have, the more detailed your enriched dataset can be. For examples, organizations can be further divided by type (commercial, educational, etc.), and thus more labels can be associated with the dataset in processing. Also, the more entity names the dataset contains, the better the ultimate performance results for the task of extraction. The extracted labels can then become input into other more complex predictions, as we will later see.

Thus, the quantity of data, the models used for labeling at different stages in the pipeline, and their quality (in terms of, e.g., authority and expertise) are what drive good machine learning outcomes. The recent explosion of data online and the ability that tech giants have to scrape, crawl, or simply collect and use their users' data for AI training—is what is making the 2020s the age of AI.

7.1.2 AI Goals in Solving the Mis- and Disinformation Problem

This segment considers specifically what AI-based systems aim to accomplish to fight mis- and disinformation. What may this *fight* entail? The concrete goals are to *detect* (i.e., identify) varieties of fakes by distinguishing them from authentic, truthful, accurate, and trustworthy content. It is at the very minimum *a binary classification problem* for AI (e.g., "good desirable" content | "bad undesirable" content of some kind).[4] The idea is that revealing "fakes" will alert users of online information to the deceptive intentions of others. If a hypothetic AI system runs checks on input data from social media regularly, much like spam filters run on all incoming e-mails, it can be used to *minimize users' exposure* to mis- and disinformation.

AI-based detectors can only *assist* users with their ultimate decisions regarding the content. (See also Chap. 3 on how we make trust decisions, and Chap. 2 on variety of psychological factors and biases that also influence our decisions.) AI-detectors' predictions can lead to *alerting* us, *quarantining* into spam-traps, temporarily *blocking,* or permanently *filtering* out undesirable types of content. However, the stricter the measures, the more controversial automated decisions become since they take agency away from humans. The more that AI-based assistive technologies move from guidance and suggestions to policing and surveillance, the more dire the consequences of incorrect predictions become (both false positives and false negatives).

Thus, currently, the primary task for AI systems developers is to *develop robust evidence-based detection methods* for *detecting mis- and disinformation.* They start with the proof of concept: these tasks are feasible, and they can perform with acceptable measures of success. Other tangential tasks are to visualize, explain, and provide rationale for concrete instances of the identification of problematic content. These tasks are also necessary for building trust in the systems' performance.

[4] Research can certainly look at the problem in a more nuanced multi-way classification, of course, not only binary.

AI systems should strive for easy-to-use interfaces. Users may also require training, technical support, and perhaps accompanying educational videos. Systems development is a separate task from usability and improvements in human computer interactions (HCI). HCI concerns are, for example, about UI layouts and functionality. HCI studies how users interact with system interfaces and provides recommendations so that users can interact with systems fairly intuitively, and have correct expectations about the systems' results, functions, and the various options within those functions. Since most of these technologies are now cutting-edge innovations in the realm of the state-of-the-art in AI, additional usability and HCI improvements will soon be well-justified. It is also important to educate potential users about such AI technologies so that they consider adoption in view of the AIs' benefits (usefulness, accuracy, ease-of-use, credibility, etc.) and with full awareness of their limitations (such as the error rates of automated predictions).

AI-based detectors of mis- and disinformation start from the premise of the existence of some undesirable problematic content online. This premise may seem trivial nowadays, as anyone alive and online can see the preponderance of problematic and undesirable content. Humans, with their general intelligence, can clearly see it happening. What needs to be spelled out though, in that fight with mis- and disinformation, is that AI systems cannot do everything, as is often mistakenly thought.[5] Preventing the creation of disinformation in the first place is an unrealistic goal for AI, though it is possible that AI can tip the scales one way or another. So, when we say AI fights mis- and disinformation, some of the fight in this techno-social phenomenon can only be resolved in conjunction with regulatory measures and the efforts of educational systems.

To be clear, and as I am sure most readers realize, AI offers no magic wand solution. AI-based detection models can aim to reveal potential mis- and disinformation, but the prevention and deterrence can only be partially impacted by AI research on its own. The more mis- and disinformation is blocked, filtered, or simply ignored, the more creative it will have to get to avoid AI-based detection. Mis- and disinformation prevention and deterrence *is* however within the regulatory purview of social media platforms.

AI-based detection can also be extended to predictions of future fakes that are not yet in existence, and in principle, could be accomplished with careful monitoring of daily viral events and trending stories. Speech generation NLP algorithms can be deployed at the service of "dark creativity" for profit and notoriety but, to the best of my knowledge, they are borderline illegal and disallowed on many social media platforms (as inauthentic bot accounts). In my view, systems with explicitly deceptive intentions for generating potent emotionally-loaded fakes put AI technology at a disservice to society, and they should be discouraged and actively regulated against. The detection methodology reviewed in this chapter aims to do precisely the opposite—to identify and stop the proliferation of fakes that are already in existence.

[5] See also Chap. 1, in which I use a quote by a contemporary software developer and data journalist, Meredith Broussard. She warns us in her (2019) book "Artificial Unintelligence" that we should not mistakenly believe that technology is a solution to every problem.

7.1.3 Chapter Roadmap: AI Detection by Type of Fake

This chapter overviews five large families of AI-enabled applications that can analyze natural language for mis- and dis-information detection: automated deception (falsehood) detectors (Sect. 7.2), clickbait detectors (Sect. 7.3), satire detectors (Sect. 7.4), rumor debunkers (Sect. 7.5), and computational fact-checkers (Sect. 7.6). Each technology is specific to the type of fake it attempts to reveal. I trace the evolution such state-of-the-art AI over the past 10–15 years and explain their inner workings, in simple terms that are accessible to readers without a computer science background.

Within each of the five families of detectors, there is variability in their approaches, contexts, and success rates. What unites them is the approximate process of inquiry. In brief, a typical AI system scans large amounts of data and learns patterns by observation. It applies statistical calculations to those observations about our language use. It then arrives at a model of distinguishing features and observed characteristic patterns of a particular variety of fake. This model is then used to make predictions about new instances of data never seen before. The system predicts if a certain text is a falsehood or not, satire or not, clickbait or not, rumor or not. The more data are analyzed, the more the model is fine-tuned to what constitutes that type of fake. Computational fact-checking is a challenging cumulative task that mimics the work of human fact-checkers and such AI-based applications are currently in their naissance (see Chap. 5, Sect 5.3 for what manual fact-checking involves).

7.2 Automated Deception Detectors

7.2.1 Linguistic Feature-Based Approaches to Identifying Falsehoods

In the field of automated deception detection, there have been steady efforts to identify deceptive texts[6] by comparison with truthful ones. An expertly trained eye can apply systematic procedures and consistent observations (such as described in Chap. 5, Sect. 5.2) to make a human judgment about the presence of deceptiveness or truthfulness. Automated methods use those observed correlations as the basis for deriving recognition cues from such methods.

The underlying assumption is that it is feasible to distinguish the speech of truth-tellers from that of deceivers, and that the distinction is present in linguistic cues. Successful implementations of automated deception detection methods had started demonstrating the effectiveness of linguistic cue identification as of the early 2000s

[6]Whenever referring to *texts* in this chapter, I mean *textual data* (i.e., any formats of written language such as news articles, tweets, reviews, posts, etc.). I typically do ***not*** mean text in its mundane use as an abbreviation for *text messages,* the brief SMS or MMS cell phone communications.

(Bachenko et al., 2008; Larcker & Zakolyukina, 2012; Mihalcea & Strapparava, 2009; Rubin & Conroy, 2012; e.g., Zhou et al., 2004). Thus, there is a growing consensus among AI and NLP researchers that distinguishing between truths and lies is feasible by looking at verbal cues, and many are pursuing this vein of R&D. This fact is little-known to the general public. The point is that it *can* be accomplished, in principle automatically, with some limited degree of accuracy, as you will further see in this Chapter. Some deception researchers who expect higher accuracies (high enough so that it may be reasonable to accuse a suspect of lying) remain unconvinced. For example, Hauch et al. (2015) are skeptical of computer programs being effective lie detectors because in their large synthesis of cue-based lie detection studies they found that "the [statistical] effects were not significant for many of the variables studied [i.e., verbal cues] or small in magnitude, or moderated by situational variables" (p. 330). Thus, the issue of AI-enabled cue-based lie detection remains controversial and an active niche for further research, especially when applied to solving the problem of identifying deceptive content online (i.e., deliberate disinformation). More definitive answers are yet to come as creative R&D minds in this field rise to this challenge.

Verbal Cues

By contrast with non-verbal cues such as physical appearance, gestures, gaze aversion or perspiration, *verbal cues*[7] are more pronounced and more strongly related to deception (Bond Jr. & DePaulo, 2006). Oddly enough, while police officers tend to pay more attention to non-verbal behavior rather than verbal behavior, researchers in the field have been arguing that they change that practice, suggesting that police may be better off simply listening more carefully to what suspects say (Vrij, 2008). "Deception research has revealed that many verbal cues are more diagnostic cues to deceit than nonverbal cues. Paying attention to nonverbal cues results in being less accurate in truth/lie discrimination, particularly when only visual nonverbal cues are taken into account. Also, paying attention to visual nonverbal cues leads to a stronger lie bias (i.e., indicating that someone is lying)" (Vrij, 2008, p. 1323).

Bogaard et al. (2016) illustrate verbal cues with the criterion of "quantity of detail" which refers to how rich the analyzed statement is in mentions of "places (e.g., it happened in the kitchen), times (e.g., on Sunday evening at 8 p.m.), [and] descriptions of people and objects (e.g., a tall man with bright blue eyes)" (p. 1). The relative presence or absence of certain verbal cues is indicative of either deception or truthfulness.

[7] *Verbal cues* are found in the content of textual data (i.e., the meaning and expression of the analyzed statements) and may include simple counts of words in a sentence, or rates of occurrence of, say, pronouns and adverbs per statement, or other particular references, as well as some more sophisticated measures, as described below. *Verbal cues* are sometimes referred to as *content cues* or *linguistic cues*. They become criteria of notice for systematic analysis, whether by hand (e.g., marking on piece of paper and counting cues manually), with computer assistance, or by completely automated means. In addition, the terms *cues, clues,* or *markers*, and even *features* and *predictors* are often used interchangeably in the context of AI-based deception detection.

For example, "deceit has been related to the use of fewer personal pronouns (e.g., using "the house" instead of "our house") and fewer negations (e.g., no, never, not), using less perceptual information (e.g., "I could smell the alcohol in his breath"), less details overall and shorter statements" (Amado et al., 2015; as cited in (Bogaard et al., 2016 p. 1); Hauch et al., 2015; Masip et al., 2005; Newman et al., 2003).

Verbal cues are often contrasted with the perceptions made with visual (observable sights during the communication, e.g., eye contact, body movements, facial expression) and auditory cues (the way in which words are said, e.g., voice pitch, pauses, hesitation) (Wiseman, 1995). "Of all three broad categories, verbal cues are the easiest for programs to observe and compute, making them amenable to automated analysis. Pure verbal cues include the number of words written or length of sentences. Verbal cues also tend to be more reliable (Wiseman, 1995), possibly since richer visual and auditory media detract perceivers' attention from the verbal essence (Vrij, 2004). Linguistic cues are verbal cues analyzed at lower linguistic levels (morphology, lexics, syntax) or higher ones (semantics, discourse and pragmatics)" (Rubin & Conroy, 2012).

Verbal cues and patterns are among those that objectively differentiate deceptive messages from truthful ones. For instance, deceivers tend to make more negative statements, give more indirect and less detailed answers (Granhag et al., 2004), use fewer self-references but more negative emotions (Zhou & Zhang, 2008), and produce more sense-based words (e.g., seeing, touching) (Hancock et al., 2007) than truth-tellers. Such patterns may be inaccessible to people without specialized analytical tools (Rubin & Conroy, 2012).

McGlynn and McGlone (2014) specifically isolate an increase in word quantity as the most consistent linguistic cue associated with deception: "Language and deception research typically finds that deceivers talk more than truth-tellers. This effect is particularly strong when deceivers are discussing non-verifiable opinions. For example, a person attempting to convince their conversational partner that their favorite food is pizza, rather than the truthful answer of quesadillas, is likely to expound for a longer period of time on the various reasons why pizza is their favorite food than would a truth-telling counterpart. Deceivers may talk more to manage information flow, to enhance mutuality with conversational partners, and to decrease the suspicions of conversational partners" (McGlynn & McGlone, 2014 p. 581).

Feasibility of Feature-Based Automation

While the feasibility of the automation of deception detection has been proven beyond doubt at this point, the success rates of such AI-based interventions vary from study to study, depending on the exact methodologies used and other pragmatic factors such as topics and formats of textual data (e.g., news reports, tweets, blog posts, hotel reviews, etc.). There is also little clarity on which set of cues can be called the decisive reliable predictors of deception, but the evidence-based studies are mounting in their volume.

Based on a literature search of 1945–2012 published and unpublished articles (via the Social Sciences Citation Index, PsycInfo, Dissertation Abstracts, and

Google Scholar), Hauch et al. (2015) extracted an initial list of 948 studies. Their meta-analysis resulted in a rather comprehensive set of known verbal cues, used up until 2012 for the explicit purpose of achieving lie detection with computerized methods. The 44 most relevant scientific articles yielded an initial list of 202 linguistic cues, which was pared down to the final list of 69 operational definitions. The first 38 verbal cues, expressed as percent of word total in a given statement, are shown as sorted by the meta-analytical study's six research questions of interest (Fig. 7.1) with additional miscellaneous 19 verbal cues (Fig. 7.2).

While the meta-analysis concluded with a general skepticism of computerized methods, it is rich in detail and worth further perusal, especially for those researchers that intend to use their own set of predictive cues for the purposes of identifying disinformation online. In broad strokes, the Hauch et al. (2015) study finds that, as expected, "relative to truth-tellers, liars experienced greater cognitive load, expressed more negative emotions, distanced themselves more from events, expressed fewer sensory-perceptual words, and referred less often to cognitive processes. However, liars were not more uncertain than truth-tellers. These effects were moderated by event type, involvement, emotional valence, intensity of interaction, motivation, and other moderators" (p. 309). The meta-analysis also emphasizes that, simply put, "humans and computers are best at different skills. Humans are less accurate in manual counting of specific cues or in rendering accurate judgments of complex syntactic relationships, whereas computers cannot provide subjective, gestalt-like judgments or capture the meaning or intention of what people are saying" (Hauch et al., 2015 p. 330).

Conceptual Clustering of Features

Once the number of predictive linguistic cues of deception reaches more than two dozen extracted features, researchers tend to cluster them conceptually, as was the original practice in the pioneering AI-based studies in the field. For example, Zhou et al.'s (2004) list of 27 linguistic features in eight broad conceptual clusters, used as a system for automatically classifying textual information either as deceptive or truthful. Linguistic features and simpler part-of-speech counts were collected into a set of concrete checkpoints that represent diversity, complexity, specificity, and non-immediacy of the texts that undergo systematic analyses (as shown in Fig. 7.3). The work was originally based on deriving predictors from law enforcement techniques such as Criteria-Based Content Analysis (CBCA) and Reality Monitoring (RM)[8]

[8] Note that Zhou's team (in 2004) reviewed another veracity assessment investigative technique widely used by authorities worldwide—called Scientific Content Analysis (SCAN). SCAN was originally developed by a former Israeli polygraph examiner Avinoam Sapir, based on his experience with polygraph examinees. SCAN has been found to have "no empirical support to date" (with the exception of "change in language" more characteristic of fabricated statements), concluding that SCAN as a technique, in spite of its popularity with authorities, has to overcome inherent methodological problems such as vague description of criteria and their ambiguous interpretation (Bogaard et al., 2016 p. 6).

	Linguistic cue	Final operational definition
Research Question 1: Do liars experience greater cognitive load?		
01[a]	**Word quantity** // word count // number of words // productivity	Total number of words
02	**Content word diversity** // diversity // content diversity	Total number of *different* content words divided by total number of content words, where content words express lexical meaning
03	**Type-token ratio** // unique words // lexical diversity // different words	% of distinct words divided by total number of words
04	**Six-letter words** // percentage words longer than six letters	% of words that are longer than six letters
05	**Average word length** (AWL; complexity) // lexical complexity	Total number of letters divided by the total number of words
06[a]	**Verb quantity** // verb count	Total number of verbs
07[a]	**Sentence quantity** // number of sentences	Total number of sentences
08	**Average sentence length** (complexity measure) // words per sentence	Total number of words divided by total numbers of sentences
09	**Causation**	% of words that try to assign a cause to whatever the person is describing (e.g., because, effect, hence)
10	**Exclusive**	% of words that make a distinction what is in a category and what is not (e.g., without, except, but)
11	**Writing errors** // typographical error ratio (informality) // typo ratio // misspelled words	% of writing errors or misspelled words divided by number of words
Research Question 2: Are liars less certain than truth-tellers?		
12	**Tentative**	% of tentative words (e.g., maybe, perhaps, see)
13	**Modal verbs** // uncertainty // discrepancy	% of modal verbs or auxiliary verbs or words expressing uncertainty (e.g., should, would, could)
14	**Certainty**	% of words that express certainty (e.g., always, never)
Research Question 3a: Do liars use more negations and negative emotion words?		
17	**Negations** // less positive tone // spontaneous negations // negation connectives	% of words that express negations (e.g., no, never, not)
18[b]	**Negative emotions** // negative affect // anger // anxiety, fear // sadness	% of words that express negative emotion/affect (e.g., hate, worthless, enemy) AND anger (e.g., hate, kill, annoyed) AND anxiety (e.g., worried, fearful, nervous) AND sadness (e.g., crying, grief, sad)
18.1	**Negative emotions (only)** // negative affect	% of words that express negative emotion/affect (e.g., hate, worthless, enemy)
18.2	**Anger**	% of words that express anger (e.g., hate, kill, annoyed)
18.3	**Anxiety**	% of words that express anxiety (e.g., worried, fearful, nervous)
18.4	**Sadness**	% of words that express sadness (e.g., crying, grief, sad)
Research Question 3b: Do liars use less positive emotion words?		
19[b]	**Positive emotions and feelings** // positive emotions // positive affects // positive feelings	% of words that express positive emotion/affect (e.g., happy, pretty, good) AND positive feelings (e.g., joy, love)
19.1	**Positive emotions (only)** // positive affect	% of words that express positive emotion/affect (e.g., happy, pretty, good)
19.2	**Positive feelings (only)**	% of words that express positive feelings (e.g., joy, love)
Research Question 3c: Do liars express more or less unspecified emotion words?		
15	**Emotions** // emotional / affective processes // affect (ratio) // positive and negative affect	% of words that express any type of emotions/affects (e.g., happy, ugly, bitter)
16	**Pleasantness and unpleasantness**	% of words that express pleasantness/unpleasantness
Research Question 4: Do liars distance themselves more from events?		
20	**Total pronouns** // personal pronouns	% of all personal (e.g., I, our, they) or total pronouns (e.g., that, somebody, the)
21	**First-person singular**	% of first-person singular pronouns (e.g., I, my, me)
22	**First-person plural**	% of first-person plural pronouns (e.g., we, us, our)
23	**Total first-person**	% of first-person singular and first-person plural pronouns (e.g., I, we, me)
24	**Total second-person**	% of second-person pronouns (e.g., you, you'll)
25	**Total third-person** // other references // third-person singular // third-person plural	% of third-person pronouns (e.g., she, their, them)
26	**Passive voice verbs** // verbal nonimmediacy	% of passive voice verbs (e.g., "it was searched for")
27	**Generalizing terms** // leveling terms	% of generalizing terms (e.g., everybody, all, anybody)
47	Past tense verb	% of past tense verbs (e.g., went, drove, ate)
48	Present tense verb	% of present tense verbs of all words (e.g., walk, run, cry)
Research Question 5: Do liars use fewer (sensory and contextual) details?		
28[b]	**Sensory–perceptual processes** // perceptual processes/information // perceptions and sense // sensory ratio // see // hear // feel	% of words that express sensory–perceptual processes (e.g., taste, touch, feel) AND visual (e.g., view, saw, seen) AND haptical (e.g., feels, touch) AND aural (e.g., listen, hearing) sensory–perceptual processes
28.1	**Sensory–perceptual processes (only)** // perceptual processes // perceptual information // perceptions and sense // sensory ratio	% of words that express sensory–perceptual processes (e.g., taste, touch, feel)
28.2	**Seeing**	% of words that express visual sensory–perceptual processes (e.g., view, saw, seen)

Fig. 7.1 Linguistic cues and their definitions, as synthesized for six research questions in Hauch et al. (2015)

28.3 **Feeling**	% of words that express tactile sensory–perceptual processes (e.g., feels, touch)
28.4 **Hearing**	% of words that express aural sensory–perceptual processes (e.g., listen, hearing)
29 **Time** // temporal ratio // temporal specificity // temporal cohesion	% of temporal words (e.g., hour, day, o'clock)
30 **Space** // spatial terms // spatial ratio // spatial specificity // spatial cohesion	% of spatial words (e.g., around, over, up)
31 **Temporal-spatial terms** // temporal and spatial details total // spatio-temporal information // space and time	% of temporal (e.g., hour, day, o'clock) AND spatial words (e.g., around, over, up)
32 **Prepositions**	% of prepositions (e.g., on, to, from)
33 **Numbers**	% of numbers (e.g., first, one, thousand)
34 **Quantifier**	% of quantifier (e.g., all, bit, few, less)
35 **Modifiers** (adverbs and adjectives) // rate of adjectives and adverbs (specificity and expressiveness)	% of modifier: adverbs and adjectives (e.g., here, much, few, very)
36 **Motion verbs** // motion terms	% of words that describe movements (e.g., walk, move, go)
Research Question 6: Do liars refer less often to cognitive processes?	
37 **Cognitive processes** // all connectives	% of words related to cognitive processes (e.g., cause, know, ought)
38 **Insight**	% of words related to a person's insight (e.g., think, know, consider)

Note. **Bold** font indicates the name of the linguistic cue chosen for this meta-analysis. % indicates number of specific words divided by total number of words.
[a]No ratio.
[b]Indicates that the specific linguistic cue is an umbrella term.

Fig. 7.1 (continued)

Linguistic cue	Final operational definition
39 Redundancy	Ratio of function words to number of sentences. Function words, such as articles and pronouns, are used to form grammatical relationships between other words. // The ratio of the number of function words to the number of messages // Repetitive words // Argument overlap: Explicit overlap between two sentences by tracking the common nouns in either single or plural form
40 Assent	% of words that express an assent (e.g., agree, ok, yes)
41 Articles	% of articles (e.g., a, lot, an, the)
42 Inhibition	% of words that express inhibition (e.g., block, constrain, stop)
43 Social processes	% of words that express social processes (e.g., talk, us, friend)
44 Friends	% of words that are related to friends (e.g., buddy, friend, neighbor)
45 Family	% of words that are related to family (e.g., daughter, husband, aunt)
46 Humans	% of words that are related to humans (e.g., adult, baby, boy)
49 Future tense verb	% of future tense verbs (e.g., will, going to)
50 Inclusive	% of inclusive words (e.g., with, and, include)
51 Achievement	% of words that express achievement (e.g., earn, hero, win)
52 Leisure	% of words that express leisure activities (e.g., cook, chat, movie)
53 Emotiveness	Total number of adjectives and total numbers of adverbs divided by total number of nouns and total numbers of verbs
54 Pausality	Total number of punctuation marks divided by total number of sentences
55 Swear words	% of swear words (e.g., ass, heck, shit)
56 Biology	% of words that express biological processes/states (e.g., eat, pain, wash)
57 Health	% of words that express health issues (e.g., hospital, pill, flu)
58 Sexual	% of words that express sexual activities/states (e.g., passion, rape, sex)
59 Optimism	% of words that express optimism (e.g., certainty, pride, win)
60 Communication	% of words that express communication (e.g., talk, share, converse)
61 Occupation	% of words that express occupation (e.g., work, class, boss)
62 School	% of words that express school issues (e.g., class, student, college)
63 Job/work	% of words that express job issues (e.g., employ, boss, career)
64 Home	% of words that express home issues (e.g., bed, home, room)
65 Sports	% of words that express sport (e.g., football, game, play)
66 Money	% of words that express money and financial issues (e.g., cash, taxes, income)
67 Physical	% of words that express physical states and functions (e.g., ache, breast, sleep)
68 Body	% of words that express body states and symptoms (e.g., asleep, heart, cough)
69 Eating	% of words that express eating, drinking, dieting issues (e.g., eat, swallow, taste)

Note. % indicates number of specific words divided by total number of words.

Fig. 7.2 Additional linguistic cues and their definitions in Hauch et al. (2015)

I. QUANTITY
1. Word: a written character or combination of characters representing a spoken word
2. Verb: a word that characteristically is the grammatical center of a predicate and expresses an act, occurrence, or mode of being [*do, happen, think*]
3. Noun phrase: formed by a noun, its modifiers and determiners.
4. Sentence: a (group of) word(s), clause(s), or phrase(s) forming a syntactic unit that expresses an assertion, a question, a command, a wish, an exclamation, or the performance of an action, which usually begins with a capital letter and concludes with appropriate end punctuation.
II. COMPLEXITY
5. Average number of clauses: total # of clauses / total # of sentences
6. Average sentence length: total # of words / total # of sentences
7. Average word length: total # of characters / total # of words
8. Average length of noun phrases: total # of words in noun phrases / total # of nouns phrases
9. Pausality: total # of punctuation marks / total # of sentences
III. UNCERTAINTY
10. Modifier (adjective and adverb): describes a word or makes its meaning more descriptive [*good, well*]
11. Modal verb: an auxiliary verb expressing modal modification of a verb [*may, might*]
12. Uncertainty: a word that indicates unsureness about something or someone [*possibly, perhaps*]
13. Other reference: a third person pronoun
IV. NON-IMMEIDACY
14. Passive voice: a verb form used when the subject is being acted upon rather than acting [*are sold out*]
15. Objectification: an expression given to an abstract notion, feeling, or ideal in a form that can be experienced by others, externalizing one's attitude
16. Generalizing term: refers to a person or an object as a class rather than an instance [*students, vehicles*]
17. Self-reference: a first-person singular pronoun [*I, my*]
18. Group reference: a first-person plural pronoun [*we, our, ourselves*]
V. EXPRESSIVITY
19. Emotiveness: (total # of adjectives + total # of adverbs) / (total # of nouns + total # of verbs)
VI. DIVERSITY
20. Lexical diversity: percentage of unique words in all words, calculated as (total # of different words or terms) / total # of words or terms)
21. Content word diversity: percentage of content words, i.e. those that primarily express lexical meaning, calculated as (total # of different content words) / total # of content words)
22. Redundancy: total # of function words / total # of sentences, where function words express primarily grammatical relationship
VII. INFORMALITY
23. Typographical error ratio: total # of misspelled words / total # of words [*wrods, speling*]
VIII. SPECIFICTY
24. Spatio-temporal information: a word indicating location or special arrangement of people or objects, or explicitly describing the timing or sequence of events
25. Perceptual information: indicates sensory experiences such as sound, smell, physical sensation, or visual details [*see, hear, sense*]
IX. AFFECT
26. Positive affect: a word indicating a conscious subjective aspect of a positive emotion apart from bodily changes [*thrilled, excited, happy*]
27. Negative affect: a word indicating a conscious subjective aspect of a negative emotion apart from bodily changes [*upset, sad, frustrated*]

Fig. 7.3 Zhou et al.'s (2004) linguistic features for deception detection. Twenty-seven linguistic-based features amenable to automation were grouped into nine linguistic constructs: quantity, complexity, uncertainty, non-immediacy, expressivity, diversity, informality, specificity, and affect. All the linguistic features are defined in terms of their measurable dependent variables. (Adapted from Zhou et al. (2004))

(see Chap. 5, Sect. 5.2 for details), and the Interpersonal Deception Theory (IDT) (see also Chap. 2, Sect 2.5 for psychology of deceit.)

When these verbal cues are implemented with standard classification algorithms (such as neural nets, decision trees, and logistic regression), automated methods can achieve up to 74% accuracy (Fuller et al., 2009). Existing psycholinguistic lexicons (e.g., LIWC by (Pennebaker & Francis, 1999)) were adapted to perform binary text

Class	Score	Sample words
		Deceptive Text
METAPH	1.71	god, die, sacred, mercy, sin, dead, hell, soul, lord, sins
YOU	1.53	you, thou
OTHER	1.47	she, her, they, his, them, him, herself, himself, themselves
HUMANS	1.31	person, child, human, baby, man, girl, humans, individual, male, person, adult
CERTAIN	1.24	always, all, very, truly, completely, totally
		Truthful Text
OPTIM	0.57	best, ready, hope, accepts, accept, determined, accepted, won, super
I	0.59	I, myself, mine
FRIENDS	0.63	friend, companion, body
SELF	0.64	our, myself, mine, ours
INSIGHT	0.65	believe, think, know, see, understand, found, thought, feels, admit

Fig. 7.4 Dominant word classes in deceptive texts and words exemplifying those classes. (Adapted from Mihalcea and Strapparava (2009))

classifications for truthful versus deceptive opinions, with classifiers demonstrating a 70% average accuracy rate (Mihalcea & Strapparava, 2009). Figure 7.4 shows which words and expressions were found to be associated with deceptive and truthful texts, ranked according to the dominance score calculated based on the total number of occurrences of the word within the corpus.

Figure 7.4. shows that the top 5 dominant classes were related to humans. “In deceptive texts, the human-related word classes (YOU, OTHER, HUMANS) represent detachment from the self, as if trying not to have the own self involved in the lies. Words closely connected to the self (I, FRIENDS, SELF) are lacking from deceptive text, being dominant instead in truthful statements where the speaker is comfortable with identifying themself with the statements they make. Also interesting is the fact that words related to certainty (CERTAIN) are more dominant in deceptive texts, which is probably explained by the need of the speaker to explicitly use truth-related words as a means to emphasize the (fake) “truth” and thus hide the lies. Instead, belief-oriented vocabulary (INSIGHT), such as *believe, feel, think,* is more frequently encountered in truthful statements, where the presence of the real truth does not require truth-related words for emphasis” (Mihalcea & Strapparava, 2009). Later works played with weighting mechanisms; some of those weighting schemes have been patented as inventions, (e.g., Chandramouli et al., 2012) (cited in (Hauch et al., 2015)).

Variability of Verbal Cues by Contexts

Deception detection researchers have also acknowledged some variation in predictive linguistic cues of deception across situations (Ali & Levine, 2011), across genres of communication and communicators (Burgoon et al., 2003), and across cultures (Rubin, 2014). The use of language changes under the influence of different situational factors, genre, register, speech community, text, and discourse type (Crystal, 1969). The situational contexts in which deceptive communications occur are of great importance, since different circumstances and formats of communication may impact the language used. For instance, fake dating profiles and forged scientific papers are both instances of intentionally deceptive texts, but each has to comply with genre specificity and each are bound to have their own structure, expectation for puffed up language use, or for precision in terminology and certainty of claims. It is natural to expect differences in deceptive cues across such situational contexts. In other words, the rate of occurrence of various cues may differ for truth-tellers and deceivers depending on how the language is used to accomplish their communicative goals. For example, there are differences in such content cues as *verbosity* (amount of words used) as well as *richness of detail* in police interrogation testimonies versus denial statements.

In synchronous text-based communication, another much-studied context, deceivers produced more *total words*, more *sense-based words* (e.g., seeing, touching), and used fewer *self-oriented pronouns* (e.g., I, my) but more other-oriented pronouns (e.g., he, she, them) (Hancock et al., 2007). Compared to truth-tellers, liars showed lower *cognitive complexity* and used more *negative emotion words* (Newman et al., 2003). In conference calls of financiers, Larcker and Zakolyukina (2012) found deceptive statements to have more *general knowledge references* and *extreme positive emotions*, and also fewer *self-references, extreme negative emotions*, as well as *certainty* and *hesitation words*. In police interrogations, Porter and Yuille (1996) found three significantly reliable verbal indicators of deception based on the Statement Validity Analysis (SVA) techniques used in law enforcement: *amount of detail reported, coherence,* and *admissions of lack of memory*. (Chap. 5, describes SVA (in Sect. 5.2) and other related veracity assessment procedures in the justice system).

In descriptions of mock theft experiments, Burgoon and colleagues (Burgoon et al., 2003) found deceivers' messages in their synchronous text-based chats were *briefer* (i.e., lower on quantity of language), showed less *complexity* in their choice of vocabulary and sentence structure, and *lacked specificity* or *expressiveness*. "Deceivers in synchronous environments often talk more as a result of conversational partners' prompts for more information and detail, while deceivers in asynchronous environments take advantage of the increased time to plan and edit messages afforded by asynchronous communication environments, increasing the detail and subsequent word count of their messages" (McGlynn & McGlone, 2014 p. 584).

One important variable in creating deceptive messages has to do with the timing of such speech acts. If deception unravels in real time, that is synchronously, deceivers may have little to no time to plan or rehearse their message. Asynchronous communications allow for re-reading and post-editing, with conscious efforts to take account of how the message is presented. "As a consequence, no single profile of deception language across tasks is likely to emerge. Rather, it is likely that different cue models will be required for different tasks. Consistent with interpersonal deception theory (Buller & Burgoon, 1996), deceivers may adapt their language style deliberately according to the task at hand and their interpersonal goals. If the situation does not afford adequate time for more elaborate deceits, one should expect deceivers to say less. But if time permits elaboration, and/or the situation is one in which persuasive efforts may prove beneficial, deceivers may actually produce longer messages. What may not change, however, is their ability to draw upon more complex representations of reality because they are not accessing reality. In this respect, complexity measures may prove less variant across tasks and other contextual features" (Burgoon et al., 2003 p. 6).

When predictive linguistic cues are developed based on general linguistic knowledge, some of the cues can be "ported" to somewhat similar pragmatic contexts, say, from e-mails to blogs, or from hotel reviews to book reviews. The logic is that the writing genres, communicative goals, and technological affordances[9] are similar within those two pairs. While contexts are highly specialized and different from each other, researchers should account for their specificity and formats when deciphering the predictive cues of deception (Höfer et al., 1996; Köhnken & Steller, 1988; Porter & Yuille, 1996; Steller & Köhnken, 1989).

In digital environments, especially on social media, in addition to pure content cues, users' behaviors can be traced with other digital "footprints" that users leave in their self-disclosed user profiles, past communications, or other monitored networking and engagement features. Uncovered incongruities in such "footprints," can in turn be harvested[10] as behavioral tell-tale signs of deception.

Message truthfulness (i.e., statement veracity) is but one of the desirable properties of information that people want to act on. Other factors—accuracy, completeness, credibility and expertise of sources, and propensity for spreading rumors—should also be considered. (See how credibility is evaluated and trust decisions are typically made, discussed in Chap. 3). It is safe to assume that deceptive cues may vary across various knowledge domains (e.g., law enforcement, advertising, social media, news) and formats, but some deceptive cues may be stable across them (e.g., *diversity, complexity, specificity, non-immediacy*) and can still be reused in computational algorithms to experiment with the most suitable inventory of features for the context at hand. Easily recognizable contexts and pragmatic uses of language simplify the job of narrowing down distinctive features of deception to the most salient ones. Mis- and disinformation detection tools should be specialized by contexts, until such similarities emerge. At this stage of the R&D of such AI-based tools, the inventory is comprised of variants of "fakes" (see Chap. 1) including detectors that target specific types of misleading information such as clickbait, satirical fakes, and rumors. These will be discussed next, each in turn.

---

[9] *Technological affordances* refers to what a specific type of technology offers to its users as possibilities (e.g., you can send tweets on Twitter and call on Skype) and under which constraints users find themselves (you cannot text-message on traditional phone landlines).

[10] Ethical considerations for the use of private personal data are obviously a concern for many, especially since many users may be unaware of the potential for third parties' access to their data. The mitigating factor that is often brought up is the fact that most digital media users should assume and be aware that whatever content is posted, tweeted, or otherwise published online is in the public domain. Privacy settings and disclosure agreements vary from platform to platform but are often part of the ad-based revenue model that allows companies to resell users' profile contacts, with little awareness of their users. (See also Chap. 6, Sect. 6.2 on for explanations of ad revenue and other marketing and profit models.)

7.3 Automated Clickbait Detectors

7.3.1 *Clickbait Defined*

Clickbait refers to content whose main purpose is to attract attention and encourage visitors to click on a link leading to a particular web page. The outrageous claims made in clickbait headlines or in their associated imagery often turn out to be ill-informed or misleading, as we find out when reading the complete message on the target website (Y. Chen et al., 2015). You may recognize clickbait formats by their more primitive listicle headlines such as "5 things you need to see," "14 strangely satisfying videos of melting cheese," or "10 ways to study you didn't know about." Others are more subtle and use more sophisticated attention-getting techniques. They are invariably written on "soft news" topics, often revolving around celebrity news (e.g., "8 things you don't want to ask George Clooney's new wife when they come home") or similarly light topics like pop culture, movies, weddings, dĕcor, or lifestyle (e.g., "Take years off your face in just 60 seconds").

Clickbait producers promote their content as "captivating," "engaging," and "meaningful for millennials" with overtly stated goals to "laugh, share and inspire" (Diply, 2017). Some clickbait writers are tasked with "taking something newsy and making it digestible"—in particular, borrowing trending ideas and "infusing them" with associated news content to "grab the feelings" of the audience (Diply, 2017). This pitch, unfortunately, often resonates with audiences on the prowl for "light-weight" viral content. Given the ad-revenue-driven motivation to "engage users," clickbait proliferates profusely (Brogly & Rubin, 2019).

7.3.2 *Clickbait Predictive Features and Clickbait Detection Performance Measures*

Since the early 2010s, it has been established that clickbait is distinct from other types of content and it is predictably formulaic in the linguistic forms and devices that it often uses. Thus, clickbait is identifiable by its characteristic features such as the use of suspenseful language, unresolved pronouns, a reversal narrative style, or forward referencing (Y. Chen et al., 2015). Biyani et al. (2016) collected a sizable dataset of 1349 clickbait and 2724 non-clickbait webpages, via the Yahoo homepage from multiple sources (such as *the Huffington Post, New York Times, CBS, Associated Press,* and *Forbes)* but achieved modest results in predicting clickbait (74.9% F-1 score). Other pioneers in AI-based clickbait detection, such as Chakraborty and colleagues (in 2016), used a set of characteristic features and were able to reach 93% accuracy in detecting clickbait and 89% accuracy in blocking it with a built-in Google Chrome extension.

In 2017, a Clickbait Challenge was set up as a competition of thirteen teams, described in Potthast et al. (2018a). The 2017 Clickbait Challenge used the

Webis-Clickbait-17 dataset, a corpus of 38,517 annotated Twitter tweets[11] created by Potthast et al. (2018b) with a focus on the detection of clickbait posts in social media. The best performing system, *zingel* by Zhou (2017),[12] achieved 85.8% accuracy rates (71.9% precision, 65% recall, 68.3% F-score), according to the (Potthast et al., 2018a, p. 5) summary of results.

The original 2017 Clickbait Challenge dataset was later enlarged for the validation of system results and made available to the R&D community. As per Gollub et al. (2017), it consisted of 4761 clickbait headlines and 14,777 non-clickbait headlines from Twitter, which were used as positive and negative examples for ML training. Judgments of the hyperlinked texts were crowd-sourced to five human judges who ranked each hyperlinked text, based on a four-point scale, with a 0 assigned to non-clickbait, 0.33 to slight clickbait, 0.66 to moderate clickbait, and 1 as definitively clickbait (Gollub et al., 2017). The mean score among the five judges indicated the level of clickbait and was used in the 2017 Clickbait Challenge as a metric to identify hyperlink texts that are clearly not clickbait (mean score < 0.1) and those that are likely to be clickbait (mean score > 0.6). Given that human judges' perceptions of clickbait vary from individual to individual (i.e., ground truth is difficult to determine for this task), it was reasonable to set aside the middle grounds of the human judgments for later resolution and start ML training using the clearest examples of positive and negative examples of clickbait for a binary classification task.

Chakraborty et al.'s (2016) dataset did not provide mean score metrics but relied on the trustworthiness of authoritative sources. Thirty-two thousand headlines were divided into two halves, one pre-classified as non-clickbait when drawn from *WikiNews,* and the other was deemed clickbait when coming from *Buzzfeed, Upworthy, ViralNova, Scoopwhoop,* and *ViralStories*. The 2017 Clickbait Challenge resulted in several accurately performing systems that varied in their inventories of features, details of implementation, and the nuances in performance results. A work by Rony et al. (2017) assembled an even larger and more varied training dataset which consisted of 1.67 million Facebook posts from 153 media organizations, which they used to achieve a near-perfect 98.3% accuracy.

Heavily influenced by the works of Chakraborty et al. (2016) and Potthast et al. (2016), the LiT.RL Clickbait Detector was implemented as part of the NV Browser, adapting previously known features and adding some original features (marked with a "yes" in the penultimate column in Fig. 7.5, per Brogly and Rubin et al. (2019)). Using the 38 features (in Fig. 7.5), the LiT.RL Clickbait Detector achieved an

[11] https://webis.de/data/webis-clickbait-17.html (Accessed on 28 February 2019).

[12] All leading approaches (by Zhou and four top-performing others) employed neural networks to optimize feature selection. Their methods may be more understandable to a non-tech savvy reader, after we discuss features of clickbait here first, and then word embeddings in satire detection. Consider coming back later to this explanation of the methodology. (Zhou, 2017) applied "a token-level, self-attentive mechanism on the hidden states of bi-directional Gated Recurrent Units (biGRU), which enables the model to generate tweets' task-specific vector representations by attending to important tokens. The self-attentive neural network can be trained end-to-end, without involving any manual feature engineering."

No.	Feature name	Feature description	Original	Precision	Recall	F1 score	Accuracy
1	*getPronounCount*	Number of pronouns		0.6658	0.9286	0.7756	0.7312
2	*getWord2GramsAvgLen*	Average length of bi-grams	Yes	0.6849	0.6235	0.6528	0.6683
3	*getDeterminers*	Number of determiners		0.6286	0.8106	0.7081	0.6658
4	*getWord3GramsAvgLen*	Average length of trigrams	Yes	0.681	0.5974	0.6365	0.6588
5	*startsWithNumber*	Begins with a number?		0.5871	0.982	0.7349	0.6457
6	*getAdvpCount*	Number of adverbial phrases (ADVPs) part-of-speech (POS) tags		0.5862	0.8984	0.7095	0.6321
7	*getNPsCount*	Number of NP POS tags		0.6342	0.6017	0.6175	0.6273
8	*containsTriggers*	Number of trigger words		0.6861	0.4394	0.5357	0.6192
9	*getNumbersSum*	Sum of numeric values in text		0.5671	0.9378	0.7068	0.6109
10	*getNNPLOCCount*	Number of location noun phrases (NNP–LOCs)	Yes	0.9123	0.231	0.3687	0.6044
11	*getFirstNNPPos*	Position of first NNP tag	Yes	0.581	0.7163	0.6416	0.5999
12	*getWordCount*	Number of words		0.6029	0.5779	0.5901	0.5986
13	*getVerbCount*	Number of verb POS tags		0.5565	0.7189	0.6274	0.573
14	*containsOn*	is “on” in the sentence?	Yes	0.5939	0.4467	0.5099	0.5706
15	*maxDistToNNP*	Maximum distance to proper noun	Yes	0.5629	0.6054	0.5833	0.5676
16	*lenOfLongestWord*	Length of longest word		0.5846	0.4394	0.5017	0.5635
17	*minDistToOn*	Number of characters to first “on” text	Yes	0.5976	0.3661	0.454	0.5597
18	*avgDistToNNP*	Average distance to proper noun	Yes	0.5423	0.7357	0.6244	0.5574
19	*similarityOfNouns*	Maximum similarity of NNPs and NPs	Yes	0.5793	0.416	0.4842	0.5569
20	*getEmotiveness*	(Number of adjectives (ADJs) + # of ADVPs) divided by (# of NPs + number of VBs)		0.5467	0.6417	0.5904	0.5548
21	*getAdjpCount*	Number of adjectival phrase (ADJP) POS tags		0.5624	0.4906	0.524	0.5544
22	*maxDistToQuote*	Number of characters to a quotation		0.5288	0.9741	0.6854	0.5529
23	*maxDistFromNPToNNP*	Maximum distance from noun to proper noun	Yes	0.557	0.5013	0.5277	0.5513
24	*getAcademicWordsCount*	Number of academic words (from pattern.en)		0.8573	0.1187	0.2085	0.5494
25	*minDistToNNP*	Min. distance to an NNP		0.5229	0.8364	0.6435	0.5366
26	*getNNPPERSCount*	Number of NNP-person (PERS) POS tags	Yes	0.5086	0.9797	0.6696	0.5164
27	*NNPCountOverNPCount*	Number of NNPs divided by number of NPs	Yes	0.5106	0.5824	0.5442	0.512
28	*getSwearCount*	Number of vulgar words	*omitted*				

Fig. 7.5 The LiT.RL Clickbait Detector feature performance using the support vector machine (SVM) classifier (Adapted from Brogly and Rubin et al. (2019)

29	*npsPlusNNPsOverNumbersSum*	Number of NPs + NNPs divided sum of numbers in headline	Yes	0.5086	0.6394	0.5666	0.5108
30	*firstPartContainsColon*	Do the first 15 characters have a colon?		0.7339	0.0321	0.0615	0.5102
31	*getNNPCount*	Number of NNPs		0.5079	0.5918	0.5466	0.5091
32	*getTimeWordsCount*	Number of time-related words (from pattern.en)		0.5991	0.0464	0.0861	0.5076
33	*getQuestionMarks*	Number of question marks		0.5034	0.9919	0.6679	0.5067
34	*getAtMentions*	Number of @ characters		0.7568	0.0148	0.0291	0.505
35	*maxDistToAt*	Distance to an @ character		0.7568	0.0148	0.0291	0.505
36	*getHashTagsAndRTs*	Number of hash tags and "RT"		0.5593	0.0175	0.0339	0.5018
37	*maxDistToHashTag*	Distance to a hash tag	Yes	0.5535	0.0155	0.0302	0.5015
38	*getCharLength*	Length of text		0.4555	0.4237	0.4391	0.4586

Fig. 7.5 (continued)

overall 94% accuracy in the binary text classification task (clickbait/non-clickbait) with a support vector machine, trained on 26,459 texts and tested on 5670 clickbait texts and 5671 non-clickbait texts dataset (Brogly & Rubin, 2019).

The best performing indicator of the set of 38 features is the *getPronounCount* function (or number of pronouns), which attains a 73.1% accuracy in the binary classification task (clickbait vs. not clickbait) (Fig. 7.5, the first line). The number of headlines without pronouns is almost doubled in non-clickbait, and the number of clickbait with one pronoun is almost seven times the number of non-clickbait single pronoun hyperlinks (Brogly & Rubin, 2019). An example of a clickbait containing multiple unresolved pronouns[13] included in the dataset was "*My* friend got with *her* boyfriend after *he* cheated on *my* sister should *I* snub *their* wedding?" Other significant predictors were the average length of bi-grams, defined as the sum of the character length of each *n*-gram divided by the number of *n*-grams[14] (Fig. 7.5, the second line) and the use of one or more determiners (such as *the, a, some, most, every, no, which*) that were used more often in clickbait texts than in non-clickbait texts (Fig. 7.5, the third line). The remaining 35 features are further detailed in Brogly and Rubin et al. (2019).

[13] When a referring expression (that person, this thing) or a pronoun (my, she, them) do not clearly point back to the previously mentioned entity or clearly understood entity in the world, linguists say that such a reference is *unresolved.* The process of determining an antecedent of a referring expression in the sentence (i.e., an anaphora) or the referent in the real world is called in NLP *anaphora resolution.* It is a much-needed step in the NLP pipeline to attribute appropriate meaning in "understanding" natural language. Humans naturally avoid unnecessary repetition of the same words and rely on the surrounding context and our pragmatic sense of the world to resolve anaphora. Algorithms need specific instructions and trained models to make such connections.

[14] Note that ***n*-grams** in this context are individual words; a **bigram** is an ***n*-gram** where ***n*** = 2, or simply put, a sequence of two adjacent words in the headline text.

7.3.3 Sample Clickbait Detector User Interface

At LiT.RL, we created the LiT.RL News Verification Browser, a downloadable stand-alone research tool, with our Clickbait Detector integrated as part of the NV Browser in one of its tabs. The Clickbait Detector User Interface (UI) shows a real-time automated color-coded analysis of any news website (Rubin et al., 2019) and it is worth explaining and demonstrating what this functionality may look like to the user.

As seen in Fig. 7.6, the Clickbait Detector identifies and signals the presence of clickbait using a traffic light analogy. Highly clickbaity headlines in red alert the user to stop and think, mildly clickbaity headlines are cautions in orange or yellow, and green is assigned to traditional headlines indicating it is safe to proceed (Brogly & Rubin, 2019).

At the top of the left-hand side of the screen, a listicle phrasing "*8 new flowering plants to show off in your garden...*" has some characteristic features of clickbait (in red). The second story withholds information referring to "*... some homeowners thinking about how...*" and is marked as a moderate clickbait (in orange). The two middle stories are sports news but presented in traditional headlinese (both marked in green). One story mentions the name of the Olympian competitor and declares her withdrawal, and the other story announces the trade of a named hockey defenseman. The last cookie-related headline at the bottom (in red) is from the Life section,

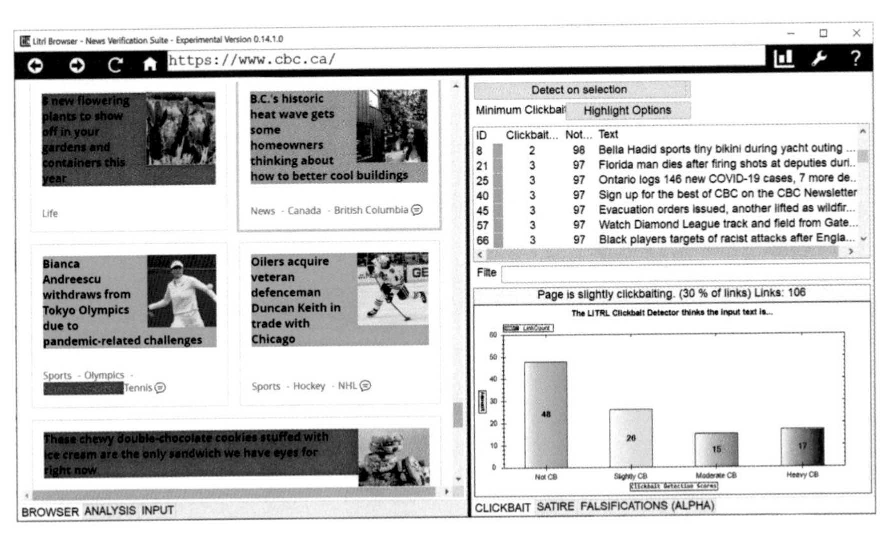

Fig. 7.6 News Verification Browser User Interface Screenshot for Clickbait Detection. The Clickbait Detector tab demonstrates the *CBC*'s front webpage (https://www.cbc.ca/) on July 13, 2021 with their five news stories from the homepage. The NV Browser's scrollable window (upper right) allows the user to examine individual headlines and scores. The bar graph (lower right) keeps track of the number of headlines considered not clickbait (green), slightly clickbait (yellow), moderately clickbait (orange), and heavy clickbait(red)

and it represents a typical "soft news." The actual article content[15] reveals a mouth-watering recipe offered by a freelance food photographer, but it can hardly be considered news. This soft-news article was ranked as clickbait by the LiT.RL Clickbait Detector based on machine learning predictions trained on the 38 characteristic features of clickbait (as discussed in (Brogly & Rubin, 2019)).

At the right-hand side of the UI, the Clickbait Detector offers analytical details for the users to explore (in Fig. 7.6). In the middle line (on the grey background), the tool keeps a tally of the total headlines analyzed (106); in this example 32 of which were heavily and moderately clickbait-ily phrased (15 and 17, respectively). Judging from the green and yellow bars in the bottom right-hand side window, 48 headlines were not clickbaity at all, and only 26 were slightly clickbaity. The tool concludes that the analyzed *CBC* home webpage on July 13, 2021 (at around 4:30 p.m.) was overall, slightly clickbaity (at about 30%, based on the ratio of heavily + moderately clickbait: slightly + non-clickbait). The top window lists each of the individual headlines and their individual likelihood of being a clickbait or not.

The effectiveness of such visual prompts, as well as the traffic light analogy, needs to be determined by further usability, HCI, and media literacy studies, but the system's capability is proven. It can clearly distinguish likely clickbait from non-clickbait with an intensity measure (as guided by the prevalence of the 38 characteristic features). Anyone who downloads, installs, and tries out this system can see its predictions and agree or disagree with its labels.

Other versions of clickbait detectors have been put forward. Some of them may lack a user interface, and others guard their algorithmic implementation details in hopes of their monetization (following the lead of the tech giants and their multiple off-shoots and start-ups), while the "know-how" has essentially been worked out and revealed in scientific R&D papers. See any of the before-mentioned works as well as other successful systems details in Elyashar et al., 2017, Grigorev, 2017, Indurthi & Oota, 2017, Papadopoulou et al., 2017, Wei & Wan, 2017, and Zhou, 2017. Examples that cover all three aspects—open access source code,[16] a UI available for experimentation, and a scientific paper with methodological explanations—are hard to come by.[17]

---

[15] See https://www.cbc.ca/life/food/salted-chocolate-chip-ice-cream-sandwiches-1.5210069 (Accessed on July 13, 2021).

[16] Open source code is typically shared via open access development repositories for non-commercial software. GitHub is an example of a collaborative distributed software development platform that allows multiple developers to access data and code for tasks including code sharing and distributed debugging. Many R&D teams use GitHub for collaborative projects; sharing internally in the development stages, and then distributing their resulting code to the broader developers' community externally (typically open access, and often linked as additional data or source code in scientific R&D papers). This popular Internet provider is known for hosting programming open-source projects and offers its basic services free of charge.

[17] The difficulty in finding such clickbait identification R&D packages may attest to my inability to find such works in languages other than English, my primary literature search language for this book.

More recent works—such as Pujahari and Sisodia (2021) —report using just eleven robust features that outperform previous results by several percentage points, inching closer to the top of the ninetieth percentile accuracy with standard ML techniques that seem to be applied with few if any significant modifications. Thus, the task can be considered technically "solved" for text-based NLP.

Research has now moved to other formats of clickbait such as images and video for an adjacent non-textual task of clickbait video identification on YouTube and other platforms. For instance, Varshney and Vishwakarma (2021) first extract audio from the videos to transform their input features to textual data, and recombine them with more features from human responses and reactions in the comments section of posted videos, as well as features of the poster's profile. Once reduced to a text-based problem and textual features, detection accuracies (of 95.5%) are similar to those of textual clickbait headlines. To detect video clickbait, the AI community uses Papadopoulou et al.'s (2018b) annotated dataset, including 380 user-generated videos, with 200 debunked and 180 verified examples and their 5195 near-duplicate reposted versions. Another useful version, the InVID Fake Video Corpus v2.0 *of* 117 fake videos and 110 real videos, with annotations and descriptions, was produced in collaboration with the AFP (France) and Centre for Research and Technology Hellas—Information Technology Institute *(CERTH-ITI, Greece)* (Papadopoulou et al., 2018a).

To sum up the state of clickbait detection, four points are important to note. First, the temptation to take the "bait" often remains irresistible (especially, if you are craving a soft ice cream chocolate chip cookie right now[18]). Second, news readers differ in their abilities and experience when distinguishing clickbait from non-clickbait, and people often disagree on what constitutes clickbait. If you were looking for a new refreshing ice cream cookie, the CBC.ca article[19] may be just what you needed to satisfy your information need. Chen and Rubin's (2017) study participants tended to base their judgments on both form and content: "In terms of form, participants identified features such as profanity, forward-referencing, and colloquial phrasing as indicators of clickbait. In terms of content, participants were more likely to rate 'soft' news headlines like entertainment/sports news and 'offbeat' stories to be clickbait" (Y. Chen & Rubin, 2017 p. 2). Third, regardless of the usability of the suggested UI, user experience, agreement, or contexts, an AI-based assistance in clickbait identification is now available. The proof-of-concept systems are now capable of objectively performing this automated task, and many of the systems are open access. The content and delivery style of clickbait is unique enough that it is recognizable with the naked eye, and clickbait has now been proven to be sufficiently identifiable with ML and NLP techniques. In practice, it is an AI success

[18] Do not follow this CBC.ca link: https://www.cbc.ca/life/food/salted-chocolate-chip-ice-cream--sandwiches-1.5210069 (Accessed on 13 July 2021). You could take a mental note of the power of suggestion: were you able to resist the temptation at the mere mention of the recipe? Would you have clicked on the link, if you simply could do so?

[19] See https://www.cbc.ca/life/food/salted-chocolate-chip-ice-cream-sandwiches-1.5210069 (Accessed on July 13, 2021).

story: nine out of ten clickbait headlines on average can be labeled appropriately by most of the competing clickbait detectors.

Lastly, the fourth point brings us to the future steps in clickbait detection. It consists of assessing the willingness of the general public to adopt such systems into their daily newsfeed reading routines. The question is about our willingness to tolerate the errors and perceived usefulness of such systems.[20] If the public perception shifts toward their adoption, clickbait filters may become just as widely and commonly used as their spam-filtering counterparts. They, in fact, may halt the incessant clickbait proliferation incentivized by businesses who are attempting to harness and monetize users' attention. (See also Chap. 6 on the ad-revenue model and other marketing techniques that fuel attention manipulation online.) Thus, clickbait detection is an important step in preventing the spread of misleading and disinformative content online as a part of ceasing mis- and disinformation.

7.4 Automated Satire Detectors

7.4.1 Satirical Fakes Defined

Much like their counterparts, outright falsehoods (Sect. 7.4 and clickbait (Sect. 7.5), satirical news pieces are implicated in contributing to mis- and disinformation online. Misunderstandings of satirical pieces often happen due to a lack of historical or cultural knowledge or hasty readings, carelessness, or lack of attention (see also Chap. 2, Sect. 2.4 on shallow reading practices). Satire is deception in plain sight, intended to be transparently understood as humorous by at least some segment of its audience, it is a type of deception "intended to be found out" (Rubin et al., 2016 p. 8).

When satire is mistaken for authentic news and unwittingly spread via networks, it contributes toward misinforming others, albeit unintentionally. On the other hand, when readers are clearly aware of the humorous intent, they may no longer be predisposed to take the information at face value.

The genre of news satire sites or news parody is exemplified by satirical fakes such as those written by *The Onion* or recorded by the *CBC's This is That*, and they used to be called simply *fake news* prior to the 2016 U.S. Presidential Election altering the term's meaning (see Chap. 1 on the weaponization of the term *fake news* and alternative terms currently favored in research literature). Satirical fakes aim to present news "in a format typical of mainstream journalism but rely heavily on irony and deadpan humor to emulate a genuine news source, mimicking credible news sources and stories, and often achieving wide distribution" (Rubin et al., 2015 p. 3). News satire exists in two slightly different forms nowadays: one is satirical

[20] Most of us—those who have ever had to "fish out" non-spam from spam-filters and who still wonder how some spam gets into our mailbox—are acutely aware of the false positives and false negatives of AI.

commentary or sketch comedy to comment on real-world news events, and the other churns out "wholly fictionalized news stories" (Editors of Wikipedia, 2021).

Both satirical commentary and humorous fictionalized stories should be distinguished from serious (non-satirical) fabrications. "Technology can identify humor and prominently display originating sources (e.g., *The Onion*) to alert users, especially in decontextualized news aggregators/platforms" (Rubin et al., 2015 p. 3).

Humoring its targets and providing a spectacle is just one element of satire. Satire is not simply just for laughs or lols on social media. Satire as a genre must also aspire "to cure folly and to punish evil" (Highet, 1972 p. 156); it requires certain writing skills and social sensibilities. Mocking the target is not the goal in and of itself, instead satirical pieces as a literary form ought to carry a form of critique and a call to action to rectify the mocked situation. "This 'element of censoriousness' or 'ethically critical edge' (Condren, 2012 p. 378) supplies the social commentary that separates satire from mere invective. However, the receptiveness of an audience to satire's message depends upon a level of 'common agreement' (Frye, 1944 p. 76) between the writer and the reader that the target is worthy of both disapproval and ridicule. This is [another] way that satire may miss its mark with some readers: they might recognize the satirist's critique, but simply disagree with his or her position" (Rubin et al., 2016 p. 8). "As a further confounding factor, satire does not speak its message plainly, and hides its critique in irony and double-meanings. Though satire aims to attack the folly and evil of others, it also serves to highlight the intelligence and wit of its author. Satire makes use of opposing scripts, text that is compatible with two different readings (Simpson, 2003 p. 30), to achieve this effect. The incongruity between the two scripts is part of what makes satire funny (e.g., when Stephen Colbert, states 'I give people the truth, unfiltered by rational argument'[21]), but readers who fail to grasp the humor become, themselves, part of the joke" (Rubin et al., 2016 p. 8).

7.4.2 *Methodologies for Computing Satire*

The idea of reducing the art of satirical writing to predictive formula may be preposterous in itself, nonetheless it has been proven feasible to articulate some of the more formulaic elements that, in combination, can predict which news are likely to be satirical and which are not. To accomplish this goal, a team at LiT.RL undertook the task of studying comparative pairs of fake and legitimate news, matched on a wide range of topics, and representative of the scope of US and Canadian national newspapers. Two diverse datasets were collected and analyzed, with 360 news articles in total.[22] The first set, studied most carefully, was collected from 2 satirical

[21] 2006 White House Correspondents' Dinner (Accessed at https://youtu.be/2X93u3anTco on 15 July 2021).

[22] The Satirical Fake and Legitimate News Datasets (S-n-L News DB2015–2016) and their documentation are available by request from the LiT.RL website via "Data-To-Go" link: https://victoriarubin.fims.uwo.ca/news-verification/data-to-go/

news sites (*The Onion* and *The Beaverton*) and 2 legitimate news sources (*The Toronto Star* and *The New York Times*) in 2015. The 240 articles were aggregated by a 2 × 2 × 4 × 3 design (US/Canadian; satirical/legitimate online news; varying across 4 domains (civics, science, business, and "soft" news) with 3 distinct topics within each of the 4 domains. As a reliability check for our manual findings within the first 2015 set, in 2016, we drew another 120 articles from additional six legitimate and six satirical English-speaking online news sources for the second dataset. The description below summarizes the predictive features of satirical fakes and the methodology for their binary classification (satire/non-satire), as reported in Rubin et al. (2016) and as incorporated as the Satire Detector in the NV Browser as per Rubin et al. (2019).

7.4.3 Predictive Feature Specifications

We identified five satirical news features—*Absurdity, Humor, Grammar, Negative Affect,* and *Punctuation*—based on our comparative data observations and our literature review of irony, humor, and satire salient features and detection studies.

The Absurdity Feature is characterized by a sudden turn of events toward the end of satirical writing. Computationally, we identified the unexpected introduction of new named entities (people, places, locations) within the final sentence of satirical news.

The Humorous Feature was based on the premises of opposing scripts and maximizing semantic distance between two statements, as a modification to a method of joke punchline identification by Mihalcea et al. (2010). Humor is found when two scripts overlap and oppose. As a joke narration evolves, some "latent" terms are gradually introduced, which set the joke on a train of thought. Due to some ambiguities in the statements, the narrative advances on two or more paths of interpretation, suddenly shifting from the initial starting point. Consider an example from Attardo et al. (2002): *"Is the doctor at home?" the patient asked in a whisper. "No", the doctor's pretty wife whispered back, "Come right in."* (p. 35). The latter path gains more importance as elements are added to the current interpretation of the reader, and eventually ends up forming the punch line of the joke (Hempelmann et al., 2006).

Polanyi and Zaenen (2006) pointed out that the context of sentiment words shifts the valence of the expressed sentiment. In Rubin et al. (2016), we used the semantic relatedness of two text segments (S1 and S2), as a metric that combines the semantic relatedness of each text segment in turn with respect to the next text segment. For each word *w* in the segment S1 we identified the word in the segment S2 that has the highest semantic relatedness, as per Wu and Palmer (1994) word-to-word similarity metric. The depth between two given concepts in the WordNet[23] taxonomy, and the

[23] WordNet is a large lexical database of English, originally complied by the Princeton University Department of Psychology and currently housed in the Department of Computer Science. WordNet's structure and lexical relations among words are a standard resource used for NLP pur-

depth of the least common subsumer were combined into a similarity score (Z. Wu & Palmer, 1994). The same process was applied to find the most similar word in S1 starting with words in S2. The word similarities were weighted, summed, and normalized with the length of each text segment. The resulting relatedness scores were averaged. Less similar semantics is more likely to generate opposing scripts. Just like with a correct punchline that generates surprise in a joke, satirical fakes finishing segments tend to have a minimum relatedness with respect to the set-up.

The Grammar Feature was a vector representing a set of normalized term frequencies matched against the Linguistic Inquiry and Word Count (LIWC) dictionaries as developed by Pennebaker et al. (2015). We accounted for the number of certain parts of speech (namely, adjectives, adverbs, pronouns, conjunctions, and prepositions) and then assigned each normalized value as the element in a feature array representing grammar properties.

The *Negative Affect Feature* was set of weights representing normalized frequencies based on term-for-term comparisons of the texts with negative affect terms in the LIWC (2015) dictionaries. Finally, the *Punctuation Features* were based on the presence of periods, comma, colon, semi-colon, question marks, exclamation, and quotes, normalized by the number of sentences.

7.4.4 Methodology for Binary Classification (Satirical Fakes Vs. Legitimate News)

Consistent with a typical NLP pipeline of text processing steps (Fig. 7.7), we trained and evaluated our model using a state-of-the-art method of support vector machines (SVMs) on a 75% training and 25% test split of the dataset, and ten-fold cross-validation applied to the training vectors, and then measured our predictions' success. We first performed a topic-based classification followed by sentiment-based classification, and then experimented with the feature selection based on absurdity and humor heuristics obtained from the literature and data-based observations.

Features representing *Absurdity, Humor, Grammar, Negative Affect,* and *Punctuation* were introduced in succession to train the model, and combined overall. The predictive performance was measured at the introduction of each new feature and the best performing features were combined and compared to the overall performance of all 5.

---

poses, for example, to ascertain words' meaning. According to WordNet main webpage (Accessed at https://wordnet.princeton.edu/ on 15 July 2021), WordNet contains English "nouns, verbs, adjectives and adverbs are grouped into sets of cognitive synonyms (synsets), each expressing a distinct concept. Synsets are interlinked by means of conceptual-semantic and lexical relations. The resulting network of meaningfully related words and concepts can be navigated with the browser (http://wordnetweb.princeton.edu/perl/webwn) and freely and publicly available for download (https://wordnet.princeton.edu/download)."

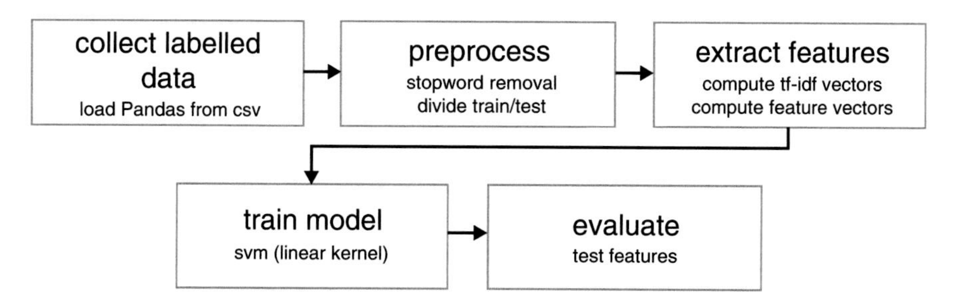

Fig. 7.7 News satire detection pipeline for distinguishing satirical from legitimate news. (Redrawn from (Rubin et al., 2016))

Table 7.1 Satirical news detection evaluation results with ten-fold cross-validation. (Adapted from (Rubin et al., 2016))

Features	Precision	Recall	F-Score	Confidence
Baseline (*tf-idf* topic vector)	78	87	82	85
Baseline + absurdity feature [0,1]	85	89	87	83
Baseline + humor feature [0,1]	80	87	83	83
Baseline + grammar features (pronoun, preposition, adjective, adverb, conjunction)	93	82	87	82
Baseline + negative affect feature	81	84	83	84
Baseline + punctuation feature	93	82	87	87
Baseline + grammar + punctuation + absurdity features (combined)	90	84	87	84
Baseline + all of the above options	88	82	87	87

We trained our binary classification model (satirical news vs. legitimate news) on 270 articles and tested it on a set of 90 news articles, with equal proportions of satirical and legitimate news. Table 7.1 shows the evaluation measures of precision, recall, and F-score with associated ten-fold cross-validation confidence result. We established a baseline model in the form of *tf-idf* feature weights, while measuring the net effects of additional features against the baseline. Our baseline model performed with 82% accuracy, an improvement on Mihalcea et al. (2010) who achieved a 65% baseline score with *tf-idf* method on similar input data. The F-score was maximized in the case when *Grammar, Punctuation,* and *Absurdity* features were used. Precision was highest when *Punctuation* and *Grammar* were included. *Absurdity* showed the highest recall performance.

As the overall result in Rubin et al. (2016), we were able to achieve relatively high accuracy rates (90% precision, 84% recall, 87% F-score), translating the theories of humor, irony, and satire into a predictive method for satire detection. We integrated word-level features using an established machine learning approach in text classification, SVM (Burfoot & Baldwin, 2009), to derive empirical cues indicative of satirical fakes. Contrary to our expectations, we discovered that individual textual features of shallow syntax (parts of speech) and punctuation marks are highly indicative of the presence of satire, producing a detection improvement of 5% (87% F-score). These findings suggested that the rhetorical component of satire may provide reliable cues to its identification. Based on our manual analysis, this finding may be due to the presence of more complex sentence structures (i.e., higher prevalence of dependent clauses) in satirical content, and the strategic use of run-on sentences for comedic effect. However, this pattern did not translate to longer sentences per se, since our average words-per-sentence feature did not increase predictive accuracy. Also contrary to our expectation, markers such as profanity and slang, word-level features deemed significant by Burfoot and Baldwin (2009), produced no measurable improvements in our trials of satirical fakes in news which may not have used many profanities or slang to start with.

7.4.5 *Neural Network Methodologies for Computing Satire*

Recent works have experimented with embedding linguistic features (at several levels—within a character, a word, a sentence, a paragraph, and a document). Such embeddings are used as input representations for neural network classifiers. Various features capture the differences between satirical and legitimate news, and such layered information is used by neural networks to learn some defining features from the labeled datasets. Such neural networks can be thought of as learning on their own, based on complicated but recurrent patterns beyond the grasp of humans, in either a supervised ML fashion (from labeled data instances) or in an unsupervised fashion (where similarities and dissimilarities are observed).

Psycholinguistic features, stylistic features, structural features, and readability features can be embedded in neural networks. Yang et al. (2017) proposed such multi-level neural networks as hierarchies—at the character-word-paragraph-document levels—and utilized an attention mechanism (by Bahdanau et al., 2016) to reveal the relative differences between satire and legitimate news with best performing accuracies at 98.4% (93.5% precision, 89.5% recall, 91.5% F-score).

The attention mechanism (Yang et al., 2017) is exemplified with the score on the right in Fig. 7.8. "The high attended paragraphs are longer and have more capital letters as they are referring [to] different entities. They have more double quotes, as multiple conversations are involved. Moreover, we subjectively feel the attended paragraph with score 0.98 has a sense of humor while the paragraph with score 0.86 has a sense of sarcasm, which are common in satire. The paragraph with score 1.0 presents controversial topics, which could be misleading if the reader cannot understand the satire" (Yang et al., 2017, p. 1986).

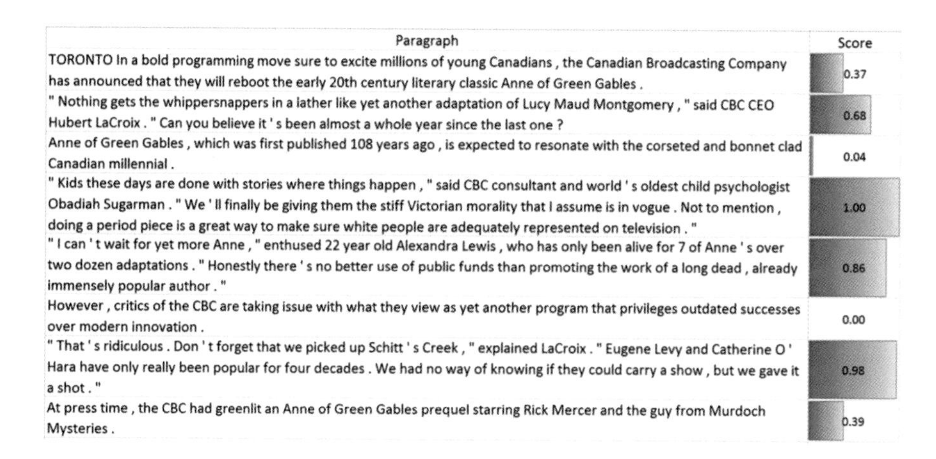

Paragraph	Score
TORONTO In a bold programming move sure to excite millions of young Canadians , the Canadian Broadcasting Company has announced that they will reboot the early 20th century literary classic Anne of Green Gables .	0.37
" Nothing gets the whippersnappers in a lather like yet another adaptation of Lucy Maud Montgomery , " said CBC CEO Hubert LaCroix . " Can you believe it ' s been almost a whole year since the last one ?	0.68
Anne of Green Gables , which was first published 108 years ago , is expected to resonate with the corseted and bonnet clad Canadian millennial .	0.04
" Kids these days are done with stories where things happen , " said CBC consultant and world ' s oldest child psychologist Obadiah Sugarman . " We ' ll finally be giving them the stiff Victorian morality that I assume is in vogue . Not to mention , doing a period piece is a great way to make sure white people are adequately represented on television . "	1.00
" I can ' t wait for yet more Anne , " enthused 22 year old Alexandra Lewis , who has only been alive for 7 of Anne ' s over two dozen adaptations . " Honestly there ' s no better use of public funds than promoting the work of a long dead , already immensely popular author . "	0.86
However , critics of the CBC are taking issue with what they view as yet another program that privileges outdated successes over modern innovation .	0.00
" That ' s ridiculous . Don ' t forget that we picked up Schitt ' s Creek , " explained LaCroix . " Eugene Levy and Catherine O ' Hara have only really been popular for four decades . We had no way of knowing if they could carry a show , but we gave it a shot . "	0.98
At press time , the CBC had greenlit an Anne of Green Gables prequel starring Rick Mercer and the guy from Murdoch Mysteries .	0.39

Fig. 7.8 An example of satirical news that exemplifies attention score, where the higher the score, the more indicative of satire the segment is (Adapted from Yang et al. (2017))

To explain word embeddings, let us look in more detail at the work by De Sarkar et al. (2018). They used previously pre-trained word embeddings such as Glove[24] (Pennington et al., 2014) and fastText[25] (Bojanowski et al., 2016) that already contain syntactic information at the word level (Baccianella et al.,2010; Miller, 1995), combined them with named-entity features (such as location, organization, GPS) and 16 values obtained from SentiWordNet features. These word embeddings (concatenated with syntax information) are multiplied with a weight matrixWemb (learned) to produce a final word embedding that summarizes the required semantics of the word for capturing satire. The sentence module (S) takes a sequence of word embeddings as input and produces sentence embeddings. Next, sentence embeddings are used as input for the document embedding module (D). Various implementations of neural networks learn correct weights based on weighted averages of these inputs and make binary predictions (satire or not). They identify which specific neural network versions perform best, with the best results achieved with implementation of the Glove embedding and syntactic information with very high 98.6% accuracy (92% precision, 91.2% recall, 91.6% F-score) and emphasize verbosity as a defining feature of clusters of satirical sentences. In Rubin et al. (2016), we previously elaborated on the specific mechanism of verbosity, calling it "sentence complexity" which is a combination of multiple clauses and extending sentence length to underscore the comedic effect. Such characteristics informed our development of the *Punctuation* and *Grammar Features*, now confirmed by De Sarkar et al. (2018). To make it clearer, let us consider two examples and compare the first excerpt (A) from *The New York Times* with its paired-up article from *The Onion (B)*, which contains not just a quote which is three times longer, but also adds flare to Dr. Berger's motivation and emotional state:

[24] https://nlp.stanford.edu/projects/glove

[25] https://github.com/facebookresearch/fastText/blob/master/pretrained-vectors.md

(A.) "With almost every bone in the body represented multiple times, Homo naledi is already practically the best-known fossil member of our lineage," Dr. Berger said." - The New York Times "Homo Naledi, New Species in Human Lineage, Is Found in South African Cave".

(B.) "Not too long ago, these early people were alive and going about their normal daily lives, but sadly, by the time we scaled down the narrow 90-meter chute leading into the cave, they'd already been dead for at least 10,000 decades," said visibly upset University of the Witwatersrand paleoanthropologist Lee R. Berger, bemoaning the fact that they could have saved the group of human predecessors if they had just reached the Rising Star cave system during the Pleistocene epoch." - The Onion "Tearful Anthropologists Discover Dead Ancestor of Humans 100,000 Years Too Late".

De Sarkar et al. (2018) also concluded that "the last sentence of a news article is a key feature for detecting satire" (p. 3378), as was previously revealed with manually-engineered features in punchline detection. De Sarkar et al. (2018) achieved "slightly superior" results that are still "comparable to the existing state-of-the-art models" which used manually-engineered linguistic features reflecting satire (p. 3379). While the manual engineering of features is surpassed by ML techniques, the results do not have as intuitive an explanation, since conceptually, satire is not typically a sentence-level phenomenon.

Yang et al.'s (2017) analysis contributed to the importance of readability, imagination, and emotionality features, and reaffirmed that psycholinguistic, stylistic features, and structural features are beneficial at the paragraph level.

McHardy et al. (2019) followed this lineage of work by Yang et al. (2017), De Sarkar et al. (2018), and the like and used word embeddings as input representations of news in German, successfully applied neural networks, and thus proved the language independence of these methods.

7.4.6 Sample Satire Detection UIs and Suggested Improvements

In an unpublished follow-up 2017 usability study of Satire Detection, we used a simple mobile UI (Fig. 7.9.) to elicit study participants' responses. The study asked its participants to contribute a news headline and body text without identifying its source. The Satire Detector tool's stated task was to label any piece of news (under 1000 words) as either satire or legitimate news. The system rendered its decision and, in a cursory way, mentioned its top predicting features. The study looked for additional human confirmation of the tool's predictions and additional human insight. Instead, we stumbled across the main difficulty for humans interacting with the tool. The Satire detector did not provide sufficient detailed explanations that users could understand and trust. Thus, our main recommendation for future R&D is not about increasing the performance of the system capabilities, since even the most sophisticated neural nets have only marginal gains to make in the accuracy metric. It is rather important to add to the explanatory powers which are in principle easier to achieve with hand-engineered features and speech generation (than less transparent "black box" ML).

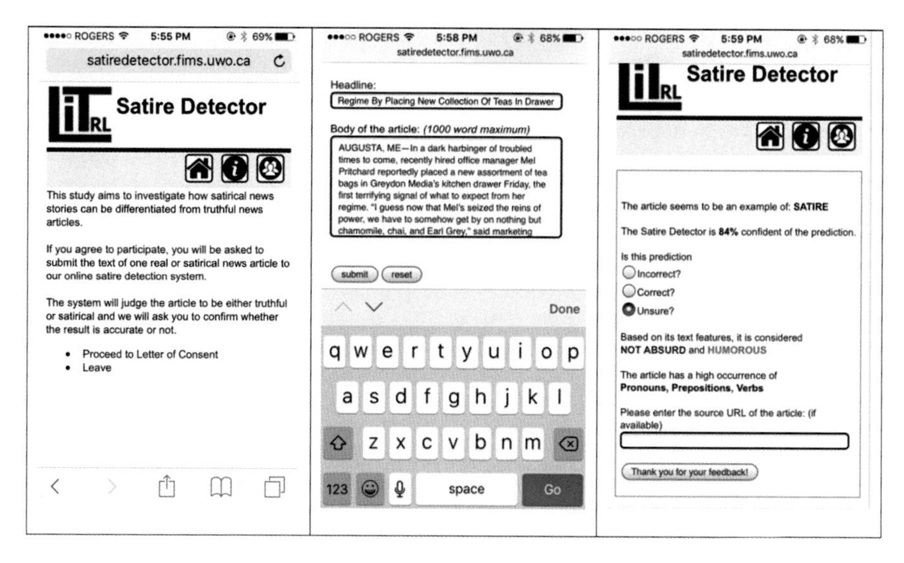

Fig. 7.9 Mobile UI screenshots for a usability (unpublished) study of the Satire Detector (based on (Rubin et al., 2016)). The UI elicits study participants' article submissions and their responses to system predictions

When explaining predictive features, it helps to show sample articles on the same topic, side-by-side. For example, multiple clauses and excessive punctuation were used by the Onion[26]: "'Proxima b is located one star away from our solar system, or just about $50 billion outside of our current budget,' said NASA administrator Charles Bolden, who explained that the terrestrial planet, which may possess the right conditions for life to exist, is situated more than two and a half times the agency's present fiscal allocation away." This satirical fake news demonstrates increased sentence complexity in terms of clauses and prevalence of punctuation. By comparison, a *New York Times* news piece similarly reports a previously unidentified planet is circling the second nearest star to Earth, and cites an astronomer who likened this fact to a flashing neon sign. Notice how much simpler the introduction of the astronomer is: "I'm the nearest star, and I have a potentially habitable planet!" said R. Paul Butler, an astronomer at the Carnegie Institution for Science and a member of the team that made the discovery."[27]

In principle, a Satire Detector UI should be able to clearly label each of the articles. In addition, digital media literacy efforts can point out satirical features and their subtlety, and particularly the effectiveness of the satirical genre as contrasted with straightforward criticism. Some systems-centered papers overemphasize either the harm or the entertainment value of satire but overlook its important role of social

[26] See https://www.theonion.com/nasa-discovers-distant-planet-located-outside-funding-c-1819579186 (Accessed on July 15, 2021).

[27] See https://www.nytimes.com/2016/08/25/science/earth-planet-proxima-centauri.html (Accessed on July 15, 2021).

critique. Creators of fake news often hide their digital content behind the term "satire," saying that they were "simply kidding." Yet, made-up disinformative content does not qualify as satire. Satire, by definition, has to offer a critical sight, and it cannot be outright deceptive without the intent to be found out.

To demonstrate a contrast between a satirical and non-satirical take on the news, let us consider an example of the news on the 2015 FDA (the U.S. Food and Drug Administration) approval of a controversial new drug, flibanserin or "female Viagra." Pharmaceutical lobbyists for the drug manufacturer attempted to make the drug's release an issue of gender-equality, despite the FDA's concerns being focused on side-effects. *The Toronto Star* writes: "The drug has overcome numerous hurdles, opposition and criticism. It was rejected by the FDA as a desire drug in 2010 and 2013. Then Sprout and other drug companies with sexual pharmaceuticals in the pipeline sponsored the campaign Even the Score. It resembled a grassroots, social media campaign accusing the FDA of sexism for not having approved this drug for women, while claiming men had similar drugs to help them treat male sexual dysfunction."[28]

The Beaverton parodies the news with "a happy customer" remark: "With this new drug, my husband and I are intimate anywhere between one and one-and-a-half times more per month. Sure, I'm throwing up every three hours, but our love life is saved!"[29] Between the two examples of the news, it is a toss-up in terms of which piece is more effective in making its point.

The 2019 NV Browser used the 2016 satire identification mechanism as one of its integrated parts—the Satire Detector tab. Since the focus at the time was on the proof of concept of an integrated UI for a suite of tools, the UI improvements were not sufficiently explored. It simply color-coded the entire text of the news as either orange (for satire, Fig. 7.10) or blue (for legitimate news, Fig. 7.11).

Explaining the rationale for predictions seems to be key to revealing either the humorous or critical intent of the genre and is currently lacking in most implementations. Explaining AI decisions is the next step in user-centered systems development. It is wise to draw on human insights and experience on the satirical literary genre and its use of humor, irony, and sarcasm.

In sum, satirical fakes are a clear specific literary genre used in a news-like way and spread widely over social media. Satire has a set of specific features that are identifiable with the naked eye and AI. These linguistic and stylistic features are entirely different from other misleading digital content such as clickbait. Satire aims to critique social ailments in witty, ironic, humorous, or sarcastic ways. While overt labeling of satirical fakes may "kill the fun" for those news readers who are "in the know," those who are not may benefit from explicit assistance and real-time education. This specifically concerns the segment of the population who struggle with interpreting sarcasm or irony in deadpan humor due to their second language use, or lack of cultural or historical contextual knowledge. AI-assistance can be of use in

[28] https://www.thestar.com/life/2015/08/19/female-viagra-okayed-in-us-may-come-to-canada.html (Accessed on July 15, 2021).

[29] https://www.thebeaverton.com/2015/06/female-viagra-allows-women-with-sexual-dysfunction-to-add-fainting-nausea-and-dizziness-to-their-symptoms/ (Accessed on July 15, 2021).

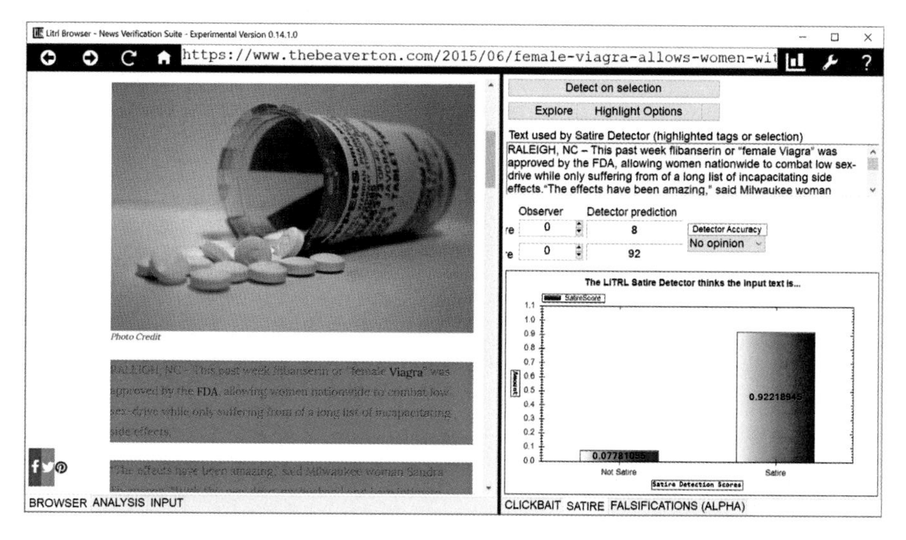

Fig. 7.10 LiT.RL News Verification Browser ((Rubin et al., 2019) Satire Detector UI, as a proof of concept, color-codes an individual news articles in orange-red when it is predicted to be a satirical fake, without any knowledge of its source

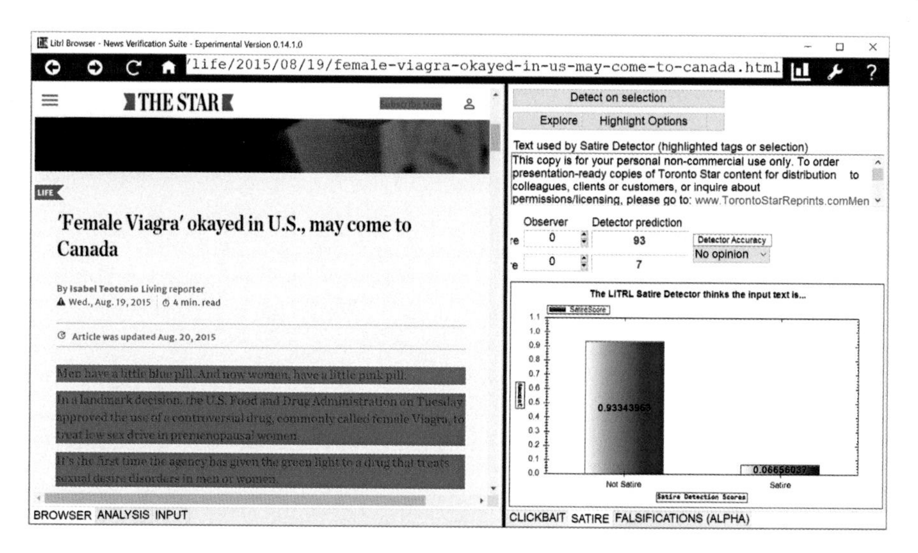

Fig. 7.11 LiT.RL News Verification Browser ((Rubin et al., 2019) Satire Detector UI, as a proof of concept, color-codes an individual news article in blue-purple when it is predicted to be an authentic news, without any knowledge of its source

multilingual information access, that is, when people access resources across language boundaries. The main effort in the identification of satire should now shift from marginal improvements in already high-performing systems to first, better justifying these sophisticated predictions to the users, and second, improving the usability of assistive or educational UIs to accomplish the goal of revealing the deception that was intended to be discovered in the first place.

7.5 Automated Rumor Debunkers

Rumors are passed around from person to person by word-of-mouth with social media allowing rumors to reach wider audiences faster. Rumors lack definitive verifiability at the time of their dissemination. As defined in standard dictionaries, *a rumor* is "a statement or report currently without known authority for its truth" or "talk or opinion widely disseminated with no discernible source" (Editors of Merriam-Webster's Dictionary, 2021). Rumors are distinctly different from mis- or disinformation in the sense that some rumors turn out to be true, while others turn out to be false. Thus, the focus of rumor debunking is to distinguish the two.

All rumors cannot be simply equated to deceptive messages, even though most people would agree rumors are harmful (Rubin, 2017). False rumors can cause negative impacts or elicit undesirable responses including defamation, protests, the destruction of property, and the spread of fear, hate, or euphoria (Matthews, 2013). Twitter rumors, for instance, are notorious for influencing the stock market. "Perhaps, one of the most infamous cases is of the hacked AP account tweeting a rumor that Barack Obama had been injured in an explosion at the White House. The tweet caused the S&P to decline and wipe $130 Billion in stock value in a matter of seconds" (X. Liu et al., 2015).

Rumor detection and debunking studies explain the spread and resolution of rumors online. For example, on Twitter, a rumor may consist of a collection of tweets with the same or similar unverified assertions, most likely worded differently from each other, yet recognizably offering versions of the same theme or topic. Such tweets propagate through the network in a multitude of cascades, and will eventually "resolve" as either true (factual) or false (non-factual), or remain "unresolved" (Vosoughi, 2015). Resolving one version of the rumor about the same topic may also offer concrete resolution for other versions of the same rumor.

In the pre-digital news era, rumors were typically resolved by either common sense judgments or with further investigations by professionals. It has been argued that resolving rumors algorithmically can speed up the process of human verification, especially when fact-checkers have access to such systems operating in real-time (X. Liu et al., 2015).

7.5.1 *Rumor Detection Stages or Sub-Tasks*

To resolve rumors algorithmically, rumor detection systems may consist of up to four separate tasks. These tasks are aligned with what a human would do to verify a rumor, although a typical system design can contain them as individual stages (in Fig. 7.12.).

In the sample pipeline, the first step is *Rumour Detection* which identifies posts on social media that can constitute an emerging rumor, if a particular rumor is not known a priori. Then, a *Rumor Tracking* component collects, monitors, and filters

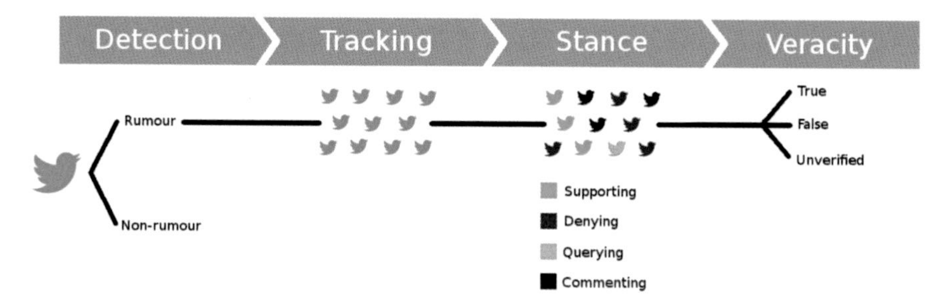

Fig. 7.12 A typical architecture for a rumor classification system. (Adapted from Zubiaga et al. (2018 p. 13))

other posts that mention the identified information based on mentions of a set of relevant keywords in the sentences. From a dataset of all the collected posts related to a specific rumor, the *Stance Classification* component determines how each post orients itself on the topic of the rumor's veracity. Each related post is labeled with a stance, usually from a predefined set of stances (e.g., agrees, disputes, supports, disproves, etc.) In some cases, this step relies solely on input from experts or on validation from authoritative sources. Finally, the *Veracity Classification* component resolves the rumor as true or false, using the input from all previous components (associated posts and their stance labels) and optimally, additional data from external sources such as news media, or other websites and databases. The final system output may consist of just the predicted truth value, or ideally the supporting contexts such as the URLs or data sources so that users can assess the system's reliability by double-checking the relevant sources (Zubiaga et al., 2018 p. 13).

Rumor debunking systems have a relatively short history, as the first attempts date back to 2010s algorithms for debunking rumors (Castillo et al., 2011; e.g., by Mendoza et al., 2010) which varied in the data they used and some features they extracted (K. Wu et al., 2015; Yang et al., 2012). Qazvinian et al. (2011) focused on rumor-related tweets to match a certain regular expression of the keyword query and the users' stance about those rumor-related tweets (that they called "believing behavior") and released an annotated dataset of Twitter "microblog[30]" for rumor detection (still currently in use by other researchers). Mendoza et al. (2010) led the way in analyzing user behavior through observing and analyzing tweets during the 2010 Chilean earthquake by elaborating on the topology network created by users' retweeting and how the rumor diffusion patterns on Twitter differed from those of traditional news platforms.

[30] Some of the early 2010s terminology is lingering in R&D papers, for example, tweets are referred to as "Twitter microblogs," and "blogs" are sometimes called "weblogs" prior to the shortening of the term.

7.5.2 Rumor Features

Liu and his colleagues from the Thompson Reuters R&D group (Liu et al., 2015) proposed a method to automatically debunk rumors on Twitter using six types of features in their real-time analysis (Table 7.2). Their proposed categories of verification features are largely inspired by journalistic insights. (See also, Chap. 5, Sect. 5.3. on "Investigative Reporting and Fact-Checking in Journalism".)

Yang et al. (2012) proposed extra features (Table 7.3) for their *Rumor Buster* for Sina Weibo, China's leading micro-blogging service provider that functions like a Facebook-Twitter hybrid. They used five broad types of evidence as the basis for their predicted resolutions: content, client, account, propagation, and location features were extracted and analyzed from Sina Weibo to make a binary classification prediction (a rumor or not).

7.5.3 Other Methodologies for Rumor Debunking

Some literature reviews have lumped rumor detection systems with others that assess the credibility of online information. For example, Saquete et al. (2020) include works on trust and the credibility analysis of tweets referencing suspicious images and videos by Middleton (2015) who extracted NLP evidence from tweets in the form of fake and genuine claims attributed to trusted and untrusted sources. Middleton (2015) reported "fake" tweet 94%–100% precision and 43%–72% recall, and "real" tweet 74%–78% precision and 51%–74% recall, but conceptually he did not declare his multimedia task as rumor detection.

Table 7.2 Six verification feature for rumor debunking on Twitter. (Adapted from (Liu et al., 2015))

Category	Feature name
Source credibility	Is trusted/satirical news account Has trusted/satirical news URL Profile has URL from top domains Client application name
Source identity	Profile has person name Profile has location Profile includes profession information
Source diversity	Has multiple news/non-news URLs after *dedup* Deduped tweets' texts are dissimilar
Source Location & Witness	If tweet location matches event location If profile location matches event location Has witness phrases, i.e., *"I see," "I hear"*
Message belief	Is support, negation, question, or neutrality
Event propagation	Event topic Retweet, mention, hashtag *h*-index Max reply/retweet *graph4* size/depth

Table 7.3 Rumor busting predictive features for Sina Weibo microblogs and their descriptions (Adapted from (Yang et al., 2012))

Category	Features	Description
Content	Has multimedia	Whether the microblog contains pictures, videos, or audios
	Sentiment	The number of positive/negative emoticons used
	Has URL	URL points to external source
	Time span	Time interval between the posting and user registration
Client	Client program used	The type of client such as web- or mobile-based client
Account	Is verified	Whether the user identity is verified by Sina Weibo
	Has description	Whether the user has personal descriptions
	Gender of user	The user's gender
	User avatar type	Personal, organization, and others
	Number of followers	The number of the user's followers
	Number of friends	The number of followers with a mutual follower relationship
	Number of microblogs posted	The number of microblogs posted by the user
	Registration time	The actual time of user registration
	User name type	Personal real time, organization name, etc.
	Registering place	The location information takes at user's registration
Location	Event location	The location in which the event happened as mentioned by rumor-related microblogs
Propagation	Is retweeted	Whether the microblog is original or a retweet
	Number of comments	The number of comments on the microblog
	Number of retweets	The number of retweets of the microblog

Liu et al. (2016) offered a representation learning method, Information Credibility Evaluation, that learns and models *who, what, when,* and *how* as key elements of user credibility, behavior types, temporal properties, and comment attitudes, respectively. Based on their ICE model and the Sina Weibo dataset, their web-based system Network Information Credibility Evaluation (NICE) automatically crawled online information from Sina Weibo and labeled it as either rumor or non-rumor based on the predicted credibility of joint key elements. (See also Chap. 3 on conceptual nuances in human credibility assessments and trust judgments.)

Hamidian and Diab (2019) decoupled rumor detection from the classification task with their system achieving 82% and 85% f-measures on a mixed and topic specific (e.g., "Is Barack Obama Muslim?") data sets, respectively. Hamidian and Diab (2019) employed novel metalinguistic features such as replay time (network-based), time of posting the tweet (regular day or busy day), and three novel content features (named entity recognition, emoticon, and sentiment). Some of the more up-to-date works have increased the number of predictive features to over a hundred, as reviewed in Reshi and Ali (2019). Sahana et al. (2016) studied rumored and non-rumored tweets posted during the London Riots in 2011 and stressed the importance of content-based features achieving 87.9% accuracy with their J48 algorithm[31] approach.

[31] J48 is a machine learning decision tree classifier, an algorithm developed by Ross Quinlan (1993). The book carries the title of "C4.5," as it is an extension of Quinlan's earlier algorithm.

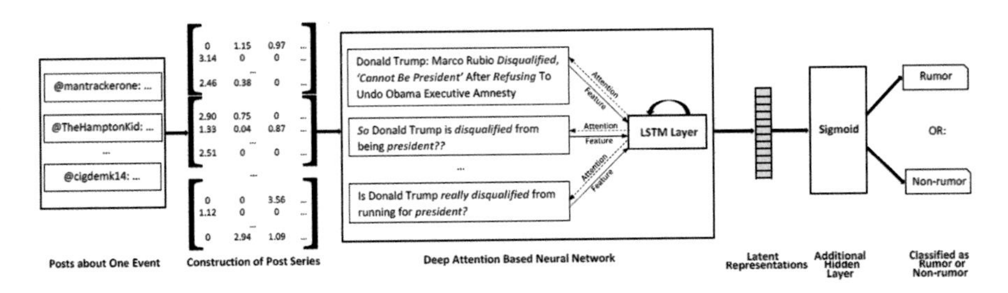

Fig. 7.13 Deep learning attentional model for rumor detection (T. Chen et al., 2017)

In addition to feature engineering, "deep learning has proven to be very advantageous over traditional machine learning in various problems owing to the fact that it is almost immune to the feature selection problem.[32] Deep neural networks need no less features to work efficiently and rather can perform well on unsifted features" (Reshi & Ali, 2019, p. 1158).

A good example of the latest developments in rumor detection with deep learning approaches includes Chen et al. (2017) who used a deep attention model on a base of recurrent neural networks (RNN) to learn temporal hidden representations of sequential posts for identifying rumors. Chen et al.'s (2017) model "delves soft-attention into the recurrence to simultaneously pool out distinct features with particular focus and produced hidden representations that capture contextual variations of relevant posts over time." Fig. 7.13 gives a schematic overview of Chen et al.'s (2017) framework. For each event, posts in sequence are collected and transformed into their *tf-idf* vectors. Then, deep recurrent neural networks augmented with a soft-attention mechanism are deployed to derive temporal latent representations by capturing long-term dependency among the post series and by selectively focusing on important relevance. An additional layer is added onto the learned representations to determine the event to be rumor/non-rumor.

Figure 7.14 illustrates how data are collected for each event and manually verified through authoritative news verification services (such as Snopes). Suitable keywords were manually extracted for each event to ensure a precise search result of only relevant posts. After that, posts were crawled with a query search and each collected data point was stored in a table with events labeled rumor [1] or normal events [0].

Attentional deep learning may be difficult to grasp conceptually[33] but Chen et al. (2017)) found good ways to visually exemplify their methodology. Lastly, see

---

[32] "The feature selection is the problem of choosing a small subset of features that ideally is necessary and sufficient to describe the target concept. For many real-world problems, which possibly involve much feature interaction, we need a reliable and practically efficient method to eliminate irrelevant features." (Kira & Rendell, 1992, p. 129).

[33] See also explanation of attentional mechanism in the previous segment on Neural Network Methodologies for Computing Satire.

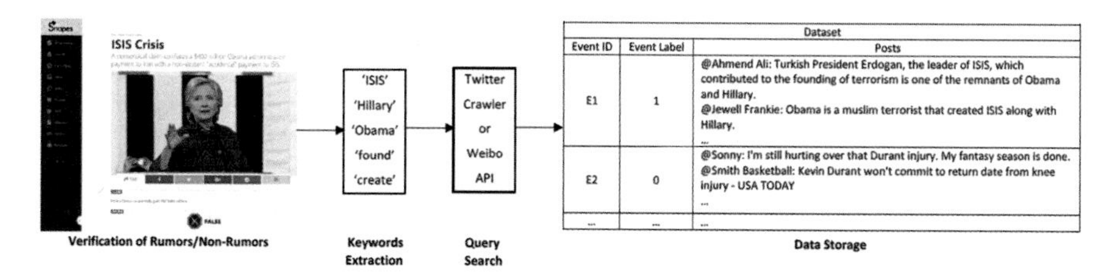

Fig. 7.14 Data collection, manual verification, and labeling [rumor = 1, normal event = 0]. (Adapted from Chen et al.'s (2017))

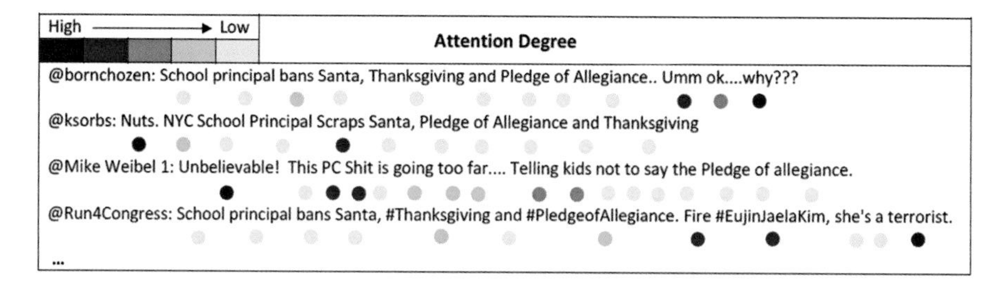

Fig. 7.15 Visualization of varied attention on a detected rumor related to the claim "School Principal Eujin Jaela Kim banned the Pledge of Allegiance, Santa and Thanksgiving." Red color and its intensity reflects higher degree of attention paid to certain words and expressions in posts that are more useful in rumor detection, and low attention to words irrelevant to the task. (Adapted from Chen et al.'s (2017))

Fig. 7.15 which illustrates the intermediate attention results on different words within a detected rumor event.

The dots' colors and their intensity in Fig. 7.15 reflect the various degrees of attention which the system paid to each word in a post. In the rumor "School Principal Eujin Jaela Kim banned the Pledge of Allegiance, Santa and Thanksgiving," most of the vocabularies closely connected with the event itself are given less attention weight than the words expressing users' doubting, enquiring, and expressions of anger caused by the rumor. The system learns to focus on useful words while ignoring unrelated words as irrelevant features. Thus, the feature selection problem (Kira & Rendell, 1992) (see footnote above) is resolved. In terms of success metrics, Chen and colleagues' CallAtRumors method was reported to outperform competitors "by achieving the precision, recall, and F-measure of 88.63%, 85.71%, and 0.869 respectively on a Twitter dataset. The same result can be seen on the Weibo dataset, where CallAtRumors achieved the precision, recall, and F-measure of 87.10%, 86.34%, and 0.8672, respectively" (T. Chen et al., 2017 p. 7).

In summary, in the past 10 years or so, the sub-field of rumor debunking has made huge strides in methodological improvements and achieved fairly accurate results. This is promising since it sets up the basis for further steps in computational fact-checking and content verification. Having identified no clear UIs of any of the recent systems, I conclude that users are still rarely introduced to such systems with readily available UIs.[34] Thus, just as with clickbait detectors and satire detectors, rumor debunkers are lagging in usability studies and user-centered UI development, and require more efforts to educate potential end-system users and encourage them to adopt rumor debunkers in their social media use routines.

7.6 Automated Fact-Checkers

The technology for content verification has moved from identification in the general sense—of deceit, emotions, credibility, or subjectivity—to the state-of-the-art computational verification of particular claims (Rubin, In Press). Computational fact-checking is an automated way to mimic the work of human fact-checkers in investigative journalism and reporting (see also Chap. 5, Sect. 5.3).

Fact-checking is a labor-intensive, information-heavy, and time-consuming process with some prescribed best journalistic practices, as described in Kovach & Rosenstiel (2010).

To mimic human fact-checkers' steps in verification, computational fact-checking systems need to perform several sub-tasks, some of which we have already seen developed for rumor debunking and detection of falsehoods. These tasks include:

- Identifying and extracting candidate claims suspected of being false, rumored, satirical, clickbaity, or otherwise inaccurate or misleading, that are worth fact-checking.
- Gathering evidence and singling out the incorrect information in the suspected content (i.e., specifying what's incorrect in the claim).
- Verifying sources' credibility.
- Rendering a verdict on the claims' veracity.
- Providing a summarized verbal justification, visualization, and/or rationale for the verdict.
- Providing ways for future human tracing and verification by pointing back to the verified evidence by credible sources.

To meet the burden of proof and be trustworthy to its end users, the system has to balance the value of transparency with the risk of overburdening its users with too much information. Such systems require layered approaches which often clog user interfaces with too much detail. Users' attention may wander at any moment while systems take extra time to compute their intermediate steps, thus the speed of processing is as crucial as the accuracy and justification process. Thus, issues of UI

[34] I suspect that some newsrooms have access to such systems, do in house development, or outsource their R&D for off-the-shelf applications, but their adoption by giant tech companies is painfully slow as of mid-2021, where users are given very little control over the verification of social media content. See Chap. 5 on the ad-revenue model and other incentives for online manipulation as potential reasons for such slow adoption in giant tech.

usability have come up recurrently, across all of the previously described systems aimed at detecting falsehoods, clickbait, satire, and rumors. What is surprising is that R&D teams seem to undervalue the role of making such UIs available, perhaps forgetting that even a picture is worth a thousand words, let alone a working user-accessible prototype.

Another issue for making fact-checking completely automated has to do with the ultimate ground truth (see Sect 7.1.1 above). Who is to say what is true or not, and how to resolve controversial cases? (See Chap. 3 which explains that trustworthiness and expertise are both necessary for credibility and trust judgments.) Pérez-Rosas et al. (2018) pointed out that automated fact-checking approaches are "built on the premise that the information can be verified using external sources, for instance, FakeCheck.org and Snopes.com, and [that] propositions made in the news articles (e.g., "Barack Obama assumed office on a Tuesday") can be assessed for the truthfulness of their claims (Conroy et al., 2015). Knowledge databases such as DBpedia 2 have been used to query the Web in a structured manner. The results of such queries can then be used to test whether different sources also contain information confirming the news claim (e.g., that Barack Obama assumed office on a Tuesday). Other works used social network activity (e.g., tweets) on a specific news item to assess its credibility, for instance by identifying tweets voicing skepticism about the truthfulness of a claim made in a news article (Hannak et al., 2014; Jin et al., 2014)" (Pérez-Rosas et al., 2018). Researchers in some of these works conflate rumors, fake news, untruthful claims, and information lacking credibility (see also Chap. 1 on nuanced typologies of various fakes, or mis- and disinformation and other problematic content online).

For ML tasks, information extraction, retrieval, and binary classification can be achieved with or without predictive features identified a priori in supervised or unsupervised learning, respectively. Such methodologies typically start from (and in fact desperately need) a corpus of fact-checked aggregate cases as a "gold standard." Even deep learning mechanisms need instances of ground truth, just as in many other previously discussed detection techniques. The resulting input datasets are extremely valued and valuable for R&D, but are difficult to construct in large volumes without biases, stereotypes, ambiguities, and human errors.

Datasets like LIAR (Wang, 2017), a larger scale dataset based on Wikipedia articles FEVER (Thorne et al., 2018), and a multi-domain dataset MultiFC (Augenstein et al., 2019) are currently in use for the purpose of evaluating deception detection techniques, and they have proven to be useful for computational fact-checking as well. A prevalent component of existing fact-checking systems is a stance detection or textual entailment model that predicts whether a piece of evidence contradicts or supports a claim (Ma et al., 2018; Mohtarami et al., 2018; Xu et al., 2018), and more rarely, "directly optimis[ing] the selection of relevant evidence, i.e., the self-sufficient explanation for predicting the veracity label (Thorne et al., 2018; Stammbach & Neumann, 2019; Atanasova et al., 2020, p. 7352; Atanasova et al., 2019, p. 4677).

Most approaches still estimate the veracity of claims to support or refute them, based on predictors extracted from texts, images, surrounding metadata, and other digitally accessible evidence. The general process is remarkably similar to many other detection tactics, as discussed above, but certainly new methodologies and improvements are emerging every month in this active area of research. The technical literature on automated fact-checking has blossomed since around 2015 and has relatively recent extensive reviews such as those of Thorne et al. (2018), Pérez-Rosas et al. (2018), Atanasova et al. (2019), and Oshikawa et al. (2020).

How successful are fact-checking approaches overall? As of 2020–2021, the lack of accuracy in predictions is the top concern: are such systems consistently arriving at the correct verdict? Atanasova et al. (2020, p. 7352) conclude in their most up-to-date ACL review, "the task of automating fact checking remains a significant and poignant research challenge" based on "the effectiveness of state-of-the-art methods for both real-world—0.492 macro F1 score (Augenstein et al., 2019), and artificial data—68.46 FEVER score (label accuracy conditioned on evidence provided for 'supported' and 'refuted' claims)" (Stammbach & Neumann, 2019).

The (Rubin et al., 2019) NV Browser is a set of assistive technologies, built into a stand-alone browser. It displays results of automated identification of "fakes" in news, guided by the original Rubin et al. (2015) typology. Acknowledging that each type of fake has its own identifying features and that a "one-stop" engineering solution would be unimaginable, the team opted to use three separate tabs of the NV Browser to allow users to scan online content for the identification of different anomalies in content based on their corresponding salient features. The NV Browser unites discrete functionalities used for the following:

(a) Detection of clickbait headlines (as per described methodology in Brogly and Rubin et al. (2019)).
(b) Detection of satirical article content (methodologically explained in Rubin et al. (2016)).
(c) Detection of falsified news articles (based on earlier work described in Rubin and Conroy (2012) and Asubiaro and Rubin (2018)).

The NV browser does not constitute a complete unified solution to computational fact-checking, neither is it as slick as commercial start-up mobile apps. The success rates for its individual functionalities were indicative of early stages of less sophisticated feature-based ML techniques, varying from mid-90% accuracies for clickbait detection, to mid-80% to 90% for satire detection, to as low as mid-60% to mid-70% accuracies for the most challenging deception detection task.

By contrast, *ClaimPortal* is a web-based platform for the continuous monitoring, searching, checking, and analysis of factual claims on Twitter (Majithia et al., 2019) (see Fig. 7.16). It is an easily accessible visual example of an integrated automated fact-checking system, and the system's implementation details are disclosed in the peer-reviewed literature. *ClaimPortal* spots and scores a claim on the importance of it being fact-checked (i.e., check-worthiness) and the portal can be searched and filtered by keywords, Twitter dates, accounts, content, hashtags, types of claims, and their check-worthiness scores (Majithia et al., 2019). It is an exciting development in the realm of computational fact-checking which holds educational promise, if it gains trust from its users.

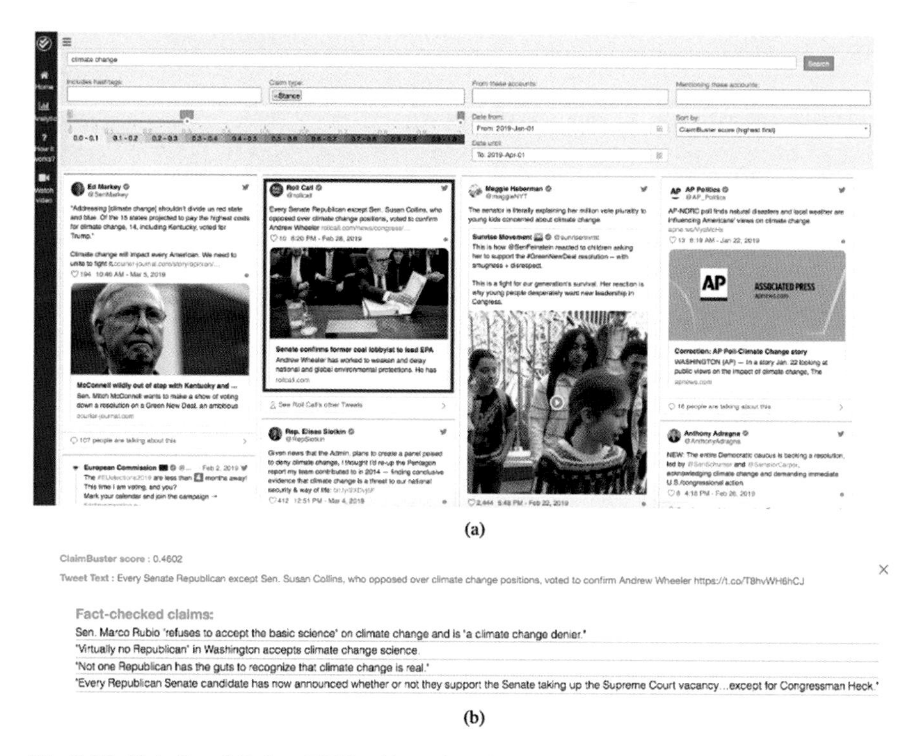

Fig. 7.16 ClaimPortal Twitter UI identifies a fact-checked claim (**a**); and (**b**) shows similar fact-checked claims related to the one highlighted in red (**a**). (Adapted from Majithia et al. (2019))

Just as with many other complex human intelligence tasks, the accuracies (even including the precision, recall, and F-score metrics) reported in computational scientific literature are not necessarily a measure of how useful, usable, and easily interpretable the systems' outputs are. The lack of explainability or lack of clarity in visual presentation can certainly impede the adoption of innovative technologies. These hurdles should be researched in HCI and usability studies, with recommendations made for developers who focus primarily on improving systems' efficiency, speed, and effectiveness. The contextual richness and broader human condition can be brought out by connecting computational experts with the more socially science-oriented disciplines. Collaborations between computationally minded disciplines and the digital humanities can increase the likelihood of success and the adoption of such pioneering and fully automated fact-checkers. User-based considerations include, for example, avoiding built-in biases and stereotypes in "gold standards" acquired during the data collection process; substantiating the quality, credibility, and suitability of external resources; and adjusting UIs to fit users' media literacy levels, tasks at hand, and levels of expertise. Thus, lots of collaborative inter-disciplinary work is still ahead.

7.6.1 Commercial Web-Based and Mobile Applications

It is understandable, on the other hand, that innovative AI technologies can generate revenue, and developers often reserve their "know-hows" and prefer to start-up enterprises of their own. For instance, *Logically* is a UK start-up run by Lyric Jain, its founder, CEO, and an alum of MIT and Cambridge. *Logically* offers services to newsrooms to verify any claims and images that the newsrooms might find dubious: "Our automated research assistant, and the world's largest team of dedicated fact-checkers, is at your service during normal times and in crisis to support topic and event-driven fact-checking."[35] If a user shares a news article with the company, its associated mobile app[36] can provide a decision on the user's submission and identify that news article as either a true news story or false news, and keep track of such verdicts in the user's news reading timeline. *Logically* also offers a Chrome browser extension[37] that is purported to work on more than 160,000 social platforms and news sites to fact-check news stories. A one-minute educational video[38] accompanies the browser extension and explains its functionality. According to a recent semi-promotional semi-innovative technology news profile in Forbes Magazine, "the company deployed first in the UK and India and then spread to the United States ahead of the 2020 election cycle to fact-check for consumers as well as government agencies." It goes on to describe that "*Logically*'s AI algorithms use natural language processing to understand and analyze text. The AI models label the credibility of the source of the content with a rating of low, medium, high, and an article as reliable or unreliable based on comparisons of similar content from more than 100,000 sources. The algorithms are checking not only content, but metadata and images too" (Marr, 2021). Several similar apps such as *Newstrition* and *NewsGuard* are discussed in Chap. 3 (Sect. 2.9. "Contemporary Solutions to Credibility Assessments and Trust Ratings of Online Content").

The Forbes Magazine article[39] also describes a collaboration among independent fact-checkers and innovative AI developers under way at *Full Fact*,[40] a non-profit company registered in England and Wales. Since 2015, *Full Fact* has been collecting and monitoring data, identifying and labeling claims, and developing automated fact-checking tools to increase their fact-checking impact (Marr, 2021). Another

---

[35] https://www.logically.ai/services (Accessed on 16 July 2021).

[36] A free downloadable app is available at https://www.logically.ai/products/app?hsCtaTracking=2bd8fffe-caed-4351-9e18-51c8e7f1f57b%7Cc6961ee1-7aa6-4cf3-a8be-1bc5245df2d0 (Accessed on 16 July 2021).

[37] https://chrome.google.com/webstore/detail/logically-browser-extensi/gadjokcojfkkjnamfjgdaacpjlnfcbdo?hl=en&authuser=0 (Accessed on 16 July 2021).

[38] See https://youtu.be/vVskLlg5O8s (Accessed on December 20, 2021).

[39] https://www.forbes.com/sites/bernardmarr/2021/01/25/fake-news-is-rampant-here-is-how-artificial-intelligence-can-help/ (Accessed on 16 July 2021).

[40] https://fullfact.org/ (Accessed on 16 July 2021).

mentioned AI company, *AdVerify,*[41] provides services to filter out inappropriate and deceiving content, spam, and fake news with their FakeRank algorithms, which Forbes Magazine claims, "helps advertisers, publishers, and ad networks to moderate content and ensure compliance with company policies to ultimately protect users and keep the reputations of brands safe" (Marr, 2021).

It is crucial that such technologies reach the public, and that AI-based apps are used more pervasively by average news readers and social media users. At the same time, AI commercialization of fact-checking, credibility assessments, and detection of other undesirable content points start-ups toward becoming corporate bodies, often run on the ad-revenue model which essentially incentivizes sensationalism and the viral spread of misleading or deceptive content. It is necessary to make sure that AI technologies do not leave out an important step of transparently providing the R&D details of their implementation to end users and sharing their know-hows via the appropriate scientific R&D channels, be that through patents or computational conferences.

7.7 Conclusions

Chapter 7 offers brief explanations for basic key terms and concepts in AI, NLP, and ML, followed by a review of several types of AI-based detectors. The technologies are grouped by the types of digital fakes they aim to detect—falsehoods, satire, clickbait, and rumors at first, with a culmination in the most challenging integrated task—that of computational fact-checking. I explain the inner workings of these state-of-the-art technologies, including their success metrics, feature selection procedures, and other, more sophisticated, ML methodologies.

Several web-based or mobile UIs are illustrated for potential users to see how these systems may look like in practice, in addition to the explanations of how they work in principle. I overview detailed features used as cues for automated detection by a kind of fake, and lay out schematically some processes in exemplified NLP pipelines. Visualizations and screenshots make each type of AI-based detector more concrete and easier to understand.

Some of the discussed systems are open access and available for download via developers' repositories such as the popular GitHub. The Open-source access movement has become instrumental for developers who want to see the details of implementation in order to replicate test results and make their own improvements to existing algorithms and trained language models. Tech-savvy or tech-adventurous users can also experiment with open-source access files that offer UIs, as well as free or commercially-available mobile apps, to see for themselves the merits and limitations of various implementations. Commercial companies (in giant tech and smaller start-ups alike) do not typically disclose their algorithmic details and

[41] https://adverifai.com/technology/(Accessed on 16 July 2021).

know-hows via open access. This makes it difficult to retrace how exactly their algorithms operate, to verify claims about their system performance, and to hold companies responsible for their systems' operations. In the scientific R&D community, by contrast to the commercial sector, scholarly literature is extensively peer-reviewed before publication, and often openly self-archived.[42]

The next challenge for AI advances in the area of automated detection of mis- and disinformation is in "AI that knows why." Accurate "explainable AI" for automated fact-checking should approach that of rigorous human investigators, lie-detectors, or fact-checkers, make the right predictions about each claim, and explain how and why the call was made.

References

Ali, M., & Levine, T. R. (2011). The language of truthful and deceptive denials and confessions. *Communication Reports, 21*(2), 82–91. https://doi.org/10.1080/08934210802381862

Amado, B. G., Arce, R., & Fariña, F. (2015). Undeutsch hypothesis and criteria based content analysis: A meta-analytic review. *The European Journal of Psychology Applied to Legal Context, 7*(1), 3–12. https://doi.org/10.1016/j.ejpal.2014.11.002

Asubiaro, T. V., & Rubin, V. L. (2018). Comparing features of fabricated and legitimate political news in digital environments (2016-2017). *Proceedings of the Association for Information Science and Technology, 55*(1), 747–750. https://doi.org/10.1002/pra2.2018.14505501100

Atanasova, P., Simonsen, J. G., Lioma, C., & Augenstein, I. (2020). Generating fact checking explanations. *Proceedings of the 58th annual meeting of the association for computational linguistics* (pp. 7352–7364). doi: https://doi.org/10.18653/v1/2020.acl-main.656.

Attardo, S., Hempelmann, C. F., & Mano, S. D. (2002). Script oppositions and logical mechanisms: Modeling incongruities and their resolutions. *Humor, 15*(1), 3–46.

Augenstein, I., Lioma, C., Wang, D., Lima, L. C., Hansen, C., Hansen, C., & Simonsen, J. (2019). MultiFC: A real-world multi-domain dataset for evidence-based fact checking of claims. *EMNLP*. https://doi.org/10.18653/v1/D19-1475

Bachenko, J., Fitzpatrick, E., & Schonwetter, M. (2008). *Verification and implementation of language-based deception indicators in civil and criminal narratives*. ACL.

Bahdanau, D., Cho, K., & Bengio, Y. (2016). Neural machine translation by jointly learning to align and translate. *ArXiv:1409.0473 [Cs, stat]*. http://arxiv.org/abs/1409.0473

Biyani, P., Tsioutsiouliklis, K., & Blackmer, J. (2016). *"8 amazing secrets for getting more clicks": Detecting Clickbaits in news streams using article informality*. AAAI.

Bogaard, G., Meijer, E. H., Vrij, A., & Merckelbach, H. (2016). Scientific content analysis (SCAN) cannot distinguish between truthful and fabricated accounts of a negative event. *Frontiers in Psychology, 7*. https://doi.org/10.3389/fpsyg.2016.00243

Bond, C. F., Jr., & DePaulo, B. M. (2006). Accuracy of deception judgments. *Personality and Social Psychology Review, 10*(3), 214–234. https://doi.org/10.1207/s15327957pspr1003_2

Brogly, C., & Rubin, V. L. (2019). Detecting clickbait: Here's how to do it/comment détecter les pièges à clic. *Canadian Journal of Information and Library Science, 42*, 154–175.

[42]For example, see such resources as the *arXiv*, a free open access self-archiving service (arXiv.org) or the *ACL Anthology*, a scholarly literature host associated with a major international scientific and professional society for people working on problems involving natural language and computation, the *Association for Computational Linguistics (ACL)* (https://aclanthology.org/) (Accessed on 17 August 2021).

Broussard, M. (2019). *Artificial unintelligence: How computers misunderstand the world.* MIT Press. Retrieved from https://mitpress.mit.edu/books/artificial-unintelligence

Buller, D. B., & Burgoon, J. K. (1996). Interpersonal deception theory. *Communication Theory, 6*(3), 203–242.

Burfoot, C., & Baldwin, T. (2009). *Automatic satire detection: Are you having a laugh?* (pp. 161–164).

Burgoon, J. K., Blair, J. P., Qin, T. T., & Nunamaker, J. F. (2003). Detecting deception through linguistic analysis. *Intelligence and Security Informatics, Proceedings, 2665*, 91–101.

Castillo, C., Mendoza, M., & Poblete, B. (2011). *Information credibility on twitter* (pp. 675–684).

Chakraborty, A., Paranjape, B., Kakarla, S., & Ganguly, N. (2016). *Stop clickbait: Detecting and preventing clickbaits in online news Media* 8. Retrieved from https://www.researchgate.net/publication/309572896_Stop_Clickbait_Detecting_and_Preventing_Clickbaits_in_Online_News_Media

Chandramouli, R., Chen, X., Subbalakshmi, K. P., Hao, P., Cheng, N., & Perera, R. (2012). Automated detection of deception in short and multilingual electronic messages. US Patent #US20150254566A1 Retrieved from: https://patents.google.com/patent/US20150254566A1/en

Chen, T., Wu, L., Li, X., Zhang, J., Yin, H., & Wang, Y. (2017). Call attention to rumors: Deep attention based recurrent neural networks for early rumor detection. *ArXiv:1704.05973 [Cs].* http://arxiv.org/abs/1704.05973

Chen, Y., Conroy, N. J., & Rubin, V. L. (2015). *Misleading online content: Recognizing Clickbait as "False News."* 15–19. doi: https://doi.org/10.1145/2823465.2823467.

Chen, Y., & Rubin, V. L. (2017, January 1). Perceptions of clickbait: A Q-methodology approach. *Proceedings of the 45th Annual Conference of The Canadian Association for Information Science/L'Association Canadienne Des Sciences de l'information (CAIS/ACSI2017).* Retrieved from https://ir.lib.uwo.ca/fimspres/44

Condren, C. (2012). Satire and definition. *Humor, 25*(4), 375. https://doi.org/10.1515/humor-2012-0019

Crystal, D. (1969). *What is linguistics* (2nd ed.). Edward Arnold.

De Sarkar, S., Yang, F., & Mukherjee, A. (2018). Attending sentences to detect satirical fake news. *Proceedings of the 27th International Conference on Computational Linguistics* (pp. 3371–3380). Retrieved from https://aclanthology.org/C18-1285

Editors of Merriam-Webster's Dictionary. (2021). *Definition of rumor.* Retrieved from https://www.merriam-webster.com/dictionary/rumor

Editors of Oxford Reference Online. (2021). GIGO: Garbage in garbage out. In *Oxford Reference.* doi: https://doi.org/10.1093/oi/authority.20110803095842747.

Editors of Wikipedia. (2021). News satire. In *Wikipedia.* Retrieved from https://en.wikipedia.org/w/index.php?title=News_satire&oldid=1000360895

Elyashar, A., Bendahan, J., & Puzis, R. (2017, October 18). *Detecting clickbait in online social media: You Won't believe how we did it.* Social and Information Networks; arXiv.org. Retrieved from https://arxiv.org/abs/1710.06699v1

Frye, N. (1944). The nature of satire. *University of Toronto Quarterly, 14*(1), 75–89.

Fuller, C. M., Biros, D. P., & Wilson, R. L. (2009). Decision support for determining veracity via linguistic-based cues. *Decision Support Systems, 46*(3), 695–703.

Gollub, T., Potthast, M., Hagen, M., & Stein, B. (2017). *Clickbait challenge 2017.* Retrieved from http://www.clickbait-challenge.org/

Granhag, P. A., Andersson, L. O., Strömwall, L. A., & Hartwig, M. (2004). Imprisoned knowledge: Criminals beliefs about deception. *Legal and Criminological Psychology, 9*(1), 103.

Grigorev, A. (2017, October 1). *Identifying clickbait posts on social media with an ensemble of linear models.* Information Retrieval; arXiv.org.

Hamidian, S., & Diab, M. T. (2019). Rumor detection and classification for twitter data. *ArXiv.*

Hancock, J. T., Curry, L. E., Goorha, S., & Woodworth, M. (2007). On lying and being lied to: A linguistic analysis of deception in computer-mediated communication. *Discourse Processes, 45*(1), 1–23. https://doi.org/10.1080/01638530701739181

Hauch, V., Blandón-Gitlin, I., Masip, J., & Sporer, S. L. (2015). Are computers effective lie detectors? A meta-analysis of linguistic cues to deception. *Personality and Social Psychology Review, 19*(4), 307–342. https://doi.org/10.1177/1088868314556539

Hempelmann, C., Raskin, V., & Triezenberg, K. E. (2006). Computer, tell me a joke... But please make it funny: Computational humor with ontological semantics. In G. Sutcliffe & R. Goebel (Eds.), *Proceedings of the nineteenth international Florida artificial intelligence research society conference* (Vol. 13, pp. 746–751). AAAI Press.

Highet, G. (1972). *The anatomy of satire*. Princeton University Press.

Höfer, E., Akehurst, L., & Metzger, G. (1996). Reality monitoring: A chance for further development of CBCA. *Annual meeting of the European Association on Psychology and Law, Sienna, Italy*.

Indurthi, V., & Oota, S. R. (2017). Clickbait detection using word embeddings. *Computation and Language; arXiv.org*. Retrieved from https://arxiv.org/abs/1710.02861

Kira, K., & Rendell, L. A. (1992). The feature selection problem: Traditional methods and a new algorithm. *Proceedings of the tenth national conference on artificial intelligence* (pp. 129–134).

Köhnken, G., & Steller, M. (1988). The evaluation of the credibility of child witness statements in the German procedural system. In *The child witness: Do the courts abuse children?* (pp. 37–45). British Psychological Society.

Larcker, D. F., & Zakolyukina, A. A. (2012). Detecting deceptive discussions in conference calls. *Journal of Accounting Research, 50*(2), 495–540. https://doi.org/10.1111/j.1475-679X.2012.00450.x

Liddy, E. (2001). Natural language processing. In M. A. Drake (Ed.), *Encyclopedia of library and information science* (2nd ed.). Marcel Dekker. Retrieved from http://surface.syr.edu/cnlp/11

Liu, Q., Wu, S., Yu, F., Wang, L., & Tan, T. (2016). ICE: Information credibility evaluation on social media via representation learning. *ArXiv:1609.09226 [Cs]*. Retrieved from http://arxiv.org/abs/1609.09226

Liu, X., Nourbakhsh, A., Li, Q., Fang, R., & Shah, S. (2015). *Real-time rumor debunking on twitter*. 1867–1870. doi: https://doi.org/10.1145/2806416.2806651.

Lynch, S. (2017, March 11). Andrew Ng: Why AI is the new electricity. A computer scientist discusses artificial intelligence's promise, hype, and biggest obstacles. *Stanford Graduate School of Business*. Retrieved from https://www.gsb.stanford.edu/insights/andrew-ng-why-ai-new-electricity

Majithia, S., Arslan, F., Lubal, S., Jimenez, D., Arora, P., Caraballo, J., & Li, C. (2019). ClaimPortal: Integrated monitoring, searching, checking, and analytics of factual claims on twitter. *Proceedings of the 57th annual meeting of the association for computational linguistics: system demonstrations* (pp. 153–158). doi: https://doi.org/10.18653/v1/P19-3026.

Marr, B. (2021, January 25). Fake News Is Rampant, Here Is How Artificial Intelligence Can Help. *Forbes*. Retrieved from https://www.forbes.com/sites/bernardmarr/2021/01/25/fake-news-is-rampant-here-is-how-artificial-intelligence-can-help/.

Masip, J., Sporer, S. L., Garrido, E., & Herrero, C. (2005). The detection of deception with the reality monitoring approach: A review of the empirical evidence. *Psychology Crime & Law, 11*(1), 99–122. https://doi.org/10.1080/10683160410001726356

Matthews, C. (2013, April 24). How does one fake tweet cause a stock market crash. *Time*. Retrieved from http://business.time.com/2013/04/24/how-does-one-fake-tweet-cause-a-stock-market-crash/

McGlynn, J., & McGlone, M. S. (2014). Language. In T. Levine (Ed.), *Encyclopedia of deception*. SAGE Publications. https://doi.org/10.4135/9781483306902.n219

McHardy, R., Adel, H., & Klinger, R. (2019). Adversarial training for satire detection: Controlling for confounding variables. *Proceedings of the 2019 conference of the North American chapter of the association for computational linguistics: Human language technologies, Volume 1 (Long and Short Papers)* (pp. 660–665). doi: https://doi.org/10.18653/v1/N19-1069.

Mendoza, M., Poblete, B., & Castillo, C. (2010). *Twitter under crisis: Can we trust what we RT?* (pp. 71–79).
Middleton, S. (2015, September 15). *Extracting attributed verification and debunking reports from social media: MediaEval-2015 trust and credibility analysis of image and video*. MediaEval 2015. Retrieved from https://eprints.soton.ac.uk/382360/
Mihalcea, R., & Strapparava, C. (2009). The lie detector: Explorations in the automatic recognition of deceptive language (pp. 309–312).
Mihalcea, R., Strapparava, C., & Pulman, S. (2010). *Computational models for incongruity detection in humour (A. Gelbukh, Ed.; pp. 364–374)*. Springer.
Newman, M. L., Pennebaker, J. W., Berry, D. S., & Richards, J. M. (2003). Lying words: Predicting deception from linguistic styles. *Personality and Social Psychology Bulletin, 29*(5), 665–675.
Ng, A. (2017, February 2). *Artificial intelligence is the new electricity (lecture)* [Stanford Graduate School of Business]. Retrieved from https://www.youtube.com/watch?v=21EiKfQYZXc
Oshikawa, R., Qian, J., & Wang, W. Y. (2020). A survey on natural language processing for fake news detection. In *Proceedings of the 12th conference on language resources and evaluation (LREC)* (pp. 6086–6093). Retrieved from https://arxiv.org/pdf/1811.00770.pdf
Papadopoulou, O., Zampoglou, M., Papadopoulos, S., & Kompatsiaris, I. (2017, October 23). *A two-level classification approach for detecting clickbait posts using text-based features*. Social and Information Networks; arXiv.org. Retrieved from https://arxiv.org/abs/1710.08528
Papadopoulou, O., Zampoglou, M., Papadopoulos, S., & Kompatsiaris, I. (2018a). A corpus of debunked and verified user-generated videos. *Online Information Review, 43*(1), 72–88. https://doi.org/10.1108/OIR-03-2018-0101
Papadopoulou, O., Zampoglou, M., Papadopoulos, S., Kompatsiaris, Y., & Teyssou, D. (2018b). *InVID fake video corpus v2.0* [Data set]. Zenodo. doi: https://doi.org/10.5281/zenodo.1147958.
Pennebaker, J. W., Boyd, R. L., Jordan, K., & Blackburn, K. (2015). The development and psychometric properties of LIWC2015. In *UT Faculty/researcher works* (September 15, 2015). University of Texas at Austin. Retrieved from http://hdl.handle.net/2152/31333
Pennebaker, J. W., & Francis, M. E. (1999). *Linguistic inquiry and word count: LIWC*. Erlbaum Publishers.
Pérez-Rosas, V., Kleinberg, B., Lefevre, A., & Mihalcea, R. (2018). Automatic detection of fake news. *Proceedings of the 27th international conference on computational linguistics* (pp. 3391–3401). Retrieved from https://aclanthology.org/C18-1287
Polanyi, L., & Zaenen, A. (2006). Contextual valence shifters. In J. G. Shanahan, Y. Qu, & J. Wiebe (Eds.), *Computing attitude and affect in text: Theory and applications* (1st ed., pp. 1–10). Springer. https://doi.org/10.1007/1-4020-4102-0_1
Porter, S., & Yuille, J. C. (1996). The language of deceit: An investigation of the verbal clues to deception in the interrogation context. *Law and Human Behavior, 20*(4), 443–458.
Potthast, M., Gollub, T., Hagen, M., & Stein, B. (2018a). The clickbait challenge 2017: Towards a regression model for clickbait strength. *ArXiv:1812.10847 [Cs]*. Retrieved from http://arxiv.org/abs/1812.10847
Potthast, M., Gollub, T., Komlossy, K., Schuster, S., Wiegmann, M., Garces Fernandez, E. P., Hagen, M., & Stein, B. (2018b). Crowdsourcing a large corpus of clickbait on twitter. *Proceedings of the 27th international conference on computational linguistics* (pp. 1498–1507). Retrieved from https://aclanthology.org/C18-1127
Potthast, M., Köpsel, S., Stein, B., & Hagen, M. (2016). Clickbait detection. In N. Ferro, F. Crestani, M.-F. Moens, J. Mothe, F. Silvestri, G. M. Di Nunzio, C. Hauff, & G. Silvello (Eds.), *Advances in information retrieval. 38th European conference on IR research (ECIR 16)* (Vol. 9626, pp. 810–817). Springer. https://doi.org/10.1007/978-3-319-30671-1_72
Pujahari, A., & Sisodia, D. S. (2021). Clickbait detection using multiple categorisation techniques. *Journal of Information Science, 47*(1), 118–128. https://doi.org/10.1177/0165551519871822
Qazvinian, V., Rosengren, E., Radev, D. R., & Mei, Q. (2011). *Rumor has it: Identifying misinformation in microblogs*. 1589–1599.
Quinlan, J. R. (1993). *C4.5: Programs for machine learning*. Morgan Kaufmann.

Reshi, J. A., & Ali, R. (2019). Rumor proliferation and detection in social media: A review. In *2019 5th international conference on advanced computing communication systems (ICACCS)* (pp. 1156–1160). https://doi.org/10.1109/ICACCS.2019.8728321

Rony, M. M. U., Hassan, N., & Yousuf, M. (2017, March 28). *Diving deep into Clickbaits: Who use them to what extents in which topics with what effects?* Social and Information Networks; arXiv.org. Retrieved from https://arxiv.org/abs/1703.09400v1

Rubin, V. L. (2014). Pragmatic and cultural considerations for deception detection in Asian languages. Guest editorial commentary. *TALIP Perspectives in the Journal of the ACM Transactions on Asian Language Information Processing (TALIP), 13*(2), 1–8. https://doi.org/10.1145/2605292

Rubin, V. L. (2017). Deception detection and rumor debunking for social media. In Sloan, L. & Quan-Haase, A. (Eds.) *The SAGE Handbook of Social Media Research Methods*, London: SAGE: (pp. 342–364). https://uk.sagepub.com/en-gb/eur/the-sage-handbook-of-social-media-research-methods/book245370

Rubin, V. L. (In press). Content verification for social media: From deception detection to automated fact-checking. In L. Sloan & A. Quan-Haase (Eds.), *The SAGE handbook of social media research methods* (2nd ed.). SAGE.

Rubin, V. L., Brogly, C., Conroy, N., Chen, Y., Cornwell, S. E., & Asubiaro, T. V. (2019). A news verification browser for the detection of clickbait, satire, and falsified news. *Journal of Open Source Software, 4*(35), 1208. https://doi.org/10.21105/joss.01208

Rubin, V. L., Chen, Y., & Conroy, N. J. (2015). *Deception detection for news: Three types of fakes* (p. 83).

Rubin, V. L., & Conroy, N. (2012). Discerning truth from deception: Human judgments and automation efforts. *First Monday, 17*(3). Retrieved from http://firstmonday.org/ojs/index.php/fm/article/view/3933/3170

Rubin, V. L., Conroy, N. J., Chen, Y., & Cornwell, S. (2016). Fake news or truth? Using satirical cues to detect potentially misleading news. (pp. 7–17). Retrieved from http://aclweb.org/anthology/W/W16/W16-0800.pdf.

Sahana, V. P., Pias, A. R., Shastri, R., & Mandloi, S. (2016). *Automatic detection of rumoured tweets and finding its origin*. Retrieved from https://idr.nitk.ac.in/jspui/handle/123456789/7425.

Saquete, E., Tomás, D., Moreda, P., Martínez-Barco, P., & Palomar, M. (2020). Fighting post-truth using natural language processing: A review and open challenges. *Expert Systems with Applications, 141*, 112,943. https://doi.org/10.1016/j.eswa.2019.112943

Simpson, P. (2003). *On the discourse of satire*. John Benjamins Publishing Company. https://doi.org/10.1075/lal.2

Stammbach, D., & Neumann, G. (2019). Team DOMLIN: Exploiting evidence enhancement for the FEVER shared task. *Proceedings of the Second Workshop on Fact Extraction and VERification (FEVER)* (pp. 105–109). https://doi.org/10.18653/v1/D19-6616.

Steller, M., & Köhnken, G. (1989). Criteria-based statement analysis: Credibility assessment of children's testimonies in sexual abuse cases. In D. Raskin (Ed.), *Psychological methods for investigation and evidence*. Springer.

Thorne, J., Vlachos, A., Christodoulopoulos, C., & Mittal, A. (2018). FEVER: A large-scale dataset for fact extraction and verification. *ArXiv:1803.05355 [Cs]*. Retrieved from http://arxiv.org/abs/1803.05355

Varshney, D., & Vishwakarma, D. K. (2021). A unified approach for detection of clickbait videos on YouTube using cognitive evidences. *Applied Intelligence, 51*(7), 4214–4235. https://doi.org/10.1007/s10489-020-02057-9

Vosoughi, S. (2015). Automatic detection and verification of rumors on twitter. In *Program in media arts and sciences: Vol.* Doctor of *Philosophy*. Massachusetts Institute of Technology.

Vrij, A. (2004). Why professionals fail to catch liars and how they can improve. *Legal and Criminological Psychology, 9*(2), 159–181. https://doi.org/10.1348/1355325041719356

Vrij, A. (2008). Nonverbal dominance versus verbal accuracy in lie detection: A plea to change police practice. *Criminal Justice and Behavior, 35*(10), 1323–1336. https://doi.org/10.1177/0093854808321530

Wei, W., & Wan, X. (2017). *Learning to identify ambiguous and misleading news headlines*. 1705.06031, 7.

Wiseman, R. (1995). The megalab truth test. *Nature, 373*, 391.

Wu, K., Yang, S., & Zhu, K. Q. (2015). False rumors detection on Sina Weibo by propagation structures. *IEEE International Conference on Data Engineering, ICDE.*

Wu, Z., & Palmer, M. (1994). Verbs semantics and lexical selection. *Proceedings of the 32nd annual meeting on association for computational linguistics* (pp. 133–138). https://doi.org/10.3115/981732.981751.

Yang, F., Liu, Y., Yu, X., & Yang, M. (2012). Automatic detection of rumor on Sina Weibo (p. 13).

Yang, F., Mukherjee, A., & Dragut, E. (2017). Satirical news detection and analysis using attention mechanism and linguistic features. *Proceedings of the 2017 conference on empirical methods in natural language processing*, 1979–1989. https://doi.org/10.18653/v1/D17-1211

Zhou, L., Burgoon, J. K., Nunamaker, J. F., & Twitchell, D. (2004). Automating linguistics-based cues for detecting deception in text-based asynchronous computer-mediated communications. *Group Decision and Negotiation, 13*(1), 81–106.

Zhou, L., & Zhang, D. (2008). Following linguistic footprints: Automatic deception detection in online communication. *Communications of the ACM, 51*(9), 119–122. https://doi.org/10.1145/1378727.1389972

Zhou, Y. (2017). Clickbait detection in tweets using self-attentive network. *Computation and Language; arXiv.org*. Retrieved from http://arXiv.org/abs/1710.05364v1

Zubiaga, A., Aker, A., Bontcheva, K., Liakata, M., & Procter, R. (2018). Detection and resolution of rumours in social media: A survey. *ACM Computing Surveys, 51*(2), 1–36. https://doi.org/10.1145/3161603

Chapter 8
Conclusions: Lessons for Infodemic Control and Future of Digital Verification

If only we had a vaccine against B.S.
(Don Lemon, CNN News, 2020)

Abstract The complexity in finding solutions to the socio-technological problem of mis- and disinformation lies in our human nature. The mind requires practical skills and digital literacy in order to overcome this problem. Socio-political and economic systems incentivize the spread of the infodemic across toxic digital media environments and require the public's effort in legislating and regulating appropriate controls. **Chapter 8** concludes this book by aggregating the key arguments and main claims about the use of automated ways of detecting various types of fakes online. Ten recommendations are put forward for educational, AI-based, and regulatory interventions, articulated summatively, as a package of countermeasures to mis- and disinformation.

Keywords Misinformation and disinformation interventions · Infodemic countermeasures · Infodemic · Causal factors · Interactions · Infodemiological framework · How-to fight "fake news" · Education · Digital literacy · Source verification · Discipline of the mind · Rationality · Critical mindset · Science-informed decision-making · Inoculation · Data deficits · Debunking · Pre-bunking · Explainers · Fact-checking · Content verification · Credibility assessment · Trust judgments · Social media · Labeling · Warning systems · Automation · AI · Artificial intelligence · Computational fact-checking · Automated deception detection · Automated lie-detection · Falsehood detectors · Satire detectors · Clickbait detectors · Rumor debunkers · Rumor busters · Automated fact-checkers · "Fake-fighters" · AI limitations · Awareness · AI bias · AI performance · Assistive technologies · Black-box · AI transparency · Explainable AI · Moral compass · Adoption · Content take-downs · Advertising transparency · AI transparency · Recommendations · Accountability · Regulation

V. L. Rubin, *Misinformation and Disinformation*,
https://doi.org/10.1007/978-3-030-95656-1_8

8.1 The Main Premise of the Book

Social and other digital media are awash in deceptive, inaccurate, and misleading information. When content is intentionally deceptive it is referred to as disinformation, while the unintentional transfer of erroneous information is considered misinformation. As of the early 2020s, solutions to the problem of the proliferation of such content remain elusive. To contribute meaningfully to potential solutions, I combine psychological, philosophical, and linguistic insights on the nature of truth and deception, trust and credibility, cognitive biases, and logical fallacies. Such background knowledge is necessary for the professional practices of many expert lie detectors and truth seekers, whose routines can, to an extent, be captured and automated to enhance our human intelligence tasks with AI.

8.1.1 Socio-Technological Solutions to Mis- and Disinformation

To control the spread of mis- and disinformation online it is necessary to use both human and artificial intelligence. Given the scale and persistence of the problem (Chap. 1, Sect. 1.1), some technological assistance is inevitable and likely to come from AI-enabled applications using the latest NLP and machine learning (ML) methodologies (Chap. 1, Sects. 1.5–1.6, and in much more details in Chap. 7). AI can deal with the problem's scale, while the final decision-making is in the human mind. Given the incessant nature of disinformation attacks, some automation to counteract them is inevitable. We need at the very least some initial assistance from systematic analysis (for example, see Chap. 1, Sect. 1.3) to sift through newsfeeds, posts, tweets, and the like.

Most of us encounter a lot of information in our daily digital media routines; the original sources of information may be obscured when mediated by social media and news aggregators in which all items look alike (Chap. 2, Sect. 1.4). The ad revenue model underlying social media platforms (Chap. 5, Sect. 5.2) motivates content producers to harness maximum user engagement and strive for virality (Chap. 6, Sect. 6.5. 5, Sect. 5.5).

Empirical social science researchers who monitor social media trends in the USA and worldwide have clearly indicated that fakes are more likely to achieve virality than well-researched authoritative content. “Falsehood diffused significantly farther, faster, deeper, and more broadly than the truth in all categories of information, and the effects were more pronounced for false political news than for false news about terrorism, natural disasters, science, urban legends, or financial information.” (Vosoughi et al., 2018 p. 1146). The proportion of lay people concerned about false information online has been steadily increasing. “Roughly half of U.S. adults (48%) now say the government should take steps to restrict false information, even if it means losing some freedom to access and publish content, according to the survey of 11,178 adults conducted July 26–August 8, 2021. That is up from 39% in 2018” (Mitchell & Walker, 2021). Moreover, at least among Americans, "a majority of adults (59%) continue to say technology companies should take steps to restrict misinformation online, even if it puts some restrictions on Americans' ability to access and publish content" (Mitchell & Walker, 2021) The question is not only in whose responsibility it is, but also in how to do it effectively.

Many, if not most computer users, are already familiar with assistive natural language processing (NLP) and machine learning (ML) technologies such as spell-checkers, spam filters, and ad blockers. Some of us are disgruntled by the instant predictive search strings filling in our search boxes, while others may enjoy the insights into human conformity in thought and speech patterns on display in such data-driven approaches. Many complain about having to fish out non-spam items from their spam filters, yet perhaps they hope similar technologies could be unleashed on other media scamming and fraud attempts. The reality is that in current computer-mediated communication (CMC) environments, NLP and ML assistive technologies are here to stay, together with the aches and pains of their technological advancement. Innovative technologies may perform at sub-optimal levels of accuracy and the general public debates their ultimate utility, ethical data use, and the day-to-day implications of AI-driven decisions. Silicon Valley has a mentality of overconfidence in the righteousness of the position that anything engineered in innovative and profitable ways is justifiable, and for the most part, beneficial to mankind (see Chap. 6, Sect. 6.7 for discussion of McNamee's (2019) criticism of Facebook's operations). Big tech companies are reluctant to take on the ultimate responsibility for rolling out automated solutions to aid in fighting mis- and disinformation. The complications start from the many ways in which people can be duped (Chap. 2, Sect. 2.3), and a lack of societal consensus on what we see as being deceptive and disinformative (Chap. 1, Sects. 1.4. and 1.6). Other difficulties are rooted in our entirely human ways of interacting with information. Such notorious terms as "fake news" have been "irredeemably polarized;" co-opted by unscrupulous politicians to refer to any news put out by sources that do not support their partisan positions (Vosoughi et al., 2018 p. 1146). (See also Chap. 1, Sect. 1.4)

Complex information-related behaviors, external circumstances, psychological reasons, and individual traits are responsible for why we may not be able to recognize mis- or disinformation (Chap. 2, Sect. 2.4). Humans are often mistaken when separating lies from truths to start with. Psychologists agree that if not alerted to the presence of deception, lay people are poor lie detectors (Twitchell et al., 2004). Unfortunately, even with training, professional lie-catchers such as law enforcement personnel and witness credibility assessment experts improve only slightly over the detection success rate of untrained people (Frank & Feeley, 2003) (Chap. 2, Sect. 2.5)

Some social media users are indiscriminate about their information resources. They may incorrectly assess their sources' expertise and trustworthiness, if they assess it at all, which leads to erroneous trust judgments (Chap. 3 on credibility assessments). Some social media users question the existence of objective reality and truth altogether, confound fact with fiction, or are simply unable to distinguish well-justified beliefs from what sounds intuitive and familiar (Chap. 4 on philosophies of truth we live by). Thus, the task of identifying mis- or disinformation, even when alerted to its presence, is—by any standard, more complicated than simpler AI-enabled tasks with NLP and ML. Spell-checkers have authoritative dictionary entries against which to verify word spelling and little, if any, controversy. Even when your search entries are filled in, the next word sequence uses AI, or specifically some ML-trained models of common patterns in similar searches found in large training datasets, also commonly known as "big data." Since there is no such thing as one agreed-upon "big dataset" of truths, creating such an authoritative resource, as a paragon for AI to rely on, can be tricky (see, for example, (Rubin & Conroy, 2012)).

8.1.2 *Infodemiological Framework for Causes and Interventions*

This book frames the problem of mis- and disinformation as a socio-technological phenomenon that is caused by at least three interacting factors—users as suspectable hosts, various types of fakes as virulent pathogens, and digital platforms as conducive environments that enable the infodemic (see the Mis- and Disinformation Triangle Model (Rubin, 2019) in Chap. 1, Sect. 1.2). I have been advocating for three types of countermeasures to disrupt the interaction of these three factors: educating susceptible minds, detecting virulent fakes at large scale, and regulating toxic environments nationally and internationally (Fig. 8.1., in red "stop signs").

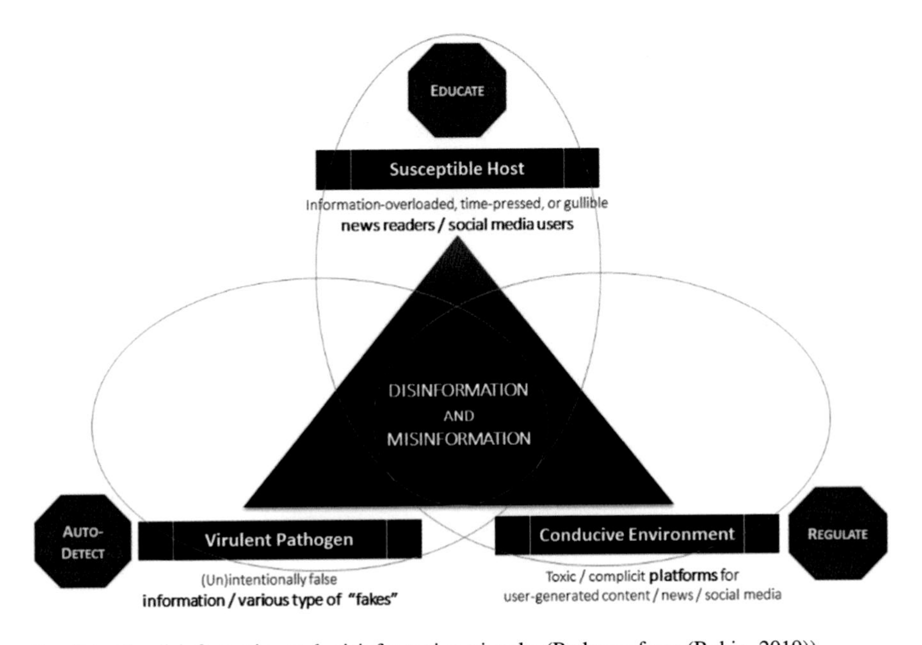

Fig. 8.1 The disinformation and misinformation triangle. (Redrawn from (Rubin, 2019))

These measures—education, automation, and regulation—are to be put in place simultaneously and pursued relentlessly. Below are some concrete recommendations as pointers toward their practical application.

8.2 Concrete Recommendations for Educational Interventions

The educational system for any democratic society should aim to teach citizens from a young age a rational way of thinking which can benefit society as a whole and each individual within it. At each age level, there are ways to explain how to use a critical mindset when faced with new information online, whichever format or platform it appears in. Digital media users should be acutely aware of the ways in which they tend to interact with information and other people through information communication technologies (ICTs). The onus is on various educational institutions to reach out to their country's population with consistent messaging, yet the final decision making will be in the minds of the individual news readers, social media users, ad consumers, or various other digital service subscribers. If a critical mindset is instilled and applied throughout public education, the public will benefit from learning to interact with online information in more thoughtfully discerning and less rushed ways. Educational campaigns, targeted literacy classes, and self-directed learning resources should be offered at various institutions such as libraries and information centers to specifically instruct people on how to identify multiple varieties of fakes and how to properly verify information.

There have been a few large-scale trials and several initiatives to include specific short-term campaigns or a more permanent embedding of various literacies (media, news, information, and digital literacy) into educational curricula, from early grade school through to graduate studies. Specific lesson plans are outside of the scope of this book, but I offer concrete ideas and recommendations on the key principles to be taught, leaving the work of curriculum development to educators. The following recommendations are for in-class and online educational efforts as well as targeted literacy campaigns.

8.2.1 Teaching to Verify your Sources

A simple piece of advice that is consistently given across reputable organizations in many fields such as health informatics, librarianship, and journalism is to verify the source, date, and author by yourself, or to consult experts or reliable fact-checking websites to make sure it is not a joke or a rumor. Chapter 1 offers sample iconographics by the World Health Organization, International Federation of Library Associations and Institutions, and Canada's National Observer (Sect. 1.3).

8.2.2 Authoritative Fact-Checking and Debunking

Learning about credible authoritative experts and fact-checker organizations and relying on their expertise is as important as realizing a certain lack of knowledge in your own expertise. "The real challenge isn't being right but knowing how wrong you might be" and it is irrational "to insist on answering difficult questions yourself when you'd be better off trusting the expert consensus" (Rothman, 2021). Non-profit initiatives, specifically dedicated fact-checking associations, or networks of independent journalists, for the most part, rely on impartial news verification practices rooted in the thorough vetting of information in the high-quality journalistic tradition (Chap. 2, see Sect. 2.6.1 and Chap. 5, Sect. 5.3) As a variation on fact-checking, "debunking" refers to exposing falseness and conspiracies, cataloging evidence of false information, or showing that something is less important, less good, or less true than it has been made to appear (Pamment & Lindwall, 2021).

8.2.3 Relying on the Scientific Consensus

The scientific endeavor, as one of the most successful ways of systematic inquiry known to humanity, has a lot to teach us about how to think critically and rationally, as opposed to going with your own gut feeling and deciding emotionally on what feels right. "Science is a determination of what is most likely to be correct at the current time with the evidence at our disposal… The scientific method, it could be said, is a way of learning or a process of using comparative critical thinking" (McLelland, 2021 p. 1). Chapter 5 offers a review of the underlying principles in scientific inquiry and the lessons science teaches us such as methodical observation and questioning (Sect. 5.4). Scientific thinking deserves promotion in classrooms with simpler teachable moments for younger digital users and more extended conversation with older children and adults. It is important to emphasize being flexible, open-minded, and willing to refine our original predictions if they do not work out. Scientists, as a group, deserve our societal trust and respect for their contributions to establishing facts and advancing reliable knowledge. (See also the distinctions between truth, reality, facts, and knowledge in Chap. 4, Sect. 4.4). Rationalists call us to monitor our own thinking, or in other words, be "metacognitively" aware of what we know and what we do not know. "Rationality is one of humanity's superpowers" (Rothman, 2021).

8.2.4 Relying on Professional Verification Experts and Distrusting Manipulators

Professionally trained experts such as detectives and journalists have worked out specific methodologies for arriving at valid conclusions by thinking critically. Each best practice in Chap. 5 is armed with various systematic analyses for purposeful investigations to answer well-formulated questions, based on experts' knowledge and the observable evidence (see Sects. 5.2 and 5.3) What unites the professional practices of detectives, journalists, and scientists is the inquisitive nature of their practices, a critical mindset, rationalism, systematicity, and evidence-based conclusions.

By contrast, Chap. 6 is about the professional manipulative practices in marketing and advertising, mass persuasion tactics, and propaganda in public relations. Propagandists, advertisers, and public relations experts are particularly well-versed in appealing to emotions, using wit and humor, and finetuning who delivers the message as a relatable credible source. To resist the powers of these professional manipulations, digital media users need to recognize a manipulative technique when they see it, as many of them are illegal, marginally legal, or at the very least immoral. When we are consciously aware of the potential tricks, a warning system is in place. Any magic trick loosens its grip on its target if the mark knows exactly where to look, what the illusion is, and how they are being conned. Once again, a piece of advice from 1930s anti-propaganda campaigns is as good as newly written for the digital age: "education is a powerful antidote to propaganda" (Hobbs & McGee, 2014).

8.2.5 Inoculating and Preemptive Messaging

Well-suited for short-term, domain-specific campaigns targeting a specific misconception (such as the efficacy of fad diets or symptoms of COVID-19) are inoculations, or a set of measures that forewarn people and help them to develop attitudinal resistance. Inoculations activate people's awareness of currently circulating false claims, stimulating their willingness to think more critically about these claims by refuting potential counterarguments to counterbalance misinformation. For example, Van der Linden et al.' (2017) study found that "communicating the scientific consensus on human-caused climate change significantly increases public perception of the expert consensus" (p. 6). (See also Chap. 2, Sect. 2.6.2).

8.2.6 Identifying and Filling Niches in Public Awareness

Monitoring public debates to predict and identify topics that are in high demand but which lack a supply of credible information is a strategy of preemptive research that can help to counter specific rumors, speculation, and mass disinformation. Such novel highly specialized "niches" have been called "data deficits" (Cubbon, 2020) but simply refer to the acutely felt lack of public awareness on a specific topic. Niches may be created by news of disasters, critical incidents, or major developments in politics, technology, health, economics, or any other social realm. People may have a number of legitimate questions, for example, about mRNA (messenger ribonucleic acid), vaccine technology, and VDPV (vaccine-derived poliovirus).

Such niche topics are typically politically salient, emotionally charged, and have a clear potential to fit into pre-existing, long-standing disinformation narratives (Cubbon, 2020). To counter manipulations around these niches, "institutions have to engage in pre-emptive research that can inform proactive messaging, such as 'pre-bunks' and explainer pieces" (First Draft, 2021). The idea of preemptive and proactive messaging is similar to inoculation, but the focus is on monitoring potentially exploitable topics and predicting future trends in public discourse. When ignored, niches can be filled with inaccuracies, controversies, and outright falsehoods that result in influencing the public mind to form erroneous public beliefs (see Chap. 6, Sect. 6.6).

8.2.7 Self-Directed Coaching and Literacy Resources

Coaching, especially through online resources and self-directed study, has been shown effective in reducing the likelihood of resharing "fake news" (Dyakon, 2020). Chapter 2 provides examples of programs and resources including an e-learning platform, an app, a podcast, shareable tips and tools, and news literacy events (in Sect. 2.6.4)

8.2.8 Making Sense of Credibility Indicators, Warning and Labeling Systems

It is important to instruct citizens about indicators of credibility and how trust decisions are made. Attempts to vet information should not be entirely delegated to tech companies. Authoritative bodies such as reputable professional associations, libraries, and governments have used external certification processes, ranking systems, and seals of approval. Chapter 3 discusses the fiascos of previous attempts at seals and stamps of approval. Nonetheless, credibility indicators persist in our assessments. Recent creative solutions have been aggregating scores of factuality, readability, virality, emotionality, opinion, controversy, authority, technicality, and topicality as a valid cumulative credibility measure. Chapter 3 provides examples of visualizations in the form of "information nutrition labels" for digital content, mimicking the concise way nutrition labels are displayed on food products. Understanding how such labels are computed and how valid they are is part of digital literacy: some are simply measures of crowdsourced popularity or user engagement under the guise of being a trust measure, while others more legitimately combine manual fact-checking and automation (Chap. 3, see Sect. 3.9.2 and Chap. 7, Sect. 7.6).

Many other warnings and labeling systems are in use by social media or individual platforms. What does each label mean? How was it decided on? Was it attached to a tweet or a post by a fact-checker or a crowd-sourced algorithm?

Teaching users to have an up-to-date mental model of the terrain is important. Does a labeling system offer a feedback loop, alerts, filters, or other checks and balances? Chapter 2 shows samples of the *PolitiFact* warning system and the mid-2021 Twitter color-coded categories to address these questions (in Sect. 2.6.3).

8.2.9 The Broader Argument for Informed Rationality and Systematic Training

This book encourages cultivating rational thinking in our education systems, no matter the grade, age, or literacy level, and reinforcing this critical mindset with corresponding public policies at institutional and governmental levels. The book encourages the adoption and explicit teaching of systematic ways of deciding whether something is true or not. The question "do I believe it or not?" should be answered based on evidence and the expertise of trusted authoritative sources. Chapter 4 overviews various philosophies of truth and their contributions to how we interpret reality. A disciplined mind knows how to apply logic, think through ideas critically, and avoid logical fallacies (Chap. 4). A critical mindset requires practice and can become more skilled with time, but it is attainable through training. Education is but one, albeit the most important one, of the three interventions in the Rubin (2019) conceptual model: the other two are automation and regulation, summarized here next, in turn.

8.3 Automated Interventions

8.3.1 AI-Based Tools for Combatting Mis- and Disinformation

Automated solutions to mis- and disinformation are likely to come from a variety of NLP and ML methods such as the following five types of AI tools:

- Automated Deception Detectors (Chap. 7, Sect. 7.2)
- Automated Clickbait Detectors (Chap. 7, Sect. 7.3)
- Automated Satirical Fake Detectors (Chap. 7, Sect. 7.4)
- Automated Rumor Debunkers (Chap. 7, Sect. 7.5)
- Automated Fact-Checkers (Chap. 7, Sect. 7.6)

Automated Deception Detectors are particularly indebted to the foundational works in psychology and communication (Chap. 2) since these AI-based methods were informed and inspired by the well-known professional practices in lie detection in the justice system (see, for example, Statement Validity Assessment (SVA), Content-based Criteria Analysis (CBCA), and Reality Monitoring (RM) in Sect. 5.2). Other kinds of detectors take advantage of linguistic patterns in a variety of fakes (clickbait, satire, and rumors) and the regularities and patterns of human behavior as captured by large datasets, or big data. I trace the evolution of these five types of technologies over the past 10–15 years, explaining how they operate in principle, and providing examples of algorithms, system designs, or feature selection, and at time UI visualization in web-based and mobile apps (in Chap. 7). A primer for non-technical readers opens the detailed discussion of AI-enabled technologies (in Sect. 7.1.1).

There are several AI-related themes woven through the chapters of this book that are worth aggregating here. They weigh the benefits and limitations in the use of AI. You will find more refined arguments and detailed discussions of these claims by going into the suggested book sections, which in turn contain extensive references from a number of social science and computational disciplines.

8.3.2 *Claim 1: AI Is Inherently Tied to Human Practices*

AI has an aura of being removed from the human experience. In fact, NLP methodologies are based on patterns and regularities observed in the user-generated content often used as ML training input. A common assumption among AI experts, but rarely spelled out to less tech-savvy audiences, is simply that AI uses a wealth of "human smarts," tricks, and captured experiences, as well as knowledge from different applied professions and fields of study. In other words, AI-enabled technology for fighting the infodemic absolutely has to be largely based on well-established human practices that allow us to observe correlations in human language use and psychological behavior. The human endeavors that inform such AI tools are computational in nature, and discussing the fundamental disciplinary background in mathematics, statistics, computer, and data sciences is beyond the scope of this book. However, this book does offer some basic explanations of AI, ML, and NLP in simple language (see Chap. 1, Sect. 1.5, "Computational Fields" for these field definitions, and Chap. 7, Sect. 7.1 for other "AI Basic Terms and Concepts Demystified"). Procedurally, AI-enabled mis- and disinformation tools are informed by well-established systematic inquiries in law enforcement, journalism, and science (each discussed in turn in Chap. 5).

8.3.3 *Claim 2: Innovative AI Solutions Are Feasible and in Existence but Are Imperfect*

Most importantly, it has been proven that it is feasible to identify certain types of fakes algorithmically. Such AI-based technologies are not yet widely adopted, neither are they commonly known outside of the computer science and IT developers' circles specializing in NLP and ML. Such AI technologies are most likely being intensely developed by big tech companies in-house due to the mounting pressure from governments and the general public. Big tech algorithmic solutions are rarely shared; "know-hows" and the data collected from their users are used for internal development, and the resulting trained models and their predictions are largely kept secret from the general public's scrutiny.

There is also very little sharing of the big tech companies' analytical tools and resources with the academic researchers and the public interested in monitoring the reach of mis- and disinformation. For example, Facebook CEO Mark Zuckerberg, in his 18 August 2021 interview with *CBS This Morning's* Gayle King, refused to answer a simple statistics question about the number of Facebook accounts that were exposed to the 18 million false COVID-related posts that Facebook had removed (Ahmed, 2021). See Roger McNamee's (2019) sizzling criticism of Facebook's manipulative practices and accusations that Zuckerberg uses his company's purposeful engineering of persuasive technologies for surveillance, attention-getting, addiction, and behavioral modification (Chap. 6, Sect. 6.7.4).

Outside of the big tech duopoly, AI-based methodologies are being developed in smaller academic R&D groups, labs, and start-ups, but their data access, audience reach, and other resources are usually inferior to the large monopolies. While Google and Facebook's research groups have large research arms that attempt to engage with the scientific community through grants, they rarely share their breakthrough technologies with the wider scientific community. Such technologies are being stifled since their wide-range adoption would limit the virality of ad content, for example, which would limit the profit of the very platforms that chose to implement them. Having no viable alternatives to the ad revenue model and undervaluing the contributions of legacy institutions such as professional news reporters has led to the erosion of the industry of news verification (See also, the overview of the current state of the digital media and news reporting in Chap. 1, Sect. 1.1, and the discussion of the changes in the production and dissemination of news brought about by the digital formats, in Chap. 5, Sect. 5.4).

At the same time, while feasible and computationally sophisticated, AI-enabled detectors of mis- and disinformation may not be good enough for everyday use. Developers are reluctant to take responsibility for erroneous or controversial automated decisions, especially with "black box" solutions that cannot transparently retrace the decision rationale. Errors, opaqueness, and controversies may bring public wrath on the developers, damage companies' reputations, or drive stocks down. Computational errors compile from one step of the NLP pipeline to the next. Human biases in the input data used to train AI models are another source of concern. System-user interfaces (UIs) may also be too cumbersome, from the usability point of view: hard to integrate within digital interactions and news browsing routines. The verification process itself may be too complex, cognitively taxing, and non-transparent to users. Many of the above complications may explain a slow uptake of these technologies that in principle can "do the job."

8.3.4 *Claim 3: Explainable, Accountable, and Ethical AI as the Future of Content Verification*

Developing an AI that is transparent to its users (can explain why and how the decisions are made) and accountable to its developers is key in gaining societal trust in such systems (see also Chap. 3, Sect. 3.4 on "Trust Modeling"). A complete end-to-end explainable AI for fact-checking is currently in its nascence and social media users, for the most part, have to sift through mountains of information unassisted, relying primarily on their intuition. There are multiple ways to acquire basic skills in digital literacy and more advanced awareness and expertise with more in-depth training. In the absence of such in-depth expertise in content verification, and when hesitant to make a judgment call, social media users can defer to authoritative credible sources and well-established professional practices such as trusted legacy journalism (Chap. 5, Sect. 5.3). Recognizing ulterior commercial or political motives (Chap. 2, Sect. 2.6 Sect. 2.2.3) ; Chap. 6 "*Cui bono?*") is important to contrast with the impartial due process of the scientific inquiry (Chap. 5, Sect. 5.4).

Reimagining our online experiences altogether may be a far-fetched goal, but perhaps the future of digital communication needs to include fundamentally different goals and standards. The moral compass for AI-enabled technologies has to point away from deceptive intentions, away from commercially and politically motivated distortions in mass communication, and away from manipulations of public perceptions through ICTs. Exactly how it might look is up for debate by philosophers, ethicists, tech-savvy sociologists, IT experts, policymakers, legislators, and society at large. What is clear to me is that new visions for digital environments have to place a greater value on morality and pro-social behaviors (see Chap. 4, Sect. 4.2).

8.4 Regulatory and Legislative Interventions

The third important intervention in curbing the infodemic is to establish and enforce a set of effective regulatory and legislative measures. Governments, legal experts, academics, policymakers, and the general public worldwide are considering and debating the potential trade-offs and stumbling blocks of countermeasures to disinformation (see Chap. 6, Sect. 6.8 for details). Some governments have taken concrete steps and enacted policies of content take-downs, advertising regulation, criminalization of content, and monitoring and reporting on citizens' public discourse, to name a few. Chapter 6 discusses "Four Types of Regulations and Legislation by Their Targets"—social media platforms, offenders, governments, and civil society.

There are governmental attempts to regulate the ICT monopolies which have historically been reluctant to assume responsibility for the spread of mis- and disinformation. Pressure is being put on the duopoly (Google and Facebook), as well as others, to actively moderate their content and mount effective and fair ways of labeling content (e.g., Rosen, 2020).

The onus seems to be heavily weighted toward tech companies' self-regulation and content moderation, but experts warn of heavy-handed approaches that can restrict citizens' liberties. "In the current highly-politicized environment driving legal and regulatory interventions, many proposed countermeasures remain fragmentary, heavy-handed, and ill-equipped to deal with the malicious use of social media. Government regulations thus far have focused mainly on regulating speech online—through the redefinition of what constitutes harmful content, and measures that require platforms to take a more authoritative role in taking down information with limited government oversight" (Nothhaft et al., 2018 p. 12).

The tech industry may take advantage of susceptible digital media users, but it is often other major entities such as large corporations, industries, or political players that incentivize the spread of disinformation, exploit human vulnerabilities, and benefit from the promotion of their commercial, political, or personal agendas. The questions remain: who should be the arbitrator to the truth? Who actually benefits from content going viral? Who directly or indirectly funds the agents of disinformation?

Another practical concern remains about the global reach of mis- and disinformation, its cross-country and cross-language nature, and the lack of international cooperation for putting legal brakes on the system in place. The overall challenge is to design adequate oversight and regulatory frameworks for social media and user data and to address the issues of responsibility and underlying societal conflicts with due attention paid to the motivating factors.

Yet, there is a definitive need to consider tighter controls to disincentivize the creation, promotion, and spread of falsehoods online by opportunistic individuals and corporations. Legal experts seem to encourage policymakers and legislators to avoid crude measures controlling public discourse. Instead, they advocate for a focus on public education, for figuring out pro-social means of bolstering facts, and that to "reestablish trust in the basic institutions of a democratic society is critical to combat the systematic efforts being made to devalue truth" (Baron & Crootof, 2017 p. 11). "There is no simple blueprint solution to tackling the multiple challenges … Likeminded democratic governments should work together to develop global standards and best practices for data protection, algorithmic transparency, and ethical product design" (Nothhaft et al., 2018 p. 12).

8.5 Policy-Making Recommendations for Governments and Institutions

In view of the discussed interventions—in education, automation, and regulation, I offer ten actionable items for public debate and consideration for governmental legislation and institutional policy-making, as a package of countermeasures to mis- and disinformation:

1. Place emphasis on digital literacy and rational thinking in general public education.

2. Support knowledge translation efforts to explain scientific research findings and actively disseminate and promote those findings to the public to uphold the value of facts.
3. Support and empower domain-specific institutions in implementing targeted messaging and informational campaigns.
4. Support grass-root movements and non-for-profit research in monitoring areas of public discourse for public awareness niches to inform the direction of informational campaigns.
5. Support curated authoritative information resources via libraries, information centers, digital portals, websites, and databases for specific domains.
6. Support newsrooms and fact-checkers, from local to national and international outlets, to restore public trust in the value of investigative journalism and the democratic values it upholds.
7. Encourage scientific research and open-source grassroots development in the not-for-profit sector to counteract the dominance of giant tech in social media and to promote the use of AI technologies for the automation of mis- and disinformation detection.
8. Regulate non-public discourse (e.g., fraud in commercial and political campaigns) to cut financing from disinformation campaigns, enforce the disclosure of financial sources, and incentivize advertising transparency.
9. Set up working groups to develop standards and best practices for social media regulation, data protection, algorithmic transparency, and ethical product design and connect them to national and global frameworks for detecting, preventing, and deterring mis- and disinformation.
10. Set up interdisciplinary cross-sector working groups to reimagine online experiences in ICT systems, based on pro-social commercial business revenue models, enabled by ethically and morally sound AI and other technologies, and set the internet on a path to regain trust in democratic institutions.

The ten recommendations above may vary from concrete to more aspirational, but as a package they lay down a foundation for one possible way to enact the interventions. The groundwork for understanding the issues behind the three interventions—education, automation, and regulation—is addressed more in depth throughout the book chapters.

8.6 Summary

In this book, I argue that a systematic rationalist inquisitive frame of mind should be cultivated more effortfully as a cornerstone of digital literacy. With continued practice, social media users with such a mindset should be able to acquire the necessary skills for spotting varieties of mis- and disinformation online to make more informed decisions when approaching any novel online information. Certain AI technologies—such as the automated detectors of falsehoods, satire, rumors, and clickbait—can assist the human mind in the effort to curtail the proliferation of the infodemic by detecting, deterring, and preventing the various kinds of fakes online. Regulatory and legislative countermeasures are being considered and implemented around the world. While the ultimate global solution remains elusive, many countries opt to emphasize non-regulatory public campaigns around education of the mind. I offer up for debate ten associated policy-making recommendations for governments and institutions.

References

Ahmed, I. (2021, August 18). *Fact-checking facebook's latest disinformation campaign.* Tech Policy Press. Retrieved from https://techpolicy.press/fact-checking-facebooks-latest-disinformation-campaign/

Baron, S., & Crootof, R. (2017). *Fighting fake News: Workshop report.* Yale Law School Publications. Retrieved from https://law.yale.edu/isp/publications

Cubbon, S. (2020, December 15). *Identifying 'data deficits' can pre-empt the spread of disinformation.* First Draft Footnotes. Retrieved from https://medium.com/1st-draft/identifying-data-deficits-can-pre-empt-the-spread-of-disinformation-93bd6f680a4e

Don Lemon, CNN News. (2020, August 20). *Don Lemon on Fox News host: Some people can't seem to quit bad habits—CNN Video.* Retrieved from https://www.cnn.com/videos/media/2021/08/20/vaccines-accountability-hannity-take-this-dlt-vpx.cnn.

Dyakon, T. (2020, December 14). Poynter's MediaWise training significantly increases people's ability to detect disinformation, new Stanford study finds. *Poynter.* Retrieved from https://www.poynter.org/news-release/2020/poynters-mediawise-training-significantly-increases-peoples-ability-to-detect-disinformation-new-stanford-study-finds/

First Draft. (2021). Vaccine misinformation insights report: July. *First Draft News.* Retrieved from https://firstdraftnews.org:443/long-form-article/vaccine-misinformation-insights-report-july/.

Frank, M. G., & Feeley, T. H. (2003). To catch a liar: Challenges for research in lie detection training. *Journal of Applied Communication Research, 31*(1), 58–75.

Hobbs, R., & McGee, S. (2014). Teaching about propaganda: An examination of the historical roots of media literacy. *Journal of Media Literacy Education, 6*(2), 56–67. https://doi.org/10.23860/JMLE-2016-06-02-5

McLelland, C. V. (2021). *The nature of science and the scientific method.* The Geological Society of America. Retrieved from https://www.geosociety.org/documents/gsa/geoteachers/NatureScience.pdf

McNamee, R. (2019). *Zucked: Waking up to the Facebook catastrophe.* Penguin Press.

Mitchell, A., & Walker, M. (2021, August 20). More Americans now say government should take steps to restrict false information online than in 2018. Pew Research Center. Retrieved from https://www.pewresearch.org/fact-tank/2021/08/18/more-americans-now-say-government-should-take-steps-to-restrict-false-information-online-than-in-2018/

Nothhaft, H., Bradshaw, S., & Neudert, L.-M. (2018). *Government responses to malicious use of social media* (p. 19). NATO Strategic Communications Centre of Excellence. Retrieved from https://stratcomcoe.org/publications/government-responses-to-malicious-use-of-social-media/125

Pamment, J., & Lindwall, A. K. (2021). *Fact-checking and debunking* (p. 51). NATO Strategic Communications Centre of Excellence. Retrieved from https://stratcomcoe.org/publications/fact-checking-and-debunking/8

Rosen, G. (2020, August 18). *Community standards enforcement report. About Facebook.* Retrieved from https://about.fb.com/news/2020/08/community-standards-enforcement-report-aug-2020/.

Rothman, J. (2021, August 13). Thinking it through. Why is it so hard to be rational? The real challenge isn't being right but knowing how wrong you might be. *The New Yorker.* Retrieved from https://www.newyorker.com/magazine/2021/08/23/why-is-it-so-hard-to-be-rational

Rubin, V. L. (2019). Disinformation and misinformation triangle: A conceptual model for "fake news" epidemic, causal factors and interventions. *Journal of Documentation, 75*(5), 1013–1034. https://doi.org/10.1108/JD-12-2018-0209

Rubin, V. L., & Conroy, N. (2012). The art of creating an informative data collection for automated deception detection: A corpus of truths and lies. *Proceedings of the American Society for Information Science and Technology, 49*, 1–11.

Twitchell, D. P., Nunamaker, J. F., & Burgoon, J. K. (2004). Using speech act profiling for deception detection. In *Intelligence and Security Informatics* (pp. 403–410). Retrieved from http://www.springerlink.com/content/ajeauy59kxjjttry

van der Linden, S., Leiserowitz, A., Rosenthal, S., & Maibach, E. (2017). Inoculating the public against misinformation about climate change. *Global Challenges, 1*(1). https://doi.org/10.1002/gch2.201600008

Vosoughi, S., Roy, D., & Aral, S. (2018). The spread of true and false news online. *Science, 359*(6380), 1146–1151. https://doi.org/10.1126/science.aap9559

Index

V. Rubin, *Detecting Disinformation and Fakes with the Eye and AI*,
https://doi.org/10.1007/978-3-030-95656-1

Printed in the United States
by Baker & Taylor Publisher Services